The Open University

Science

Environmental Science
TOPICS 1 to 4

Block 5

Cover image An extreme environment: Antarctica.

The Open University, Walton Hall, Milton Keynes, MK7 6AA

First published 2002. Reprinted 2006.

Copyright © 2002 The Open University

All rights reserved. No part of this publication may be reproduced, stored in a retrieval system, transmitted or utilized in any form or by any means, electronic, mechanical, photocopying, recording or otherwise, without written permission from the publisher or a licence from the Copyright Licensing Agency Ltd. Details of such licences (for reprographic reproduction) may be obtained from the Copyright Licensing Agency Ltd of 90 Tottenham Court Road, London W1T 4LP.

Edited, designed and typeset by The Open University.

Printed in the United Kingdom at the University Press, Cambridge.

ISBN 0 7492 6991 X

This publication forms part of an Open University course, S216 *Environmental Science*. Details of this and other Open University courses can be obtained from the Course Information and Advice Centre, PO Box 724, The Open University, Milton Keynes MK7 6ZS, United Kingdom: tel. +44 (0)1908 653231, e-mail ces-gen@open.ac.uk

Alternatively, you may visit the Open University website at http://www.open.ac.uk where you can learn more about the wide range of courses and packs offered at all levels by The Open University.

To purchase this publication or other components of Open University courses, contact Open University Worldwide Ltd, The Open University, Walton Hall, Milton Keynes MK7 6AA, United Kingdom: tel. +44 (0)1908 858785; fax +44 (0)1908 858787; e-mail ouwenq@open.ac.uk; website http://www.ouw.co.uk

2.1

s216 block 5i2.1

TOPIC 1
EXTREME WEATHER

Ross Reynolds

1	Introduction	5
2	**Explosive mid-latitude frontal cyclones**	7
2.1	What's the recipe for a 'bomb'?	9
2.2	The forecasters' problem	10
3	**Hurricanes**	12
3.1	Origin and development	13
3.2	What do hurricanes look like?	17
3.3	What does the damage?	18
3.4	Hurricanes in the USA: a case study	20
4	**Tornadoes**	30
4.1	Origin and development	31
4.2	Incidence, longevity and intensity	33
4.3	The forecasters' problem	37
5	Summary of Topic 1	42
	Learning outcomes for Topic 1	43
	Comments on activities	44
	Answers to questions	46
	Acknowledgements	47

Introduction

The *Extreme weather* DVD activity complements this text. It is divided into three sections ('Hurricanes', 'Tornadoes' and 'Bombs') and includes several video clips and animations with supporting narration, which are referred to at the points in the text where they are particularly relevant. You could either study the DVD material in sections or work through it all in one session when you reach the end of the text.

Weather that causes problems is probably the most frequently reported type of natural hazard in the press, on the radio and on television. Although we normally hear about hazardous events in the UK, we are also of course aware that much worse conditions occur in many other parts of the world. Quite what type of weather problem affects particular areas is partly dependent on the latitude and season: hurricanes and typhoons rage more commonly in some months than others, across parts of the tropical oceans and some adjacent islands and continents. Widespread river flooding or gales in the British Isles are much more likely in winter than in summer, while tornadoes, which wreak havoc over the High Plains of the USA, normally peak in May.

Extreme weather is, by definition, unusual — it doesn't happen every day, or even every year or every decade, depending on the phenomenon and the region. In the autumn of 2000, York saw its worst floods for some 400 years (Figure 1.1). In the 'great storm' of October 1987, a good deal of southern England experienced the worst winds for well over 200 years. In late August 1992, Hurricane Andrew proved to be the most costly weather-related disaster to hit the USA in the 20th century.

Figure 1.1 Flooding of the River Ouse in York, UK, during November 2000.

The rarity of such very extreme events means that for some of them, like the October 1987 storm in the UK, the combination of atmospheric conditions that produced such devastatingly powerful winds was itself rare. It is true that this particular storm's intensity was under-predicted, but it's also true that it was not a hurricane. If it had been, the winds would have been much stronger with perhaps ten times as much rainfall, resulting in huge devastation.

Meteorologists are naturally concerned to predict these hazardous events, so that they can give reliable warnings with as much lead time as possible. One problem for them is that extreme events are produced by a combination of atmospheric circumstances that occur only rarely. The computer-forecast models used today are very reliable, but they must be able to capture those rare circumstances that lead to extremes. This means that the weather observations that form the lifeblood of such predictions must be made at the appropriate times and in the right places, so that the 'ingredients' of a potential extreme weather event are well represented.

One problem for British and western European forecast services generally is that the really severe frontal depressions run in from the North Atlantic Ocean, where the observational network is understandably somewhat sparse. Similarly, forecasters at the National Hurricane Centre (NHC) near Miami have to look out for tropical cyclones that track across the vast, empty expanse of the tropical North Atlantic. The frequent cloud images produced from weather satellite observations are critical in keeping track of troublesome weather systems over these ocean areas.

The utility of these satellite images rests partly on the fact that frontal depressions that are rapidly deepening (i.e. where the central pressure is falling well below the pressure outside the depression) present very characteristic cloud features, and these provide a very strong clue to the UK Meteorological Office forecaster as to whether the depressions are likely to produce severe winds. Similarly, NHC forecasters can, if they wish, use the characteristics of a hurricane's cloud pattern viewed from space to estimate its likely surface wind strength (Figure 1.2).

This topic focuses on three quite distinct hazardous weather phenomena that affect three different geographical locations — explosive frontal cyclones, hurricanes and tornadoes. You will learn about the 'recipes' for the events and how they are predicted, and appreciate the types of hazard they pose.

Figure 1.2 Visible image of Hurricane Andrew. Some of the damage produced by this hurricane is shown later (Figure 3.11 in Section 3.4).

Explosive mid-latitude frontal cyclones

Frontal depressions or cyclones can occasionally wreak havoc across the British Isles and other parts of northwest Europe. The term 'explosive' relates to the fact that they deepen unusually rapidly and simultaneously have intensifying winds because of the steepening horizontal pressure gradient. (Remember that 'deep' in this context refers to how low the pressure in a system is compared to its surroundings.) **Explosive frontal cyclones**, along with the more typical frontal systems, are the bearers of much of the rainfall to northwest Europe and other similar areas around the world, such as the coast of northwest USA, British Columbia, the Pacific coast of Alaska and northeastern USA. The widespread cloud they bring is dull but also life-giving, in the sense that it provides the majority of the water supply of these regions.

So, most of the time — principally, but not exclusively, during autumn and winter — these frontal systems sweep across Britain from the North Atlantic and off into the continent. Their presence is made known in the spate of gale warnings for sea areas and coastal waters, which can be heard on the radio. These systems sometimes produce a gale inland, although this is quite a rare occurrence.

Sometimes, however, frontal systems are a lot more severe in their effects; they deepen very rapidly over the North Atlantic to generate storm force (Beaufort force 10) or stronger winds that can produce swathes of structural damage and fell trees in very large numbers. In rare circumstances, they can be associated with hurricane force winds (Beaufort force 12). The term 'hurricane force' has led to confusion in the media, and thus in the eyes of the public. In the 'great storm' of October 1987 mentioned above, some of the average surface winds along the Sussex Coast, for example, reached force 12, gusting to even higher values (Figure 2.1). Hurricane force simply relates to the *strength* of the wind, not to the *nature* of the low centre that produced it; in other words, the October 1987 event was *not* a hurricane.

For those who remember, the UK Met Office forecaster Michael Fish said that there would not be a hurricane on that October day in 1987. He was quite correct, and not just on some sort of technicality either. What parts of the British Isles experienced that night was the passage of a very unusually deep mid-latitude low. It had no eye, no spiral rainbands (both features of hurricane systems, described later), less widespread intense winds, and about one-tenth of the rainfall we might expect from a hurricane!

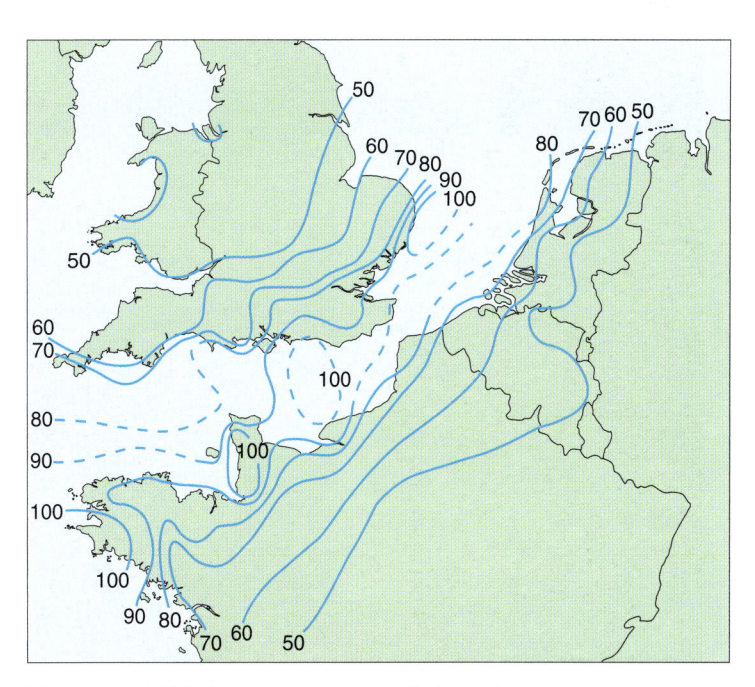

Figure 2.1 Highest reported gusts (in knots) over southern England and the near continent during the 'great storm' of 16 October 1987. (Contours shown dashed denote estimated values.)

A broadly similar, very deep frontal low battered the British Isles on Burns' Day (25 January) 1990 (see Figure 2.2). Another such event occurred on Christmas Eve 1998, manifest this time as two bands of very high gusts that rushed as distinct features from the coast of Wales, across parts of northern England and off into the North Sea. A Meteosat animation of the Burns' Day storm is in Section 3 of *Extreme weather* on the DVD.

Figure 2.2 Thermal infrared image of the Burns' Day storm, 25 January 1990.

2.1 What's the recipe for a 'bomb'?

Meteorologists have coined the term '**bomb**' for a mid-latitude frontal depression that deepens (has a falling central pressure) at a rate that is at least 24 hPa in 24 hours. Such events are rare, and have special ingredients to help them 'explode'.

All frontal depressions grow on shallow, sloping zones of relatively steep horizontal temperature and humidity contrast — the elongated frontal zone along which the frontal waves form and grow (refer back to Figure 4.10 in Block 2, Part 1). The gradual ascent of the warm, moist air over the colder, drier air is associated with the conversion of potential energy into that of motion, which is observed as the intensifying winds in the system.

If a storm becomes very windy while over the North Atlantic, there will be a lot more evaporation into it than if it were less windy. So it is often the case that the bombs are not only damagingly windy, but they can be wet enough to produce river flooding, especially if the event occurs over saturated or near-saturated soils. A rainfall of 25–40 mm in a day would be fairly typical for this type of storm, with generally higher totals in the autumn when the North Atlantic sea surface is at its warmest, or just starting to cool down from it. The warmth of the sea is an important factor in evaporation, as well as the wind speed. This is, broadly, why the records of most UK and Irish stations show October and November to be the wettest months on average.

Now, to get back to bombs. Since the great storm of 1987, meteorologists in western Europe have become aware of certain tell-tale cloud patterns on weather satellite images that provide strong clues as to whether a specific frontal system is likely to be a rapid deepener. One critical ingredient is the presence of a tongue of cloud-free air that streams along with the low: such a feature is the signature of a **dry intrusion** or **stratospheric dry intrusion**.

A (stratospheric) dry intrusion is air that has, most often quite independently of the original frontal system, subsided from the lower stratosphere right down into the troposphere — so far down that its presence in juxtaposition with the warmer, moister air of the pre-existing frontal feature enhances the thermal and moisture contrast, which tends to strengthen the wind flow.

Air from the lower stratosphere is extremely dry and very cold of course. The sinking air is warmed by adiabatic compression but it isn't moistened significantly, because no water can evaporate into the subsiding air. In addition, air from the lower stratosphere is quite stable, in contrast to the unstable air that is so common in the troposphere. (Remember that 'unstable' means that the temperature and humidity conditions are suitable for convection to occur.) As the dry intrusion moves quite rapidly down through the troposphere, it tends to act to strengthen the circulation around the pre-existing frontal cyclone (Figure 2.3).

Figure 2.3 Schematic diagram of a dry intrusion. Trajectories of air parcels from the lower stratosphere descend and fan out.

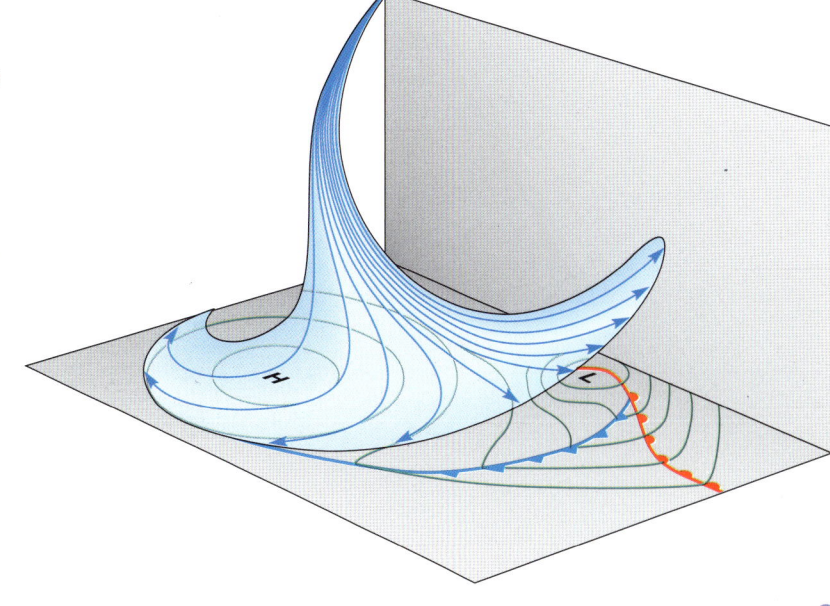

2.2 The forecasters' problem

Modern-day operational weather forecast models generally work very well indeed. Modern analysis of observed weather data is able to show clear evidence of relatively narrow tongues of dry stratospheric air being 'extruded' downwards into the troposphere. It is then possible to feed these data into computer models and forecast the consequences as dry stratospheric air moves both horizontally and down through the atmosphere.

A dry intrusion could, for example, start its descent from the lower stratosphere above Labrador, possibly a few thousand kilometres from a frontal depression in the central North Atlantic. This type of phenomenon is mapped and predicted well nowadays, for the quality of the UK Met Office forecasts of such severe events has improved since the under-prediction in October 1987. For example, both the Burns' Day storm and the Christmas Eve storm mentioned earlier were predicted a few days ahead, with warnings issued over a similar time frame.

As the computer-based forecast is issued, the operational meteorologist keeps a wary eye on rapid deepeners. Indeed, he or she can use water vapour images provided by the Meteosat satellites to map the location of the all-important dry intrusion, which shows up as a distinct feature known as a 'dry slot'. The absence of water vapour means that the satellite receives a very strong infrared signal from below, which is translated into a dark region on the image. In contrast, if water vapour is present, it absorbs much of the outgoing radiation, so that a very weak signal reaches Meteosat and the image appears white in such regions. Low water vapour concentrations are quite often associated with air that has subsided from the stratosphere, while high concentrations are characteristic of air that has ascended from the lower troposphere.

○ Study the Meteosat water vapour image (Figure 2.4) and locate a dry slot/dry intrusion.

● The dry intrusion that was associated with the surface gusts can be seen in the subtle purple feature that stretches from North Wales to Lincolnshire in Figure 2.4.

In general, the purples and blues depict dry air, while the greens and reds indicate moist air. The strongest winds are often associated with this kind of feature, running in behind the cold front. Note how there was a region of dangerously strong gusts (Figure 2.5) that swept across parts of northern England (and other areas) on Christmas Eve 1998. This relatively small expanse was directly related to the presence of, and the scale of, the downward-plunging dry air.

Such satellite images are of very great value in monitoring frontal systems, because they can be used to assess the quality of the forecasts from computer models. This is done by, for example, locating the frontal system's centre from the satellite image and comparing it with the predicted location for a particular time. In addition, and perhaps more critically, the presence of a dry slot will alert the forecaster to check whether the model is predicting a rapid deepening. Or, conversely, if the model is predicting such an eventuality, then the satellite image can be used as corroboration. This is invaluable when the system approaches from the mainly data-sparse expanses of the North Atlantic.

Explosive mid-latitude frontal cyclones

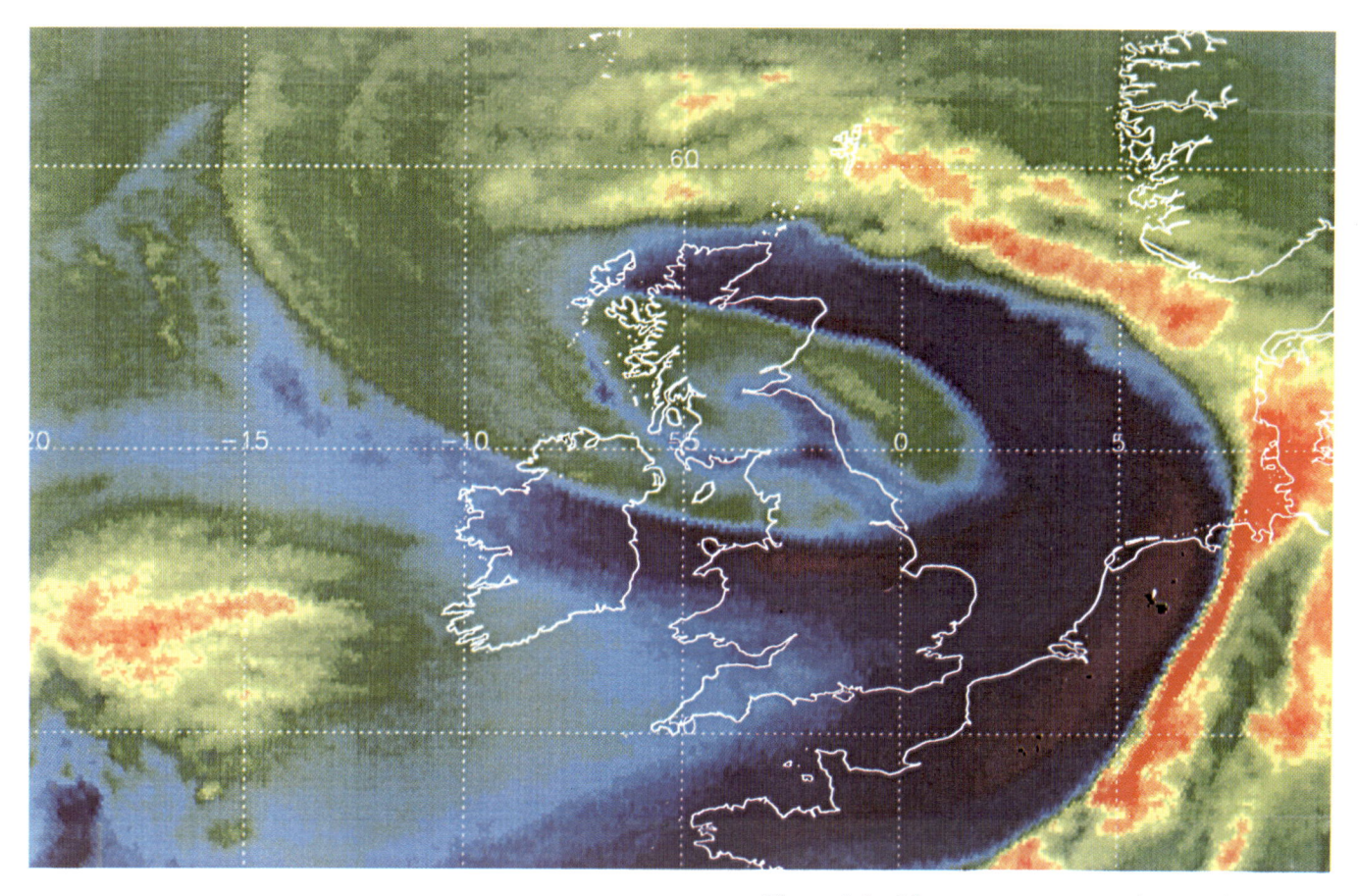

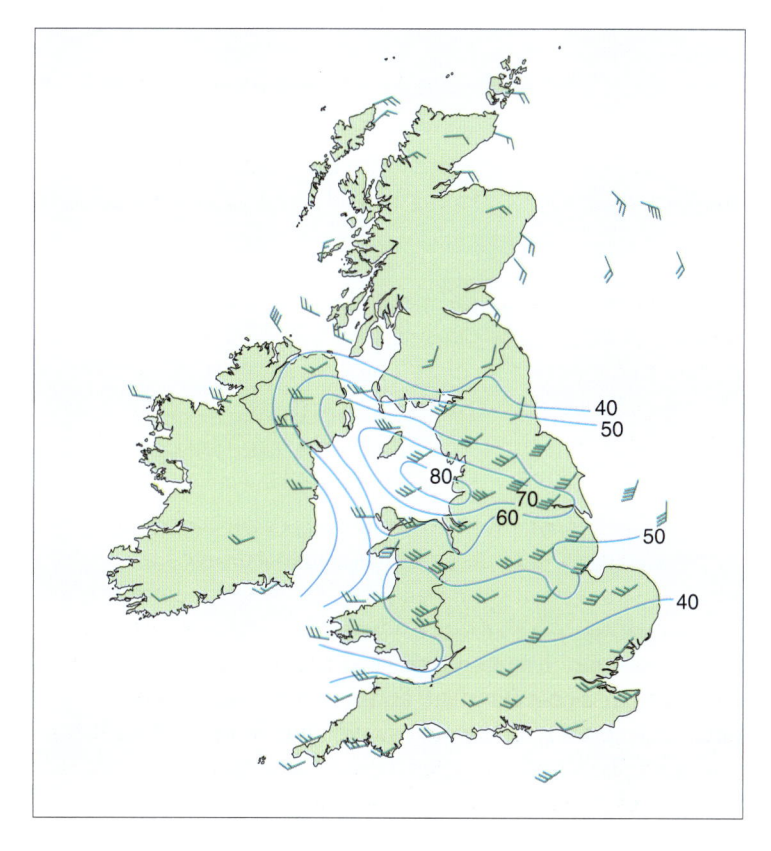

Figure 2.4 Meteosat water vapour image of a dry intrusion, Christmas Eve 1998. The purple areas are those where the air is driest and the red areas denote the most moisture-laden air.

Frontal depressions are also the source of the great majority of Britain's snowfall. While snow showers occur in the winter and spring, they are by their nature scattered and not very long-lived. Frontal systems, however, are linked to widespread snow (as they are with rain) and often strong winds. These two combined are the recipe for blizzard conditions and all the problems they bring. While widespread, persistent snow is not very common in Britain, it does create many problems when it occurs.

Figure 2.5 Contours of extreme gusts (knots) for the hour ending 21.00 UTC on Christmas Eve 1998. The average wind speed and direction at 21.00 UTC are also plotted.

11

3 Hurricanes

Hurricane is the regional term for a **tropical cyclone** that has attained a specific surface wind strength across the tropical North Atlantic and tropical North-East Pacific. The general term 'tropical cyclone' refers to a low-pressure system in tropical latitudes that is typically 500–800 km in diameter. The 'cyclone' means of course that the air circulates in towards the low centre in an anticlockwise fashion in the Northern Hemisphere (clockwise in the Southern Hemisphere); a somewhat old-fashioned but nevertheless descriptive term for a tropical cyclone is 'tropical revolving storm'.

The internationally agreed definition of the different phases and names used in the development of a tropical cyclone is based on five-minute averaged surface wind observations (except in the USA where a one-minute average is used). In its earliest phase, the system is termed a **tropical depression**, when the wind speed is 33 kn or less and a few closed isobars define the centre of the feature. The next phase is when the surface wind strengthens to 34–63 kn, when the system is termed a **tropical storm**. It is during this period that the disturbance is given a name.

When the averaged surface wind speed exceeds 63 kn, the tropical storm is upgraded to a hurricane (or **typhoon** in the North-West Pacific or cyclone in the South-West Pacific and Indian Ocean). This terminology is related to the definition of a 'hurricane' force 12 wind on the much-used Beaufort wind scale; for mariners, the description goes as follows: 'air filled with foam and spray, sea completely white with driving spray, visibility seriously affected'.

So whether the disturbance is a depression, a storm or other event rests completely on the averaged surface wind speed. The classification does not depend on the central pressure value or the amount of rain, for example. The definitions are well established, being related to different values of the Beaufort wind scale (Table 3.1).

Table 3.1 International definition of intensity grades of tropical cyclones.

Definition	Beaufort force	Wind speed/kn	Description of effects
tropical disturbance	less than 6	up to 21	—
tropical depression	6–7	22–33	whole trees in motion, inconvenience felt when walking against the wind (force 7)
tropical storm	8–11	34–63	trees uprooted, considerable structural damage
hurricane	12 and above	64 and greater	widespread damage

Something that is also occasionally listed on a table of Beaufort forces is the *dynamic pressure* that is imposed on a flat plate at right angles to the wind; dynamic pressure is that associated with the air's movement, in contrast to the barometric pressure, which is a *static* type of pressure. The dynamic pressure is usually quantified as the **wind loading** (Table 3.2). A wind loading of 1 kg m^{-2}

exerts a force equivalent to the weight of 1 kg on an object that presents a 1 m² area perpendicular to the wind direction. Since the gravitational force acts downwards and wind is usually horizontal, this wind loading does not guarantee that a 1 kg object will be lifted. The wind's ability to lift depends on the shape of the object and the frictional forces involved.

Table 3.2 Wind loading (kg m^{-2}) on a vertical surface for different Beaufort forces and the corresponding wind speeds.

Beaufort force	Upper limit of wind speed		Wind loading/kg m^{-2}
	kn	m s^{-1}	
2	6	3	0.4
3	10	5	1.4
4	16	8	3.4
5	21	11	6.4
6	27	14	11.3
7	33	17	17.7
8	40	21	26.6
9	47	24	37.9
10	55	28	51.7
11	64	33	68.9

Activity 3.1

Plot a graph of the wind loading (on the *y*-axis) against the upper limit of the Beaufort wind speed (from Table 3.2, in m s^{-1}) on the *x*-axis. Calculate the ratio of the wind loading of a force 11 tropical storm to the wind loading of a 'near gale' (force 7) wind — which occasionally occurs overland in Britain a few times in the winter. How does this value compare to the ratio of force 11 to force 7 maximum wind speeds? What would a graph of wind loading against the *square* of the wind speed look like?

Activity 3.1 indicates that the force applied by the wind on structures increases very dramatically as the wind speed increases, particularly when it is in excess of hurricane strength. If you extrapolate your graph to a wind loading of 175 kg m^{-2}, you will see that this value corresponds to a wind speed of roughly 100 m s^{-1}, which is characteristic of tornadoes, and thus illustrates why extremely heavy objects can be lifted and transported some distance by them.

3.1 Origin and development

Tropical cyclones develop from quite large areas of deep cumulus clouds, called 'cloud clusters', which occur over the tropical oceans most commonly between about 5° and 15° of latitude. One favoured region for their occurrence is off the West African coast in the vicinity of the Cape Verde Islands. The cloud clusters have often travelled westwards across sub-Saharan Africa (during the northern summer) as rain-bearing systems prior to this. In this ocean area they are embedded in the North-East Trades, which means that the cloud cluster moves westwards towards the Caribbean.

Most depressions in middle latitudes travel from west to east, though summertime tropical disturbances over West Africa travel from east to west. The direction in which pressure disturbances move is related to that of the deep flow within which they are embedded. The westerlies of middle latitudes are related to the poleward decrease of temperature there. In complete contrast, over West Africa in the summer the lower troposphere's temperature *increases* towards the North Pole, because of the intense heating over the Sahara. The result of this is an easterly flow across West Africa.

In order for a cloud cluster to evolve into an organized circulation, with the wind flowing into a low centre, it must move over ocean where the surface temperature is warmer than 26 °C to a depth of about 60 m. This means that a significant quantity of water vapour produced by evaporation from the tropical ocean can flow up into the system to later condense in the deep cloud, and to therefore release substantial amounts of latent heat. This heating leads to vertical expansion of the tropospheric air column, which is linked to a fall of surface pressure and thus increased wind flow into the low. The stronger flow means stronger evaporation, and so the process goes on to strengthen the evolving system.

○ What is the source of the tremendous energy of a tropical cyclone's winds?

● The thermal energy of the sea and the latent heat involved in evaporation.

It is not only the warmth of the sea that is critical. There must also be a pre-existent horizontal shear in the surface and low-level wind in which the disturbance will grow. This means that within the North-East Trades, for example, the wind speed should be strongest on the northern flank and weakest on the southern, with a gradual weakening in strength from north to south. Such a shear pattern promotes cyclonic motion.

In order to concentrate the latent heating in a vertical column, the lower tropospheric wind speed and direction should not be very much different from that in the upper troposphere over the central regions of the developing system. If there is a big difference in wind speed and direction through the troposphere, the heating cannot be usefully concentrated above the low centre, but instead will be dispersed. This 'ventilation' by the larger-scale environmental winds within which the cyclone is embedded would effectively suppress the system.

It is also very important that energy available through the evaporated seawater is pumped up into the disturbance to some depth. This is carried out by very deep cumulonimbus clouds, whose formation can be suppressed if the air through which they must grow, in the middle troposphere, is too dry. This form of convection does not occur over the tropical ocean if the mid-level humidity is less than 50–60%.

The final factor has to do with the Coriolis force. Because this force vanishes at the Equator, and is very small near the Equator (over about 5° of latitude either side), circular motion around isobars cannot and does not occur in that belt. This is another constraint upon the genesis of disturbances that may ultimately become hurricanes, typhoons or cyclones.

Not surprisingly, the incidence of tropical cyclones is strongly related to the geographical and seasonal distribution of the factors listed above. A study of the global location of the origins of tropical storms reflects the uneven distribution of such disturbances, including some tropical ocean regions where they simply never occur (Figure 3.1).

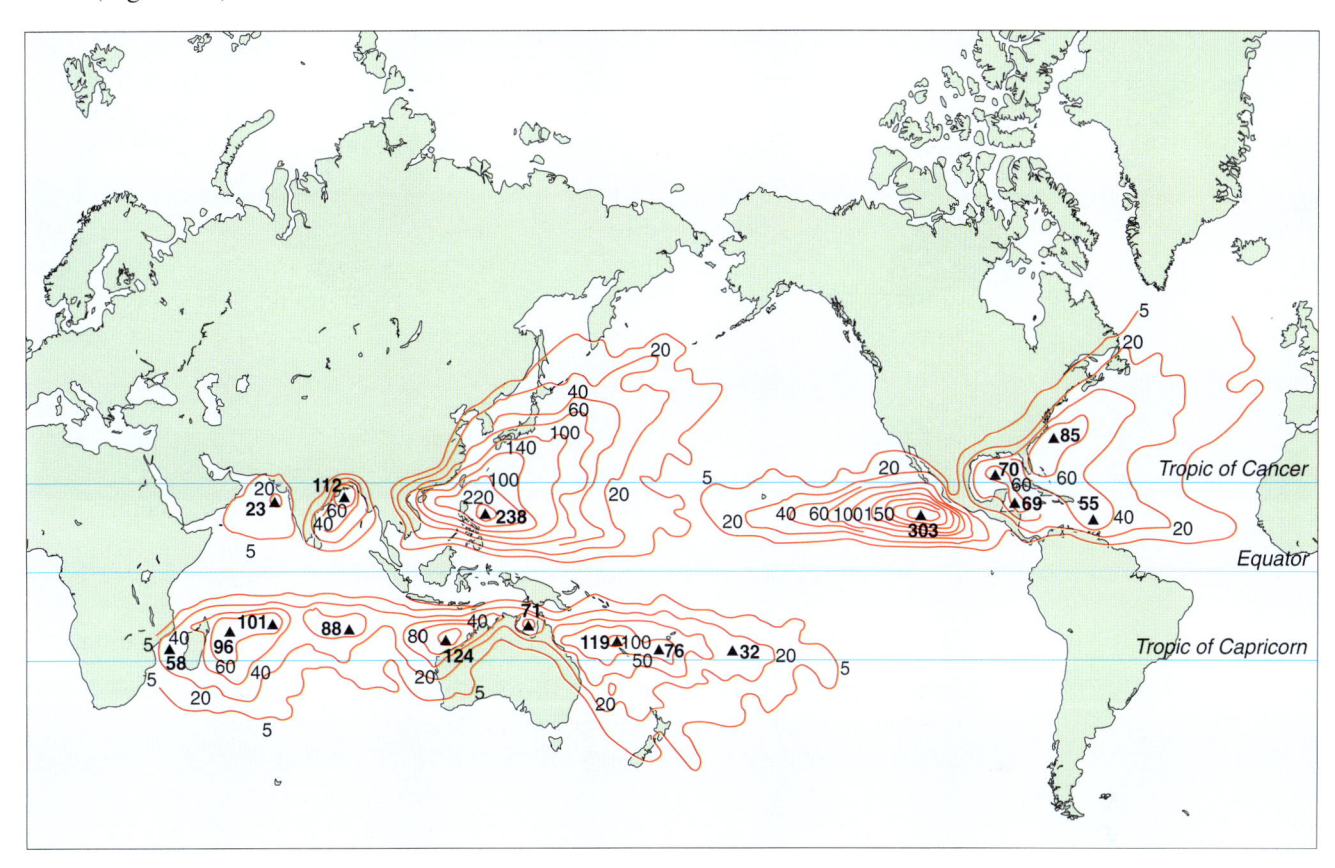

Figure 3.1 Global distribution of the frequency of tropical cyclones per 100 years within 140 km of any point. The locations of frequency maxima are shown as solid triangles with the frequency values in bold.

Question 3.1

Use the maps of global sea surface temperature and ocean currents (Figure 3.2) to help establish why hurricanes are not observed in the South Atlantic and South-East Pacific.

The average annual number of tropical storms in the Northern Hemisphere during a 20-year period was 55 while in the Southern Hemisphere it was 25 over the same period; so the average annual total was 80, with a strong bias towards the Northern Hemisphere. Looking at their location in more detail indicates that certain ocean areas are very much more active than others (Table 3.3).

Table 3.3 Average annual number (rounded) of tropical storms in different ocean areas.

Ocean area	Number of storms
NW Atlantic	9
NE Pacific	14
NW Pacific	26
N Indian	6
S Indian	9
S Pacific	10
Australasia	6

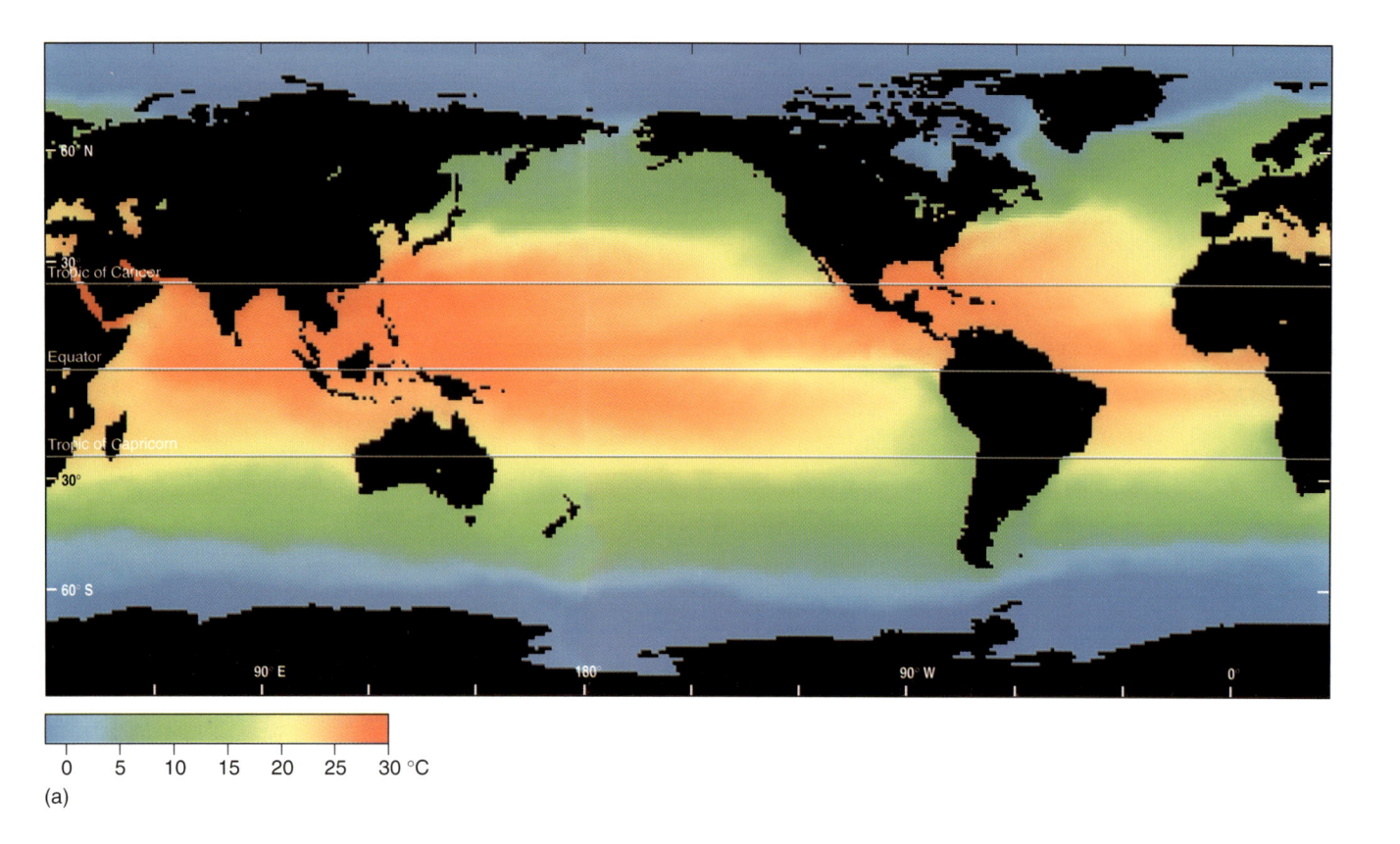

Topic 1 Extreme Weather

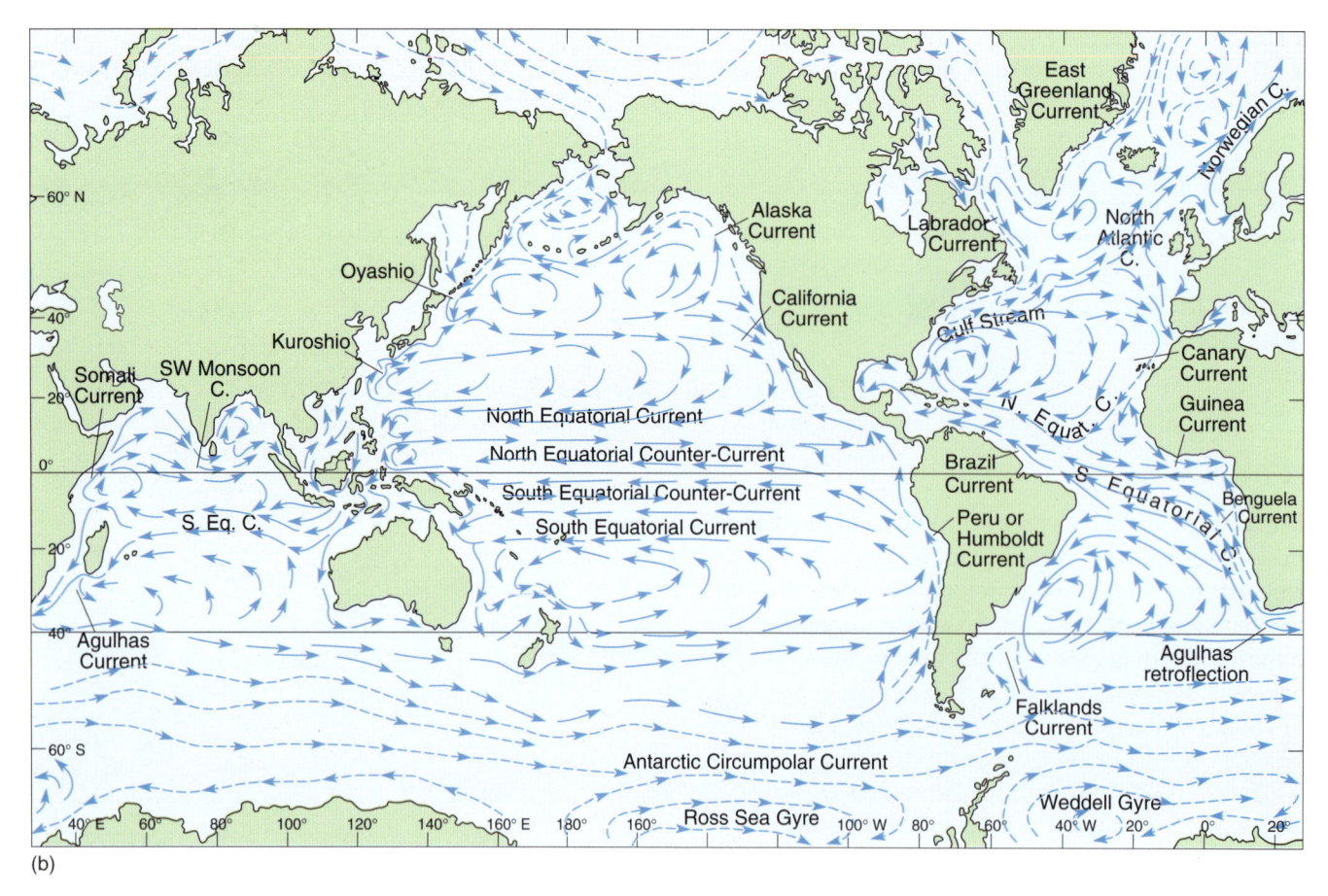

(a)

(b)

16

◀ **Figure 3.2** (a) Global sea-surface temperatures and (b) surface currents in the Southern Hemisphere summer.

The important role played by the warmth of the sea surface is illustrated by the strong seasonality in the frequency of tropical storms (Table 3.4). The sea surface in the Northern Hemisphere is warmest in September and coolest around March, while in the Southern Hemisphere, waters are warmest in February and coolest during August/September.

Table 3.4 Average monthly number (rounded) of tropical storms in each hemisphere.

Hemisphere	Jan	Feb	Mar	Apr	May	Jun	Jul	Aug	Sep	Oct	Nov	Dec
Northern	1	0	0	1	3	5	9	11	12	8	4	2
Southern	6	6	5	2	1	0	0	0	0	1	2	4

3.2 What do hurricanes look like?

When tropical storms attain hurricane, typhoon or cyclone strength, they have certain typical components to their structure. Firstly, they are vertically orientated disturbances, which means that their central axis, the **eye**, is upright. The eye is surrounded by the **eyewall cloud**, which is like an upright cylinder or throat of cloud. In gross contrast to the eye, the air in this region *ascends* at very rapid rates within the deep, vigorous cumulonimbus clouds, which can stretch up to heights of around 15 km. It is the eyewall cloud that is the location of most of the really severe winds and torrential rainfall: when we hear reports that Hurricane so-and-so had winds of 125 kn, they are referring to the winds at the surface across the eyewall cloud zone. This, like the eye, is not very wide. The band of cumulonimbus clouds is normally not more than a few tens of kilometres across and, because it is the zone of maximum destruction, poses problems for the forecast, which we'll look at a little later.

Within the eye itself, the air sinks (subsides) in great depth, from the upper to the lower troposphere. This subsiding air is compressed and warmed, making it cloud-free and warm compared to the surrounding air circulating within the hurricane. It is warmed at the dry adiabatic lapse rate in the unsaturated air of the eye, which is typically 20–50 km in diameter (Figure 3.3), being about half the size of the entire system, i.e. eye plus eyewall cloud.

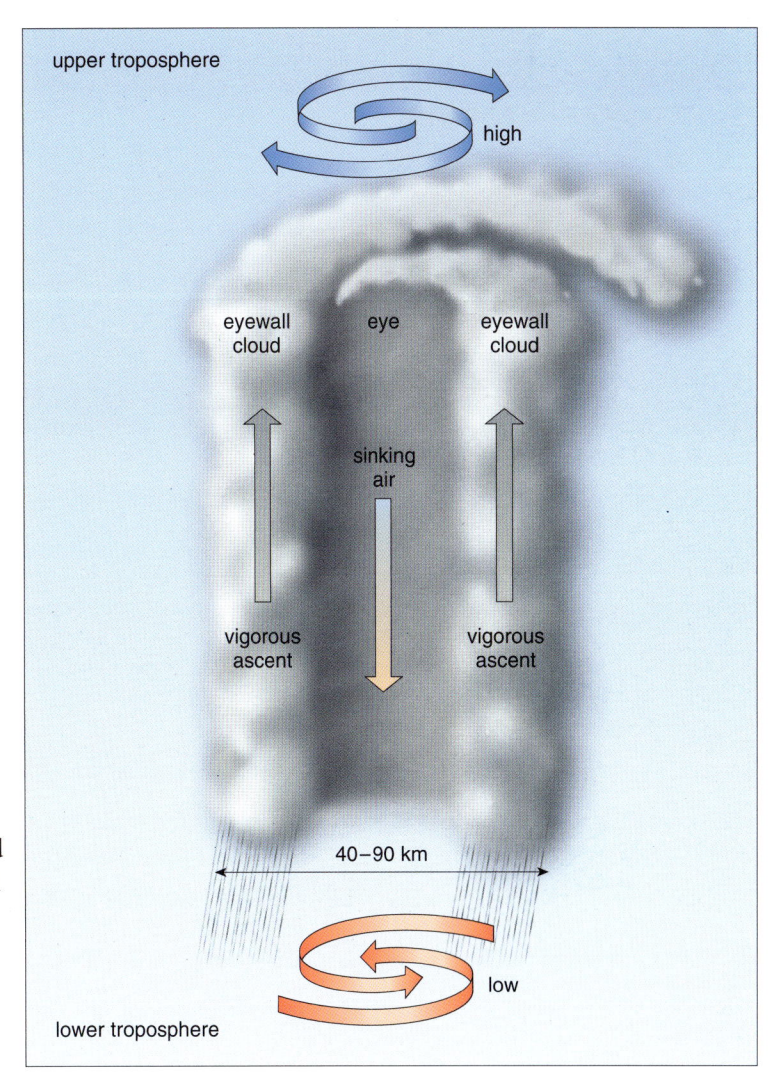

Figure 3.3 Large-scale horizontal inflow and outflow and the vertical motion within the eye region (Northern Hemisphere hurricane).

Precipitation from the cumulonimbus clouds can be as much as 25 cm h^{-1}, but more typically 25 cm d^{-1}, which is about one-third of the average annual rainfall in much of southeast England.

Another unique aspect of these systems is that they are characterized by marked cyclonic *inflow* in the lowest 1–2 km over the tropical ocean and anticyclonic *outflow* in the uppermost 3–4 km of the troposphere (Figure 3.3). Within a hurricane, this means that thousands of millions of tonnes of air spiral anticlockwise towards the low-pressure centre across the ocean surface, and thousands of millions of tonnes spiral away clockwise from the system at high altitudes. In between these layers of 'input' and 'output' of air and water vapour, the air generally ascends as part of the cycling of mass by the hurricane.

In Block 2, Part 1, you saw that barometric pressure measurement is equivalent to weighing the atmosphere. In the case of a hurricane, typhoon or cyclone, it is critical for the forecast to make an accurate assessment of the amount of mass coming in at low levels and leaving at upper levels.

Another feature of these systems is **spiral rainbands** (Figure 3.4) which, as the name suggests, are elongated, curved features that run in towards and wrap around the centre. They are composed of very deep cumulonimbus clouds in lines and are associated with torrential rain, extremely gusty conditions and the occasional tornado. The very vigorous upward motion within the clouds and the rainbands as a whole is linked to areas of deep sinking motion between them. These are regions within the hurricane's circulation that experience windy conditions and clear blue skies.

Figure 3.4 Spiral rainbands, extending out some 200–300 km from the eye of the hurricane

○ If a greater mass of air is leaving at upper levels than coming in at the lower levels, how will the surface pressure change?

● If there is more air being exported aloft than is being imported into the lower layer, then the surface pressure will fall. If, however, the flow of mass is greater at low levels, then the pressure will increase.

3.3 What does the damage?

It's true to say that most of us think 'wind' when the subject of hurricanes comes up. However, hurricanes can generate serious coastal flooding, and then river flooding from the intense rain following landfall (i.e. when they reach the land). This double source of water can make flooding events extremely serious, particularly if (as sometimes happens) the system is moving slowly.

The Saffir–Simpson scale (described in Section 3.4) indicates the scale and intensity of wind damage; as mentioned above, the worst winds are confined to the eyewall cloud zone, although winds can also be damagingly strong outside this region. The wind can be particularly severe, occurring as violent downbursts that flow out of the deep cumulonimbus clouds of the spiral rainbands. Such downbursts form within rainshafts that fall through part of the cloud. What happens is that the rain evaporates and so chills the surrounding air, which is dragged down by the falling drops. The result is that a strong pulse of air shoots down with the rain, hits the Earth's surface and fans out in all directions as an often dangerously gusty downburst.

We know that barometric pressure is related to the weight of air above a barometer. High values mean more air and lower values less air. This fact has an important consequence for the sea surface across which hurricanes (and all sorts of other pressure systems) travel. A deep low, such as a hurricane, in a sense 'sucks' the sea surface up a little; this is called a **surge**. A high pressure in a ridge or anticyclone depresses the surface somewhat. The effect is such that a 1 hPa change in the surface pressure leads to a 1 cm change in the elevation of the sea surface. So for a hurricane whose central pressure is, say, 940 hPa, the sea surface could be elevated by as much as 60 cm. This is of no consequence across the open ocean, but when a hurricane approaches and crosses a gently shelving coastline the surge can be greatly amplified (Figure 3.5).

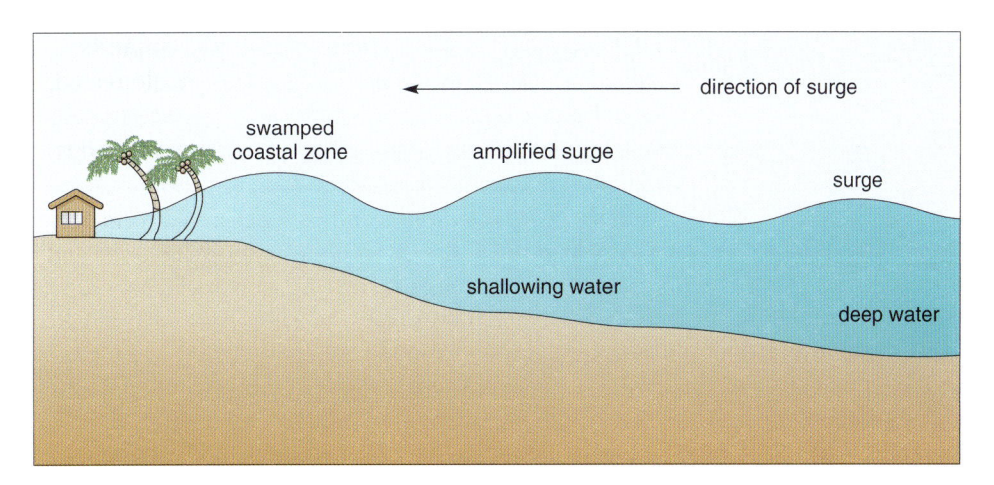

Figure 3.5 The amplification of a surge as it moves into shallower coastal waters.

The deepening of a surge occurs in much the same way as the increase in height of a wind-driven wave that approaches a beach. As a mass of water approaches a shoreline — as a surge or as very strongly driven wind waves — it is 'squeezed' vertically so that as the water depth becomes shallower, the surge becomes higher. However, the surge is not a wall of water, but a gradual increase and then decrease of water depth over quite a few hours.

The height of the storm surge is accentuated by the strong surface winds piling water up; the result is that for the more extreme hurricanes, the surge can reach 4.0–5.5 m! As a point of information, storm surges are sometimes termed *tidal* surges. This is incorrect: they have nothing to do with tidal forcing by the Sun and Moon, in much the same way that *tidal waves* (properly termed tsunamis) are unrelated to tidal effects.

An important aspect of the surge height (and of the low-level wind strength) in hurricanes, typhoons and cyclones is that there is a significant asymmetry in their pattern with respect to the line of motion of the disturbance. Hurricanes

Topic 1 Extreme Weather

move forward at a speed of 10–20 kn, buried within the larger-scale easterly tropical winds. Their low-level anticlockwise circulation means that the forward motion of the low centre is added to the winds blowing around it, which is in the same direction on the right flank of the track.

○ What is the wind speed at the edge of the hurricane in Figure 3.6 *relative to the eye of the storm*?

● The eye is moving northwards at 10 kn. The wind speed, relative to the eye, on the right flank is 110 kn − 10 kn = 100 kn to the north. The wind speed relative to the eye on the left flank gives the same value to the south: 90 kn + 10 kn = 100 kn. The winds at the edge of the hurricane are therefore 100 kn.

This means that when a hurricane makes landfall, the worst surge flooding and wind damage is to the right of the eye's track. In fact, more generally, the forward right quadrant of a hurricane or typhoon is the worst region for damage (Figure 3.6). For Southern Hemisphere cyclones, it is the forward *left* quadrant where most damage occurs, because the low-level wind flow is clockwise.

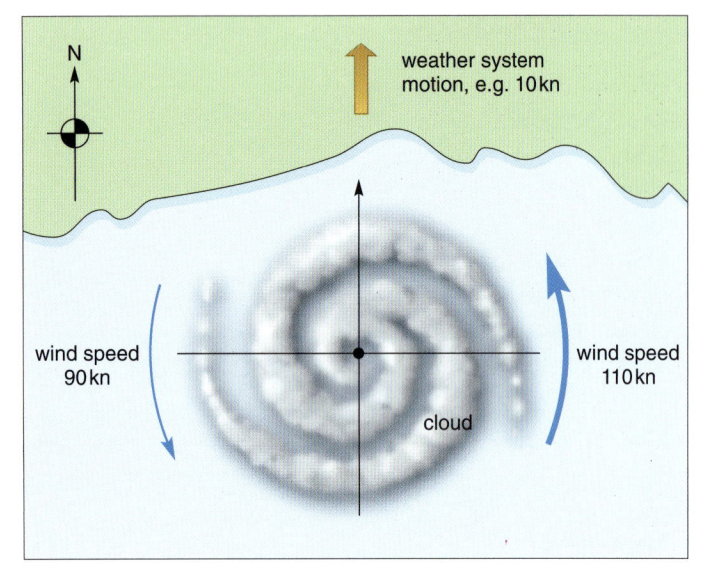

Figure 3.6 The forward right quadrant is the worst region for wind and surge damage for hurricane systems in the Northern Hemisphere.

3.4 Hurricanes in the USA: a case study

For some years now, the **Saffir–Simpson scale** (Table 3.5) has been used to express the intensity of hurricanes that threaten and actually damage the Gulf and Atlantic coasts of the USA. The two names are those of a meteorologist and a structural engineer, respectively, who teamed up to produce a meaningful, workable scale that could be easily appreciated by professionals and, perhaps more importantly, by members of the public in susceptible areas.

Table 3.5 The Saffir–Simpson scale of hurricane intensity.

Category	Description	Wind speed/kn	Surge height/m	Damage caused
1	weak hurricane	65–83	~1.5	damage mainly to trees, shrubs and unanchored mobile homes
2	moderate hurricane	83–96	~2–2.4	some trees blown down; major damage to exposed mobile homes; some damage to roofs
3	strong hurricane	96–114	~2.5–4.0	foliage removed from trees; large trees blown down; mobile homes destroyed; some structural damage to small buildings
4	very strong hurricane	114–135	~4.0–5.5	all signs blown down; extensive damage to roofs, windows and doors; complete destruction of mobile homes; flooding as far as 10 km inland; major damage to lower floors of structures near shore
5	devastating hurricane	>135	>5.5	severe damage to windows and doors; extensive damage to roofs of homes and industrial buildings; small buildings overturned and blown away; major damage to lower floors of all structures less than 4.5 m above sea-level within 500 m of the shore

Table 3.6 lists the *frequencies* of the five categories allotted to hurricanes that struck the US mainland during successive decades of the 20th century.

Table 3.6 Hurricane frequency in the USA.

Decade	Category 1	2	3	4	5	Total (1–5)	Major (3–5)
1900–1909	5	5	4	2	0	16	6
1910–1919	8	3	5	3	0	19	8
1920–1929	6	4	3	2	0	15	5
1930–1939	4	5	6	1	1	17	8
1940–1949	7	8	7	1	0	23	8
1950–1959	8	1	7	2	0	18	9
1960–1969	4	5	3	2	1	15	6
1970–1979	6	4	2	0	0	12	2
1980–1989	9	1	5	1	0	16	6
1990–1996*	0	3	3	1	0	7	4

* Note that this is a seven-year period only.

Activity 3.2   Hurricane frequency analysis

(a) Construct a bar graph with one bar (based on the *x*-axis) for each of the ten periods in Table 3.6. The *y*-axis should depict both the *total* number as the height of the bar and the number of *major* hurricanes too, represented as, say, a shaded fraction of the total bar height.

(b) Is there any trend in either total hurricane number or number of major events, or do the inter-period variations appear to be more or less random?

(c) What is the average annual frequency of (i) all hurricanes and (ii) major systems over the 97-year period?

There were only two category 5 systems in all of the 97-year period covered in Table 3.6. The most recent of them was Hurricane Camille in 1969, which produced gusts of up to 160 kn and a surge of some 7 m, which swamped large areas of the State of Mississippi. The earlier one was the Labor Day hurricane that occurred in 1935.

Hurricane Mitch was the most recent category 5 event in the North Atlantic, where it wreaked havoc across Guatemala and Honduras during the last week of October 1998 (Figure 3.7). Indeed, it was the strongest ever such system to be observed so late in the year. The estimated 11 000 deaths caused by its passage across Central America is believed to be the highest death toll due to a hurricane in that region since 1780.

Figure 3.7 Damage produced by Hurricane Mitch.

3.4.1 Hurricane birth in the North Atlantic: the forecasters' problems

Officially, the hurricane season in the North Atlantic starts on 1 May each year. Plots of the number of tropical storms, all hurricanes and major ones for the inclusive period from 1944–1996 indicate different average start dates (Figure 3.8). The very earliest in the year that one would reasonably expect to see the first named tropical storm is around mid-May, the first full-blown hurricane in early June, and the first major hurricane of the season in late July. The typical dates for the first of each type are 11 July, 14 August and 4 September, respectively. So for the North Atlantic the hurricane season doesn't take off typically till mid-August. The steepest slope for all three curves in Figure 3.8 occurs roughly between mid-August and mid-September. This is the period during which all three systems are increasing in number most rapidly. It is clear too that all three types of event do hang on into October.

○ Typically, what fraction of hurricanes have occurred by 1 October? What is this fraction for named systems (tropical storms)?

● On 1 October, the hurricane curve is at about 4.5; at 1 December, the curve flattens out to about 6, so the fraction is 4.5/6 = 0.75. For named systems the fraction is 8/10 = 0.8.

Nowadays forecasters at the National Hurricane Centre (NHC) in the USA, which is situated just outside Miami, have varied support in their task of monitoring and predicting the track, intensity changes and landfall of hurricanes that cross the Caribbean and the USA. Tracking such systems across the expanse of the tropical North Atlantic is relatively easy because the NHC receives geosynchronous weather satellite images every 30 minutes from the European

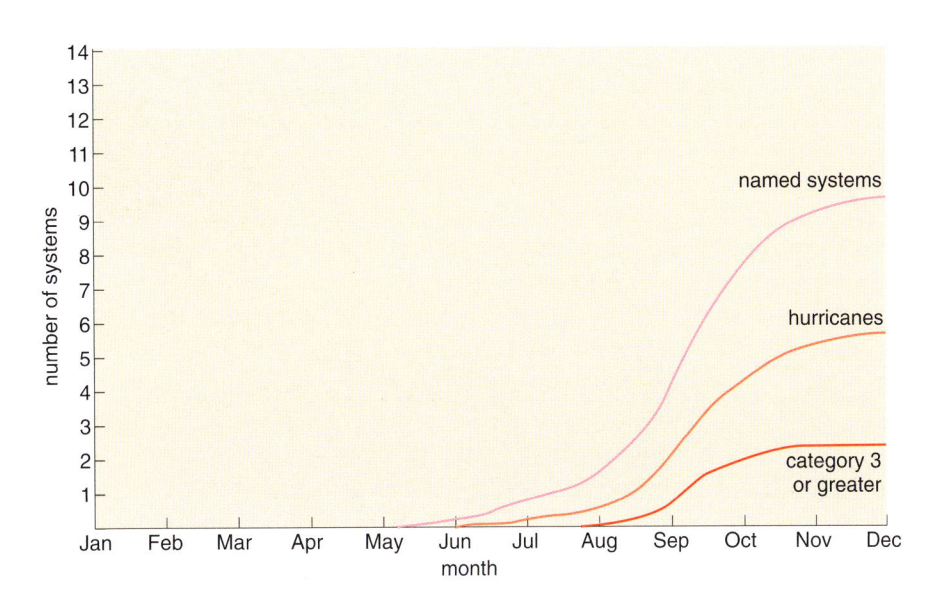

Figure 3.8 Plots of cumulative number through the year (i.e. the number that have occurred by a given date), of North Atlantic tropical storms and hurricanes against starting date. The data are averages for the period 1944–1996.

Meteosat, which covers the sea to the west of tropical Africa, and also from the American GOES-East geostationary satellite, which covers the central tropical Atlantic, Caribbean and USA.

What follows is an account based on a case study of the worst hurricane to hit the US mainland in the 20th century. Of course, damaging tropical cyclones affect many other parts of the world as well, including southeast Asia, Bangladesh, countries bordering the Arabian Sea, Madagascar, Mozambique, Australia and the islands of the southwest Pacific Ocean (Figure 3.1). Weather satellites play a universal role in monitoring the strength and track of such systems, while the various weather services issue warnings of impending danger. Quite how the respective societies respond is, however, related to such factors as: the efficacy of the public broadcast system; public awareness of both the problem and of the best course of action; the nation's infrastructure and means of transport; and the degree of organization of the emergency authorities.

It is a reasonably straightforward job for a trained analyst in any forecast office to spot tropical cyclones on satellite images. These systems present cloud features that are indicators of their status and their vigour: the presence of an eye, the existence of rainbands and how wrapped-around they are, and the extent of cirrus clouds spiralling anticyclonically in the high-level outflow. Indeed, satellite images are the best data for locating the system centre and thus its track over time. It's not surprising that we don't see direct observations from ships in the area; they're more likely to be steaming full speed away from the system, rather than towards it for the convenience of the forecasters!

So the *actual* track isn't too much of a problem. The *future* track is predicted by operational computer weather prediction models that utilize all sorts of observational data on a global scale; that is to say, the hurricane is only one small (but significant!) part of a global prediction. As far as the North Atlantic is concerned, it will be handled not only by the forecast issued by the US National Weather Service, but also (for example) by that of the European Centre for Medium-Range Weather Forecasts (ECMWF) in Reading, UK, and the UK Meteorological (Met) Office in Bracknell (in Exeter from approximately mid-to-late 2003). Such predictions are broadcast to *every* weather service, and act as critical guidance for tropical cyclone warnings, no matter where.

Table 3.7 Average locational error of Hurricane Andrew's eye in the North Atlantic derived from forecasts for increasing lengths of time ahead.

Forecast period/h	Error/km	No. of forecasts
12	61	57
24	120	35
36	196	33
48	261	31

The Miami forecasters take note of a range of global prediction models, including those from the ECMWF and the UK Met Office. In order to get the hurricane's track right, the models have to get the surrounding large-scale circulation right first, for the entire depth of the troposphere and into the stratosphere too.

The relative paucity of data in critical regions like the tropical North Atlantic means that the accuracy of the forecast of the track over such data-sparse regions (up to, say, seven days ahead) isn't as good as a forecast of the track of a frontal depression over the USA, where the data coverage is much better. The location error (the difference between the average location over *all* forecasts and the actual observed location) for the hurricane's eye tends to increase as the forecast projection time increases, and to vary between the forecast models (Table 3.7). Forecasters will therefore always use the most recent forecast, no matter what the situation, and keep a careful eye on the hurricane's track prediction out to, say, day three or four of the outlook. They will want to start issuing warnings for those US coastal States under threat, with a reasonable lead time.

○ Why does the error in the predicted location of Andrew's eye grow with time? Does the error of 120 km just one day ahead have important consequences for evacuation, given that the hurricane's eye and eyewall are about 60 km across?

● The basic reasons for the error growth are that (i) all observations used in weather forecast models are imperfect; (ii) observations are sparse over the ocean where the hurricane is tracking in from; and (iii) the physical and dynamical processes involved in its track and evolution are not perfectly represented.

Given that Andrew's eye and eyewall cloud were together some 60 km across, an error of 120 km means that even one whole day ahead there is an uncertainty regarding the location of the eye's landfall.

Currently, this level of error is insurmountable, and is handled by the forecasters in such a way that they issue *probability* forecasts for various possible landfall locations. Evacuation plans are based on these probability assessments made a day or so before the event.

Of course, it is also extremely important to get the strength of the winds right. Forecasts based on models will indicate the predicted speed and, while the system is over the data-sparse tropical North Atlantic, will act as a useful guide to the actual wind speed at landfall.

It is possible for trained analysts to make a reasonable estimate of the surface wind speed from very careful assessment of weather satellite visible images taken when the disturbance is over the data-sparse areas. Quite how powerful the wind will be depends on variables such as the sea-surface temperature and the flow patterns in the upper troposphere. Both of these have to be well represented in the model in order to get the surface pressure gradient right, and then the winds right. The satellite images act as useful checks on the quality of the forecast. At an operational weather centre, the eye centre can be located accurately from the image (to within about 10 km) while the wind speed can be estimated to within about 10 kn either side of the actual value. This 'truth' can be used by the forecaster to assess the quality of the prediction for the same time as the satellite observation. For example, he or she will be able to assess whether the prediction is pushing the system on too fast, or not in quite the right direction, or underestimating the wind speed.

In the tropical North Atlantic, direct measurements are not made until the disturbance approaches, or is within, the Caribbean Sea. The US Air Force has a fleet of Hercules aircraft (Figure 3.9) based in Biloxi, Mississippi, which are used to fly into threatening hurricanes (and tropical storms) on a routine basis. The point of these missions is to take *direct* readings of wind speed and direction, temperature, pressure, etc. diametrically across the disturbance at a variety of levels. Such data can be used to calculate the surface wind speed before landfall and thus, for one thing, to corroborate (or refute) the values predicted by the weather models. In addition, higher-flying jet aircraft occasionally cross the uppermost outflow region of these systems in order to collect useful data for the forecast model regarding the strength of the outward-spiralling flow. This is done by mapping the winds in detail and calculating the air density for the same layer. In this way, the mass outflow can be estimated. This helps to improve the accuracy of prediction of changes in surface pressure and thus, more generally, the surface intensity of the disturbance.

Figure 3.9 A Hercules aircraft of the 'Hurricane Hunters' based at Keesler Air Force Base, Biloxi, Mississippi.

Question 3.2

Why is it important for high-flying aircraft to measure the winds in the *outflow* region in the upper troposphere? Is there a method based on weather satellite observations that could help too?

3.4.2 Hurricane Andrew — the most costly of the 20th century

It all started on 14 August 1992 when a **tropical wave** crossed the West African coastline on its way over the tropical North Atlantic. Like many other such waves, it had developed a few days before, much further east over West Africa, and had skirted the southern flank of the Sahara desert before arriving at the coast. These waves are broad undulations in the generally east to west air flow and are associated with roughly north–south bands of heavy, squally showers.

Scientists at the NHC were aware of the existence of the tropical wave on that day; they keep a very careful eye on every cloudy disturbance that leaves the West African coast during the hurricane season. The cloudy wave moved westwards at around 20 kn, passing to the south of the Cape Verde Islands on the 15 August. By 18.00 UTC on the next day, it had been assigned the status of a tropical depression — not from any direct measurement of the surface wind, but on the basis of careful analysis of its cloud organization from visible images provided by the geosynchronous weather satellites.

Table 3.8 Six-hourly central pressure and maximum wind speed for Hurricane Andrew.

Date	Time (UTC)	Central pressure /hPa	Maximum wind speed/kn
22 Aug	00.00	1000	55
	06.00	994	60
	12.00	981	70
	18.00	969	80
23 Aug	00.00	961	90
	06.00	947	105
	12.00	933	120
	18.00	922	135
24 Aug	00.00	930	125
	06.00	937	120
	12.00	951	110
	18.00	947	115
25 Aug	00.00	943	115
	06.00	948	115
	12.00	946	115
	18.00	941	120
26 Aug	00.00	937	120
	06.00	955	115
	12.00	979	80
	18.00	991	50
27 Aug	00.00	995	35
	12.00	997	30

Just 18 hours later, the name Andrew was assigned to the first tropical storm of the 1992 season, somewhat later than the average date indicated in Figure 3.8. Between 17 and 20 August, Andrew moved threateningly towards the Caribbean. A reconnaissance Hercules from the USA flew into the storm on 20 August; at 15.00 UTC on that day, the aircraft measured winds of 70 kn at 457 m (1500 feet) on the storm's northeastern flank.

During 21 August, Andrew exhibited significant cloud pattern changes that indicated to the NHC analysts that it was intensifying. It was on the morning of 22 August that it became a full-blown hurricane, sporting an eye by 12.00 UTC on that day.

Activity 3.3 Hurricane Andrew's intensity

Using the data from Table 3.8, plot a graph of central pressure against maximum wind speed. You should produce a scatterplot of 22 points for the 22 pairs of values. While you have learned that it is the horizontal pressure gradient that drives the wind, so that the stronger the gradient, the stronger the wind, your graph should indicate that the value of the central pressure is also a good indicator of the surface wind strength. Why should this be?

In addition, deduce from your graph the upper and lower pressure values that correspond to the upper and lower limits of wind speed for Hurricane Andrew when it was a category 3 system (see Table 3.5).

Luckily for most of the Caribbean, Andrew tracked northwestwards far enough out over the tropical North Atlantic to not be a bother — except for the Bahamas, that is. Andrew's eye crossed land later on 23 August, swamping Eleuthera Island in the northern Bahamas, where a wind of 120 kn was logged at 21.00 UTC. Four fatalities occurred in the Bahamas as a result of Andrew.

By this time, of course, the NHC was gearing up to issue orders to State authorities that long tracts of susceptible coastline of the US mainland should be evacuated. Quite where this should occur was obviously critically dependent on the model predictions, and on the monitoring of Andrew's track over the preceding few days.

The hurricane's potency while approaching southeast Florida is quite simply expressed by what happened to the sunken *Belzona* barge that lay in 21 m of water just offshore. It was 66 m long, weighed 350 tonnes and was sunk with the help of 1000 tonnes (10^6 kg) of cement in its hold. Andrew shifted the *Belzona* about 220 m and managed to lighten its 'ballast' by 900 tonnes!

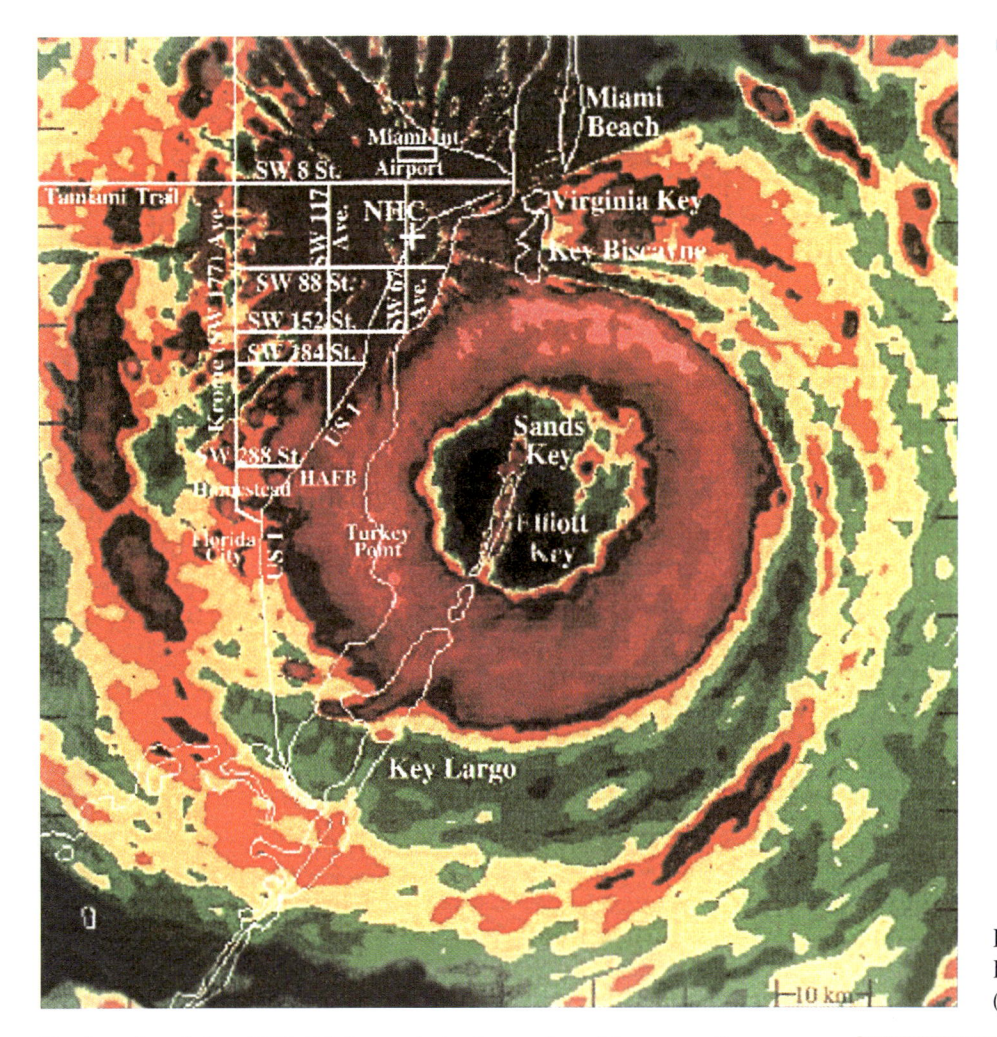

Figure 3.10 Radar rainfall map of Hurricane Andrew for 08.35 UTC (03.35 local time) on 24 August 1992.

During the night of 23–24 August, Andrew closed in on southern Florida. The watchful eye of the precipitation radar on the roof of the National Weather Service building in Miami mapped the torrential precipitation falling from the eyewall cloud and the wrapped-around spiral rainbands. Figure 3.10 is a dramatic depiction of these features and also the rain-free eye and 'lanes' between the spiral bands—the last available image before the radar antenna was destroyed by Andrew's extreme winds! The community of Homestead just to the south of Miami was one place that bore the brunt of the eyewall passage. Severe devastation was widespread there (Figure 3.11).

Figure 3.11 Damage produced by Hurricane Andrew at Homestead, Florida.

Activity 3.4 The passage of Andrew's eye region

Figure 3.10 is the radar precipitation map just before Andrew's eye crossed over the Florida mainland. Actual observations from an aircraft from Keesler Air Force Base in Mississippi indicated that the eyewall had a width of 20 km. At this time Andrew moved about due west across Homestead at 16 kn (8 m s^{-1}).

Calculate how long the torrential rain (and severe winds) of the eyewall would have lasted (to the nearest minute) in Homestead, if the eye and eyewall crossed the community along its diameter. Assuming that Andrew moved due west at this time, the first bout of extreme weather would be followed by the calm and dry conditions of the eye, followed by a second bout as the trailing eyewall ran across. You need only calculate the duration of *one* eyewall passage.

As Andrew went ashore, a number of intrepid members of the public who opted to stay in their properties — including a Mrs Hall and Mr Martin, a brother and sister living about 400 m apart — measured a now-confirmed minimum barometric pressure of 922 hPa at 09.05 UTC (04.05 local time) as the eye crossed their community. Interestingly, Mrs Hall's home was constructed in 1945 by her grandfather to be hurricane-proof, with concrete walls 64 cm (22 in) thick! She survived, as did her barometer. The wind was estimated to have gusted to 155 kn in the eyewall around this time too.

A Mr Fairbanks living in the area reported a gust of 184 kn in Andrew's northern eyewall; this value was decreased to 164 kn after the equipment used was tested at the Virginia Polytechnic Institute. Likewise, the 155 kn value quoted above was reduced to 125 kn after careful calibration tests. Wind speeds in the southern eyewall were believed to have reached a maximum of 99 kn.

- ○ Work undertaken in the 1960s indicated that the difference in the wind speeds (kn) in the northern and southern eyewalls is just about twice the forward speed (kn) of the hurricane. Why should this be so?

- ● As noted earlier (p. 20), the wind that we observe has two components. One is that related to the horizontal pressure gradient, while the other is the actual bodily motion of the hurricane (or tornado or explosive cyclone, etc.) itself. This means that in a circular system, like a hurricane travelling due west, more or less as Andrew was doing, its line of motion and the wind blowing around it at the surface act in the same direction on its *northern* flank. This means that they should be added together. In contrast, the forward motion of the system is opposed by very strong eastward winds on the *southern* side. All this results in significantly stronger winds in the north eyewall than in the south one (for westward moving hurricanes). The values outlined above confirm this.

Andrew crossed southeastern Florida quite rapidly, which meant that the rainfall from it was 180 mm over large areas. The total number of properties destroyed was 25 524, with an additional 101 241 damaged in some way; unfortunately, 44 people died as a direct or indirect result of the hurricane. There were some 725 000 insurance claims filed by Florida residents in its aftermath.

The storm surge coincided with an extremely high tide along the Atlantic coast of Florida, resulting in inundation including the flooding of the international headquarters of Burger King by a surge of 5.1 m! The predicted combination of wind, rain and flood meant that some 1 150 000 people had to be evacuated from low-lying areas of southeastern Florida. Quite where this evacuation is carried out clearly depends critically on the forecast of the precise track, extent and intensity of the system. Ideally, of course, there should be a lead time of a day or two in order to give ample time for the evacuation process. The State authorities have to consider the margin of error in the location of the eye's landfall and evacuate accordingly.

Andrew carried on across the Gulf of Mexico to effect a second landfall on US territory, when the eye crossed the coast near Morgan City in Louisiana. On its way over, it toppled 13 oil platforms, leading to seven pollution events, and destroyed five drilling wells. The warm Gulf waters added to the water content of the clouds — and to the rainfall total, such that a massive 303 mm fell at Hammond, Louisiana. The spiral rainbands' cumulonimbus clouds were tornadic in places; indeed, two people were killed and 32 injured when a funnel cloud (see later) ripped through Laplace. As in Florida, the authorities organized the evacuation of about 1 250 000 inhabitants away from the flat, low-lying coast. In total, 17 people perished in Louisiana.

Once overland, Andrew was downgraded to a tropical storm after 10 hours and then to a tropical depression after a further 12 hours. Although its vigour declined rapidly, it still produced heavy rain in some of the States it swept over.

The estimated total cost of the damage caused by Hurricane Andrew across the USA was $30 475 000 000, far outstripping the second in line, Hurricane Hugo, which devastated South Carolina in 1989 at a cost of some $8 490 000 000.

There is a video animation of Hurricane Andrew in Section 1 of *Extreme weather* on the DVD.

4 Tornadoes

Tornadoes are the most violent atmospheric phenomena ever observed. They are small-scale rotating features with a typical width of 50–100 m, although they can be as wide as 1500 m in extreme cases. On average they last for 5 minutes, move at 10 m s^{-1}, and therefore have typical damage paths that are a few kilometres long. However, the swathe can vary considerably in extent, from little more than a point to a line almost 200 km in length.

Tornadoes can produce surface wind speeds in excess of 280 kn — much stronger than those recorded in hurricanes and the like — but only across a very limited area. However, if this 'limited area' is in an urban region, then the scale and cost of the devastation can be immense. A tornado occurs when a **funnel cloud** (Figure 4.1) touches the ground. This type of cloud is the obvious visual expression of the presence of these violent phenomena. A funnel cloud can also protrude from the base of a slowly rotating cumulonimbus cloud, but not necessarily reach the ground. The condensation in the form of a funnel shape is related to the extreme pressure drop experienced by moist air flowing into the column-like upward motion in a tornado. The pressure drop from the surrounding air into the tornado wall can be as much as 200 hPa across a few tens of metres!

Tornadoes occur most notably across the High Plains of the USA (Figure 4.2), but are also observed in other parts of North America and in many other countries around the world, including the British Isles.

Figure 4.1 A tornado funnel cloud in North Dakota.

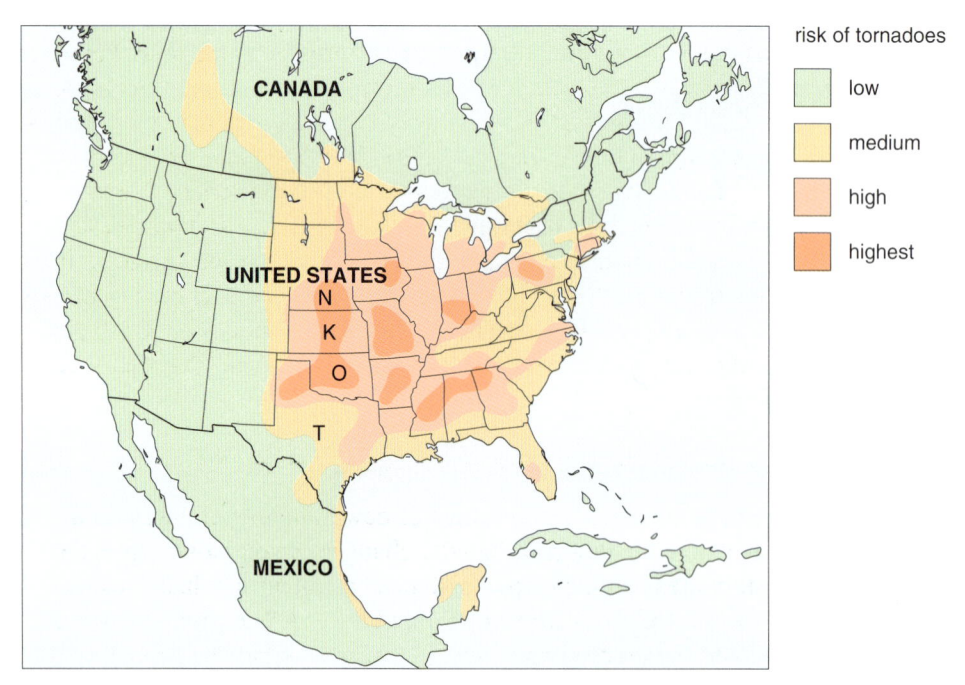

Figure 4.2 Annual mean tornado frequency in North America. (N = Nebraska; K = Kansas; O = Oklahoma; T = Texas.)

4.1 Origin and development

Tornadoes are always associated with a 'parent' cumulonimbus cloud that is itself rotating. One ingredient for the development of deep, vigorous cumulonimbus clouds is a steep decrease of temperature with height within the troposphere. This condition is most exaggerated in the spring when the surface heating of lower-level air can be strong, while the air aloft is still very cold indeed, particularly when it flows across the High Plains from much further north. Not only should the surface heating be strong, but it should also be reinforced by the northward flow of warm, moist air from the Gulf of Mexico. The high humidity of this low-level Gulf air is a critical part of the recipe for a tornado.

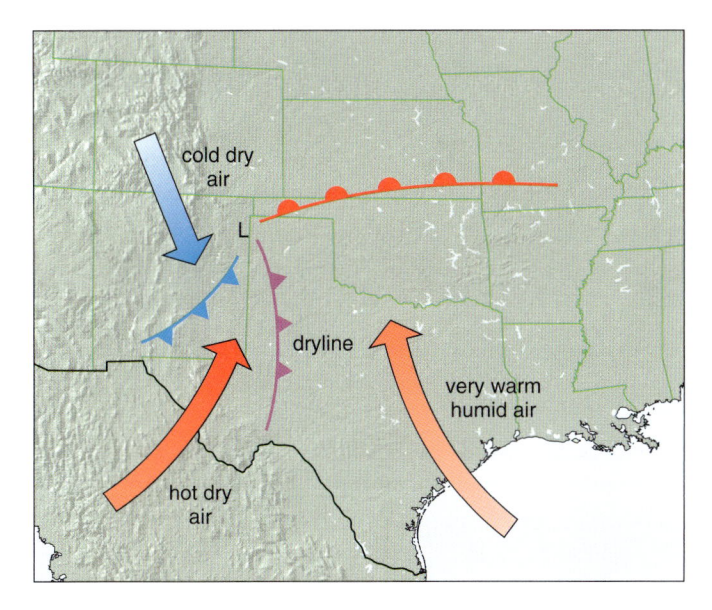

Figure 4.3 Schematic of the dryline.

The severe storms that spawn tornadoes quite often form along a feature known as the **dryline**, which occurs in the warm sector of a frontal system over the USA (Figure 4.3). The fronts are located in such a way that the warm sector air has two distinct streams: the warm and very moist Gulf air from the southeast and the hot and dry air flowing from the southwest across the interior of northern Mexico. These two air streams are juxtaposed along a line that is quite often observed across western Texas, Oklahoma and Kansas, especially in the spring and early summer. The more or less north–south orientated separation line is the dryline, so-called because there is a very rapid horizontal change in humidity across it; the dewpoint temperature can decrease by as much as 9 °C km^{-1}, moving west from the moist to the dry air.

○ How does the dryline manifest itself in the weather data in Figure 4.4a?

● It is located where the horizontal gradient of dewpoint temperature is at a maximum, i.e. where the dewpoint rapidly changes as you move across the land. The eastern areas on the map experienced moist air that had streamed northwards (from the Gulf of Mexico), while the western region was being influenced by very much drier air moving eastwards, which had come from the hot and dry stretches over New Mexico and northern Texas (see Figure 4.4b).

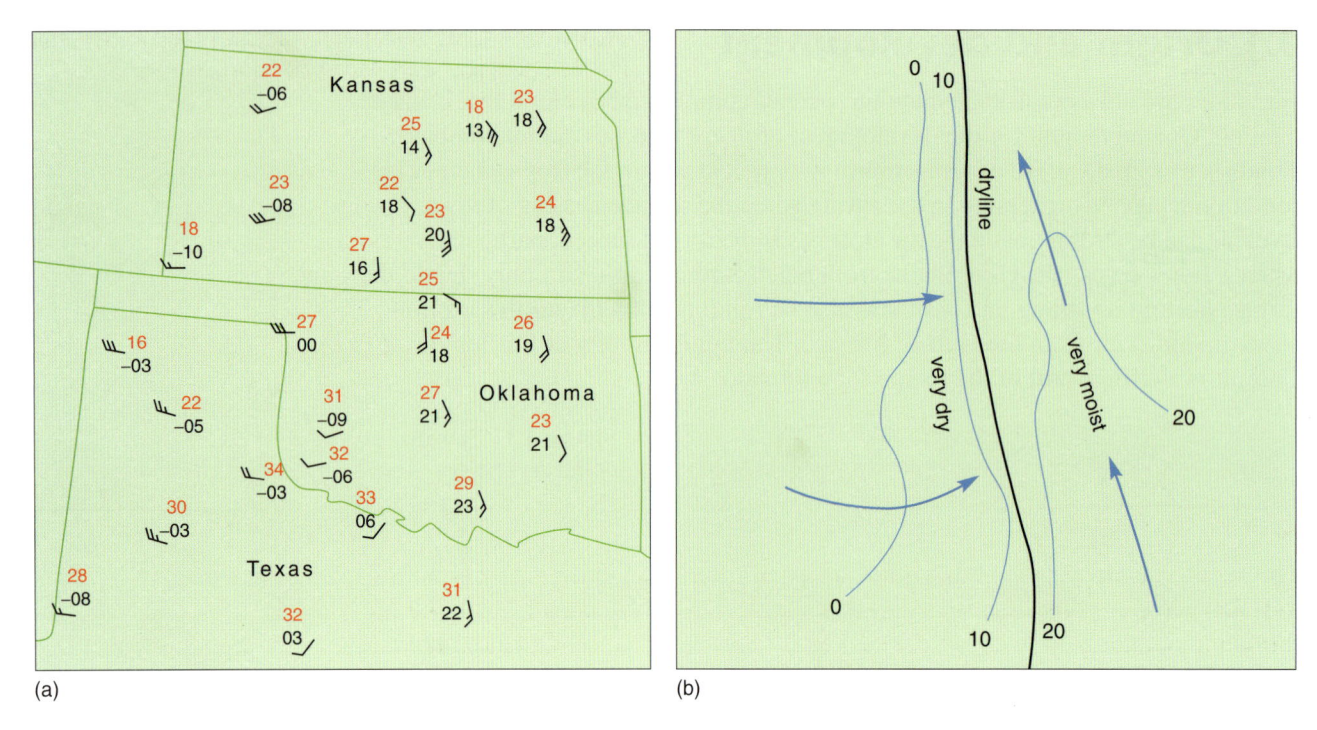

Figure 4.4 A springtime dryline case over Texas, Oklahoma and Kansas. (a) Wind speed and direction, drybulb temperature (red numbers, °C) and dewpoint temperature (black numbers, °C). (b) Schematic showing overall wind flow and dewpoint contours.

The upward slope of the High Plains towards the west means that the hot, dry air coming from that direction rides up and over the slightly cooler, damp air from the Gulf. This situation often leads to the upward burst of the overridden Gulf air, aided by the flowing together of the two airstreams. The consequence is very deep, violent thunderstorms just to the east of the dryline; they tend to run from southwest to northeast in the general deep flow from that direction in the warm sector.

These large-scale atmospheric conditions, which are known to provide the breeding grounds for tornadic storms, are quite well predicted nowadays by the operational weather forecast models. This means that forecasters in Oklahoma and Kansas, for instance, are forewarned by a day or more of the risk of severe convection in their region. But exactly where and when it will develop is another matter entirely.

Predicting the location, duration and intensity of tornadoes, even a few hours ahead, is not possible with precision. Such forecasting is unlikely to ever be possible, simply because Nature doesn't supply the clues that far ahead. It's suspected, however, that there are a few critical factors that can be monitored in order to help forecasters produce a useful risk assessment.

One factor is the existence of leading edges of low-level cold pools that are remnants of pre-existing thunderstorms, from the downdrafts and gust fronts they produced. The idea is that these regions are ones of enhanced horizontal thermal contrast that set up an invisible rotation about a horizontal axis, called a horizontal vortex roll, somewhat like a slowly turning rolling pin (Figure 4.5). Another critical ingredient of a tornado-producing cloud is the presence of a thunderstorm with an updraft supplying it — the updraft crosses the spinning tube of air and tilts it gradually into vertical alignment, rather like a top, and often intensifies the spin by vertical stretching.

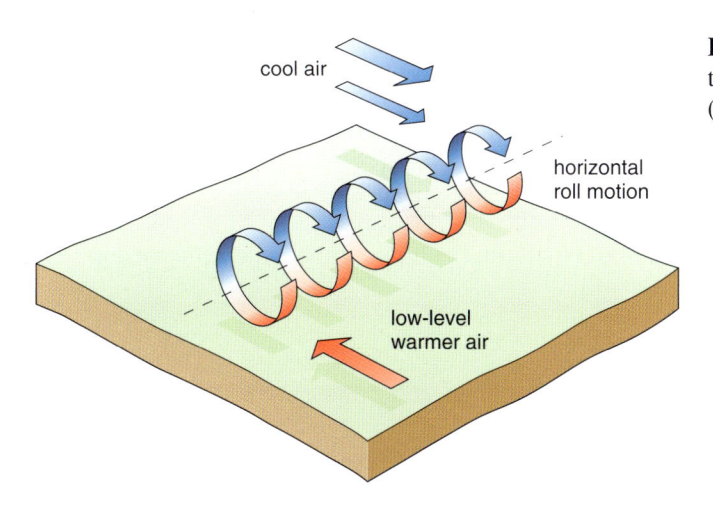

Figure 4.5 A horizontal vortex roll between a cool remnant thunderstorm outflow (above) and warmer environmental air (below).

This rotation is imparted to the whole updraft of the cloud, which can be some 3–10 km across, and is a necessary, but not wholly sufficient, condition for the formation of a tornado. It is associated with an area of low pressure below and within the cloud, known as a **mesocyclone**, into which the air swirls.

Another essential ingredient appears to be the character of a feature known as the rear-flank downdraft. This is a heavily rain-laden, descending shaft of air that sinks at the trailing edge of severe storms. Once at low levels, it is drawn into the storm's circulation in a hook shape, concentrating spin over a region that is less than a few kilometres across. The concentration of spin is linked to the formation of a funnel cloud.

4.2 Incidence, longevity and intensity

The ingredients required to produce severe storms are geographically biased, so that the distribution of tornadoes across the USA shows a strong pattern in which there is a maximum frequency that stretches north from Texas, through Oklahoma and Kansas and into Nebraska (Figure 4.2). This elongated zone is sometimes termed 'tornado alley'.

Even where there is the highest average annual frequency of tornadoes — in central Oklahoma — only nine, on average, occur in a circular area of some 110 km radius. Put slightly differently, in an 'average' year one tornado occurs in an area of some 4200 km^2, equivalent to a square with sides of 65 km. This, of course, assumes a uniform distribution in space, but it does emphasize the relative rarity of such features, even where they are most common.

Tornadoes do, however, occur in 'outbreaks', rather than one or two scattered randomly here and there. Some of these have been particularly deadly: one of the worst occurred over 3 and 4 April 1974 when, during a 16-hour period, 148 tornadoes ran through 13 States, killing 307 people and producing damage valued at $600 million at the time. The infamous 'Tri-State' outbreak of 18 March 1925 killed 695 people across Missouri, Indiana and Illinois. The 3 May 1999 outbreak in Oklahoma (and Kansas) was the worst in that State's history (of which more later).

Topic 1 Extreme Weather

The seasonal incidence of tornadoes in the USA is strongly biased towards a springtime peak, although the systems can and do occur in any month of the year. Figure 4.6 illustrates the very marked maximum during May and June, when some 41% of tornadoes occur, followed by roughly 24% in March and April. The frequency declines rapidly into the high-summer months of July and August with a much lower total of about 16%, tailing off to 8% for September and October, and 6% during both November and December, and January and February.

Figure 4.6 The mean monthly incidence of reported US tornadoes and tornado fatalities since the 1950s.

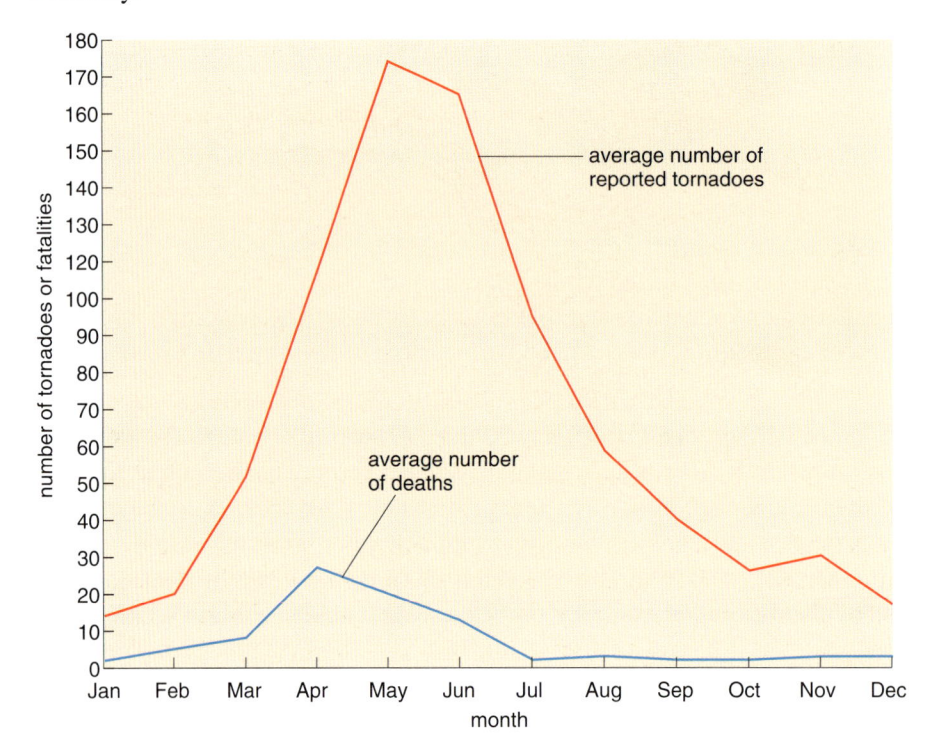

Another strong signal related to tornado frequency occurs within the day of the tornado strike itself. When a very large number of cases are analysed, by far the highest frequency of occurrence is found in the afternoon and early evening period, between roughly 13.00 and 18.00 local time. This is related to the progression of surface heating and the time taken for the deep cumulonimbus clouds to grow to great heights. The very rapid decline in occurrence after sunset is of course related to this too.

Just like the Saffir–Simpson scale for hurricanes, the **Fujita F scale** (Table 4.1) has been developed to express the intensity of a tornado, in order to convey the seriousness of such a situation to both professionals and the public in North America. The scale is not perfect, nor does it pretend to be; the numbers used express the tornado intensity on the basis of the damage caused. Forecasters use this scale to express the predicted likely intensity of tornadoes and the actual intensity once they've developed.

The Fujita F scale was designed to pick up where the Beaufort scale leaves off, at hurricane force 12 winds. One major problem for such a scale is that it has to be based on the damage caused, because direct measurement of the wind speed is not possible. This is quite simply because tornadoes rarely cross weather stations, and if they do, they destroy the instruments! So, basing the scale on

Table 4.1 The Fujita F scale.

F value	Wind speed/m s^{-1}	Description of damage caused
F0	18–33	light damage; some damage to chimneys; breaks branches off trees; pushes over shallow-rooted trees; damages signboards
F1	33–50	moderate damage; lower limit is the start of the hurricane wind speed; peels surfaces off roofs; mobile homes pushed off foundations or overturned; moving cars pushed off the road
F2	50–70	considerable damage; roofs torn off frame houses; mobile homes demolished; rail wagons pushed over; large trees snapped or uprooted; light-object missiles generated
F3	70–93	severe damage; roofs and some walls torn off well-constructed houses; trains overturned; most trees in forest uprooted; heavy cars lifted off road and thrown
F4	93–117	devastating damage; well-constructed houses levelled; structures with weak foundations blown off some distance; cars thrown and large missiles generated
F5	117–138	incredible damage; strong-frame houses lifted off foundations and carried a considerable distance to disintegrate; car-size missiles fly through the air in excess of 100 metres; trees debarked
F6	>138	maximum speeds within tornadoes are not expected to reach F6

damage is the best option, even though its reliability is limited because the structural integrity of the building affected is often not known. The damage is inflicted by the dynamic pressure of the wind, which varies with the square of the speed (Activity 3.1) and a factor that depends on the shape of the building and the direction of the wind. A further limitation of the Fujita F scale is that many tornadoes travel across regions where there are no buildings to damage.

Nevertheless, the scale has been in use since its foundation in 1971 and it is a well-established way of reporting a tornado's severity to those who need to know.

Where the fatalities tend to occur indicates the relative risk involved in living in certain types of home: some data for the period from 1985 to 1997 are collected in Table 4.2.

Table 4.2 Incidence of US tornado-related fatalities according to type of location.

Number of deaths	% of total	Location
230	38	mobile homes
162	27	'permanent' homes
66	11	vehicle
27	5	business
56	9	outdoors
15	2	school
49	8	other

Activity 4.1 The Andover tornado

26 April 1991 was a prime day for severe weather across large tracts of the USA. There were 203 reports that covered a whole range of severe weather, including 113 of hail and 56 of tornadoes that had affected States from Nebraska to Texas. There were 298 people injured and 19 died.

It was, however, the town of Andover, some 10 km east of Wichita in Kansas, that a rare F5 tornado ripped through. The damage caused across Kansas by the severe weather as a whole was costed at $272 000 000 and included the destruction of 1728 homes. Of these, 241 were fixed caravans in the Golden Spur Mobile Home Park in Andover; 13 people died there, even though a seven-minute warning was issued for the area.

Study the map of the track of the Andover tornado (Figure 4.7) and answer the following questions:

(a) What was the duration (in minutes) of the tornado's track across Kansas?

(b) Roughly what proportion of its lifetime did it spend producing severe damage or worse? (Refer to Table 4.1.)

(c) What was the approximate length of the tornado path, and its approximate average speed (m s^{-1}) over its whole lifetime?

(d) What was the approximate width of its damage swathe when it hit the Mobile Home Park (when it was estimated to have been an F5)?

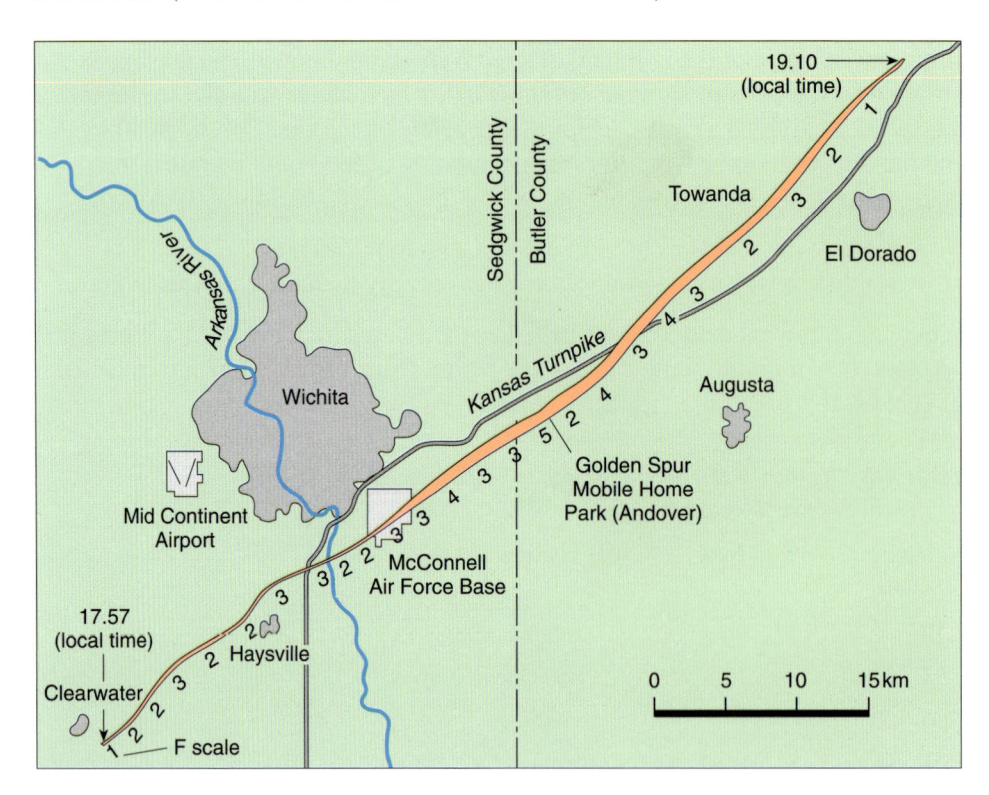

Figure 4.7 The Andover F5 tornado, which occurred on 26 April 1991. The area shaded orange represents the width of the damage swathe and the numbers on the tornado's trail denote its F-rating.

4.3 The forecasters' problem

Predicting tornadoes is an immense challenge to operational meteorology. In the USA, the National Severe Storms Forecast Centre is adjacent to the National Severe Storms Laboratory (NSSL) in Norman, Oklahoma.

As mentioned earlier, the risk of the development of severe storms is indicated by the forecast models, with a lead time of one or more days. The indication will be of a high risk over a region covering a few adjacent States, for example. This means that forecasters will obviously be on the look-out for signs of the development of very deep, rapidly growing convective cloud.

They do this by using a combination of observations. In susceptible areas like Oklahoma, there is a dense surface network of frequently reporting automatic weather stations called a Mesonet (Figure 4.8). The stations transmit their data instantaneously to a central point for display and analysis. This helps the forecasters to keep tabs on the evolution of significant features such as the dryline, by looking at both drybulb temperature and dewpoint temperature, and how they are evolving over short periods of hours.

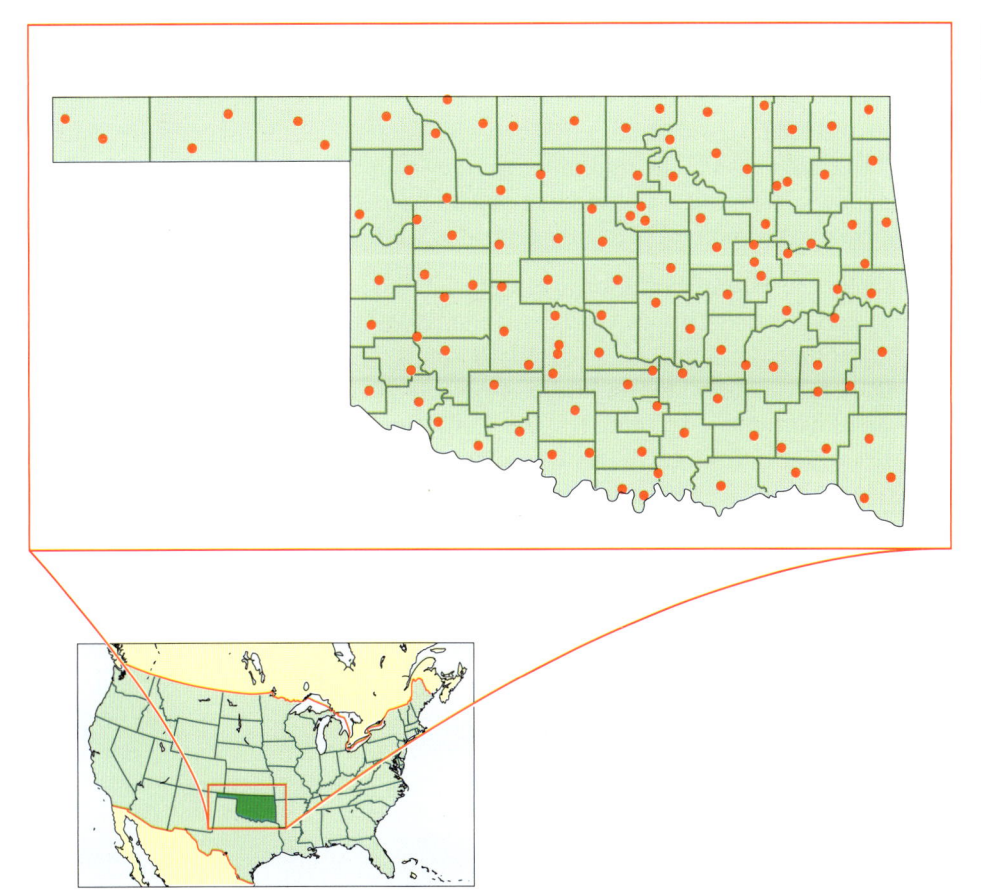

Figure 4.8 The Oklahoma Mesonet of weather stations (shown as red circles).

Forecasters can also have access to frequent, up-to-the-minute satellite images from the US GOES-East geostationary satellite. If a very high risk of tornadic storms is predicted, the forecasters in Norman can request that the satellite switches to its 'rapid scan' mode, when it images the entire USA at 7.5-minute intervals. It can focus instead on a more limited chunk of the country even more frequently. These scan modes are critical for two reasons. First, the storms that generate tornadoes 'explode' so quickly that an image every 15 or 30 minutes is really not satisfactory; hence the very frequent images. Second, the satellite cloud and water vapour maps capture the presence of storm-associated clouds before they start to precipitate. This means that they are detected by satellite imaging before they are by precipitation radar and by the surface raingauge network.

On the ground, the National Weather Service has a network of Doppler radars that map precipitation and surface, lower and mid-tropospheric convergence and rotation; both convergence and rotation are evidence of potential tornadic storms. These radars work by mapping the movement of particles that serve as tracers of the flow in which they are embedded. The particles could be precipitation but could also be dry matter, so they work as an aid even when it's not raining. The radar maps the component of flow towards or away from its antenna, and a network of such radars is used to map the three-dimensional structure of storms or, just as importantly, the pre-storm environment.

Radar networks can map zones of convergence and divergence over a wide region and in considerable depth within the troposphere. This means that they can spot areas where developments are likely to occur, for example in association with a line of convergence at and near the surface. Over the last ten years, the introduction of improved radar observations, together with the parallel training of forecasters, has led to an increase in the minimum warning time of a tornado strike from five to 12 minutes. This doesn't sound like much of an improvement, but the addition of five or so extra minutes is clearly critical in such situations.

Figure 4.9 is an example of a warning issued in the middle of March, 2001 by the National Weather Service Office in Tallahassee, Florida. This was issued at 06.04 local time (EST or Eastern Standard Time). The Tallahasee office houses a Doppler radar that provided an image of reflectivity for 06.45 EST for the duration of the tornado warning (Figure 4.10). Strong reflection of the emitted radar beam comes from relatively heavy precipitation, and weaker reflection from relatively lighter precipitation.

All these observations together, combined with radiosonde balloon data, offer forecasters the opportunity to monitor the development of weather on a day when there is a high risk of tornadoes. In areas of the US High Plains that are susceptible, there is in addition a network of voluntary storm-spotters. These people are trained individuals who report sightings of funnel clouds to the National Weather Service, for example. Within this so-called 'Project Skywarn', they report information such as time, location, track direction and intensity of any damage (if possible), so that a live picture can be built up of where the activity is occurring.

It's all very well being supplied with all these types of data, but what the forecasters must do is keep a cool appreciation of the data and monitor the outbreak in all its detail. This is no mean feat, particularly when they have to issue warnings to clusters of counties within a State towards which the tornadoes are heading. In addition to those tornadoes already in existence, they have to worry about where and when new ones might pop up. It is clear that the system of monitoring and transmitting warnings to those at risk has to run very smoothly indeed.

```
WFUS52 KTAE 151104

TORTLH
GAC087-131-151145-

BULLETIN - EAS ACTIVATION REQUESTED
TORNADO WARNING
NATIONAL WEATHER SERVICE TALLAHASSEE FL
604 AM EST THU MAR 15 2001

THE NATIONAL WEATHER SERVICE IN TALLAHASSEE HAS ISSUED A

* TORNADO WARNING FOR...
   EAST CENTRAL DECATUR COUNTY IN GEORGIA
   CENTRAL GRADY COUNTY IN GEORGIA
   THIS INCLUDES CAIRO...

* UNTIL 645 AM EST...

* AT 600 AM EST NATIONAL WEATHER SERVICE DOPPLER RADAR INDICATED A
  DEVELOPING TORNADO 13 MILES SOUTHWEST OF WHIGHAM...OR ABOUT 9 MILES
  SOUTHEAST OF BRAINBRIDGE...MOVING NORTHEAST AT 50 MPH.

* THE TORNADO WILL BE NEAR...
   WHIGHAM BY 615 AM EST...
   CAIRO BY 620 EST...

A TORNADO WATCH REMAINS IN EFFECT UNTIL 800 AM EST THURSDAY MORNING
FOR ALABAMA AND FLORIDA AND GEORGIA.

IF YOU ARE IN THE PATH OF THIS TORNADO...ABANDON YOUR MOBILE HOME FOR
A REINFORCED SHELTER.

LAT...LON 3084 8459 3075 8452 3084 8411 3107 8419

.END

12-WATSON
```

Figure 4.9 An official tornado warning.

Topic 1 Extreme Weather

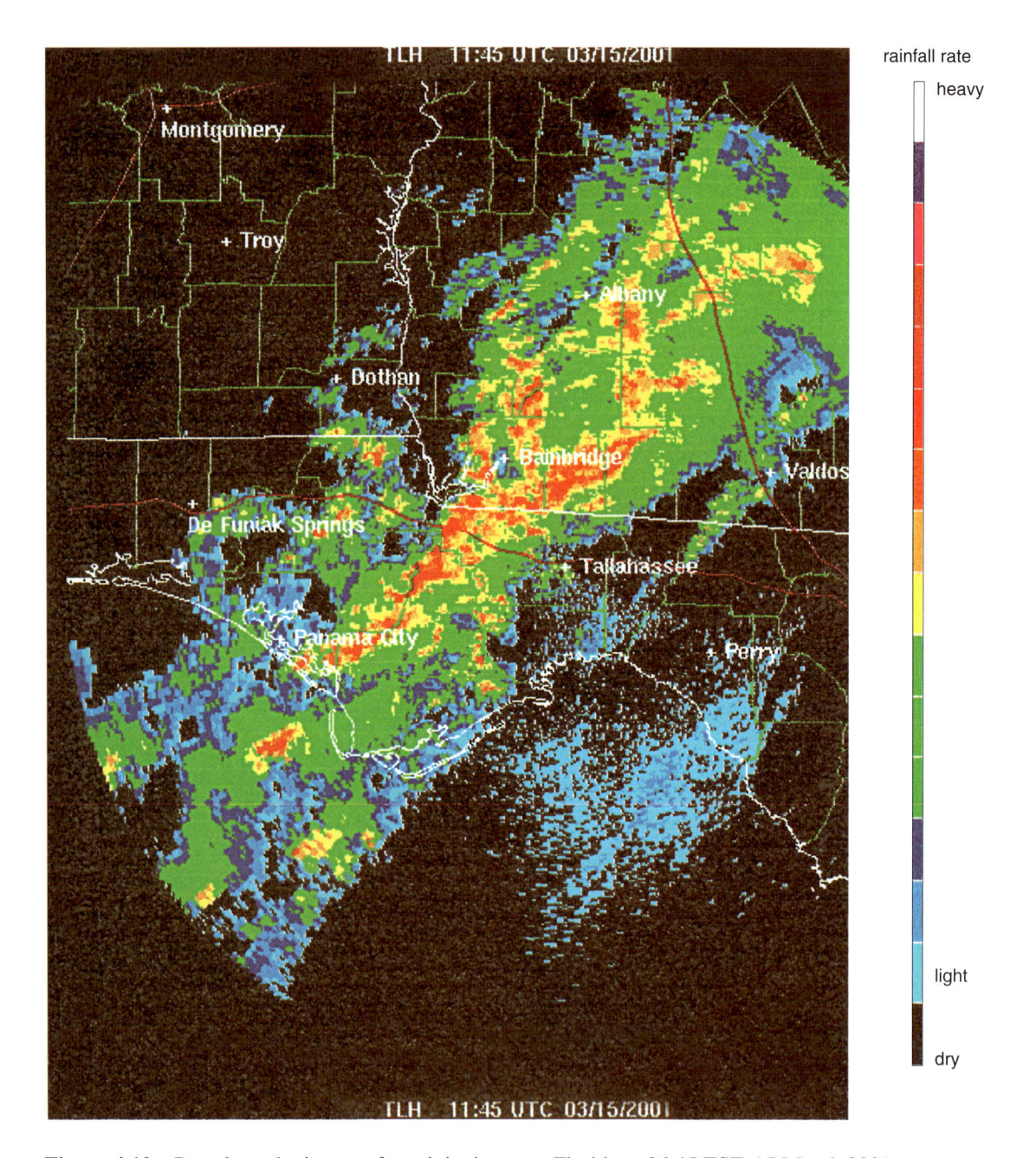

Figure 4.10 Doppler radar image of precipitation over Florida at 06.45 EST, 15 March 2001.

4.3.1 What's the damage?

Contrary to popular belief, a tornado does not do its damage because the extremely low pressure at its centre causes a property or other structure to explode. Damage to property and fatalities are caused by the extreme dynamic pressure force associated with very strong winds blowing against structures, *and* by exceedingly dangerous missiles flying fast within the tornado's circulation.

In the USA, tornadoes are ranked third with respect to the annual average number of fatalities caused by extreme weather, following floods and lightning.

Over four recent decades, the annual average number of tornado deaths was 75 (Table 4.3). Tornado forecasters have access to ever-improving surface, upper air, satellite and radar observations, which are fed to them almost instantaneously, so they can keep tabs on how severe storms are forming and evolving. The forecast models are being improved as a continuing process. For example, results of research into the role of the larger-scale environment in the genesis of severe storms are now incorporated into the models.

Not only are the monitoring, prediction and warning systems improving, but the American public are very much aware of what to do in a tornado outbreak — not by chance, but as part of a concerted federal and State effort to heighten both the awareness and preparedness of the public. The other important factor is the ease of communication within the USA; getting the warning across quickly is not a problem, unless tornadoes strike at night, which they occasionally do. There are plans to supply all households in tornado-prone regions with a special radio that can be switched on remotely (automatically) in order to broadcast warnings.

Table 4.3 Annual average US tornado-related deaths for different decades of the 20th century.

Period	Annual average US fatalities
1940–49	154
1950–59	135
1960–69	94
1970–79	99
1980–89	52
1990–99	56

○ What factors could have contributed to the significant decline in the number of tornado-related deaths in the USA?

● It is possible that there has been a decline in the number of tornadoes across the USA over this period. This is not correct, however. The decline is associated with a number of factors related to how weather prediction and society in general have changed over these many decades.

4.3.2 Oklahoma's worst tornado outbreak

On 3 May 1999, more than 50 tornadoes swept across part of the State of Oklahoma, killing 40 people that day, mainly in and to the southwest of Oklahoma City. The forecasters did well, issuing timely warnings for the pertinent counties as the massive outbreak evolved, which was mainly during the afternoon and into the evening. Unfortunately for the generally vast, thinly populated State, the major tornadoes ran through Oklahoma City and its suburbs.

The killer tornadoes grew from a massive storm cell that had earlier produced a crop of F3 tornadoes to the southwest of the State capital. One landed as an F1 near the town of Chikasha and sped northeastwards towards the large metropolitan area, intensifying to an F4, then declining to an F3 before hitting the suburb of Bridge Creek as an F5. It completely destroyed 680 homes there. During the evening, a second, 1.5 km-wide tornado was spawned near the Canadian River and smashed as an F5 through the community of Moore. 1225 houses, 274 apartments, 50 businesses, two schools and some churches were razed.

In addition, 4000–5000 buildings were damaged in an event that is estimated to have cost $1 000 000 000. The tornadoes' ferocity was such that even the most substantial buildings were destroyed; for many people, there was quite literally no safe shelter.

There is a video sequence showing the origin and nature of tornadoes and the damage they can cause in Section 2 of *Extreme weather* on the DVD.

5 Summary of Topic 1

1. Very rapid, or explosive, frontal cyclones occur occasionally in middle latitudes. The unusual intensity of the October 1987 storm that swept across Britain and Ireland, the worst for over 200 years in parts of southern and eastern England, was not well predicted. This was due in part to insufficient observations in the critical region at the critical time.

2. Since the October 1987 storm, meteorologists have gained insight into the important role played by relatively thin filaments of dry stratospheric air that become engaged with some developing frontal cyclones — over the North Atlantic, for example. Such dry intrusions plunge down from the stratosphere, enhancing horizontal pressure gradients and therefore winds at the surface. They can sometimes descend to the lower troposphere and are known to have been associated with similarly-sized areas of very strong surface gusts.

3. Meteorologists employ strict wind-speed-based criteria to define hurricanes and the like, extending also to pre-hurricane-intensity disturbances, i.e. tropical depressions and tropical storms. Additionally, they use a damage-related scale to express the intensity of a hurricane by a number from 1 to 5 (the Saffir–Simpson scale).

4. 'Hurricane', 'typhoon' and 'cyclone' are names for the same rotating tropical weather phenomenon in different parts of the world. The formation of this weather phenomenon depends in part on the sea-surface temperature (above 26 °C), the requirement to be at least some 5° of latitude away from the Equator, and the presence of moist air in the middle troposphere, which minimizes the risk of erosion of the all-important deep convective clouds.

5. Tornadoes are rapidly rotating features, typically 50–100 m across, that last for several minutes. With wind speeds of 80–140 m s^{-1}, they are the most destructive weather phenomena. Their wind speed and damage-causing ability are used to classify them on the Fujita F scale (1–5).

Learning outcomes for Topic 1

After working through this topic you should be able to:

1. Appreciate the role of stratospheric dry intrusions in the explosive development of frontal cyclones in middle latitudes, and the importance of good quality observations in predicting these and other extreme weather phenomena.

2. Identify tell-tale signs of such intrusions on a weather satellite water vapour image, and explain why the feature is associated with localized but very gusty weather at the Earth's surface.

3. Understand why frontal systems are the source of the most widespread severe weather (particularly gales and heavy, flood-producing rain) in Britain.

4. Identify and follow a developing frontal system across the North Atlantic through studying a video clip of thermal infrared images from the Meteosat geostationary satellite. (*DVD, Section 3*)

5. Define the terms 'hurricane' (typhoon or cyclone) and 'hurricane intensity' and describe the basic structure of a hurricane and the factors that promote its formation. (*Questions 3.1 and 3.2*)

6. Explain the reasons for the seasonality and geographical variability of hurricanes, and appreciate the nature and origin of the principal dangers they pose to society.

7. Appreciate the problems of forecasting the track and intensity changes of hurricanes, and the implications this has for evacuating very large numbers of people from susceptible coasts, in southeast USA, for example.

8. Given appropriate information, detect the presence (or absence) of trends in hurricane frequency in a given region. (*Activity 3.2*)

9. Using appropriate materials (text, diagrams, video, etc.) identify details of the track, evolution, rain pattern, damage production and decline of a given hurricane. (*Activities 3.3–3.4 and DVD, Section 1*)

10. Appreciate the typical scale and longevity of a tornado. (*Activity 4.1 and DVD, Section 2*)

11. Understand the way in which tornadoes are dependent on the much larger-scale atmospheric circulation and thermal structure across the High Plains of the USA.

12. Know how potential tornadic storms are monitored, both by using sophisticated space- and ground-based instrumentation and by video evidence, partly from scientific tornado-chasers; be familiar with the Fujita F scale used to classify tornadoes operationally in the USA. (*Activity 4.1 and DVD, Section 2*)

13. Describe and explain the nature and extent of damage created by a tornado, and understand how tornadoes are predicted. (*DVD, Section 2*)

14. By studying the Andover tornado in Kansas, give an account of an actual tornado outbreak and its consequences. (*Activity 4.1*)

Topic 1 Extreme Weather

Comments on activities

Activity 3.1

Your graph of wind loading versus wind speed should look like Figure 3.12. The ratio of force 11 to force 7 wind loading is 68.9/17.7 = 3.89 and the ratio of maximum wind speeds is 33/17 = 1.94. The wind loading increases dramatically with wind speed. In fact, the wind loading is proportional to the square of the wind speed. For example, $(33^2)/(17^2) = (1.94)^2 = 3.76$, which is close to the corresponding wind loading ratio of 3.89. The graph of wind loading against wind speed squared would therefore be a straight line.

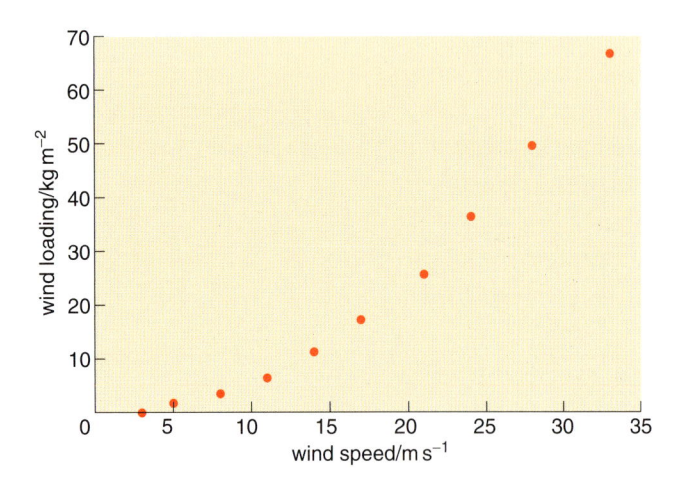

Figure 3.12 Plot of the wind loading and wind speed data in Table 3.2.

Activity 3.2

(a) Your bar graph of hurricane frequency should look like Figure 3.13.

(b) Both the total number and the number of major events show no increasing or decreasing trend over the period shown, i.e. the fluctuations in these data appear random. (Insurance companies are nevertheless concerned that, with global warming, the number of major hurricanes likely to hit the USA will increase.)

(c) (i) From Table 3.6, 157 hurricanes crossed the USA in 97 years, which is an average of about three hurricanes every two years. (ii) The number of major systems over the 97-year period was 62, i.e. roughly one every two years.

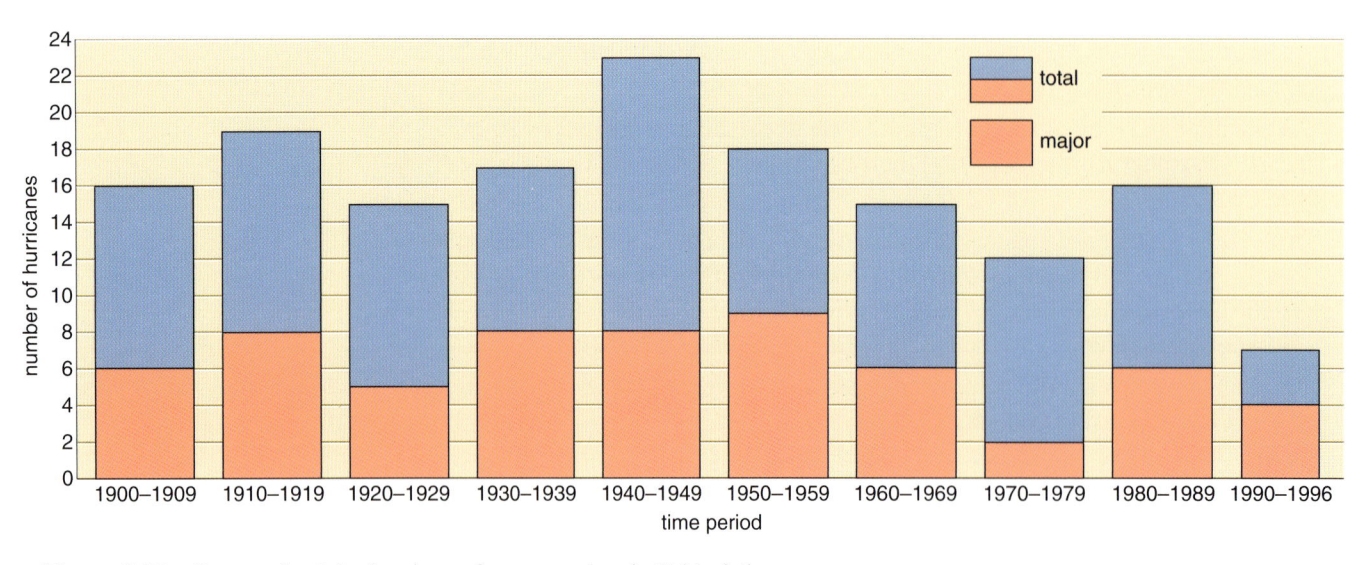

Figure 3.13 Bar graph of the hurricane frequency data in Table 3.6.

Activity 3.3

Figure 3.14 is a scatterplot of the data from Table 3.8. The lower the central pressure, the greater the difference between it and the 'normal' air pressure outside the hurricane and hence the greater the strength of the surface wind. Category 3 corresponds to wind speeds of 96–114 kn (Table 3.5), which is a central pressure range of about 940–960 hPa on Figure 3.14.

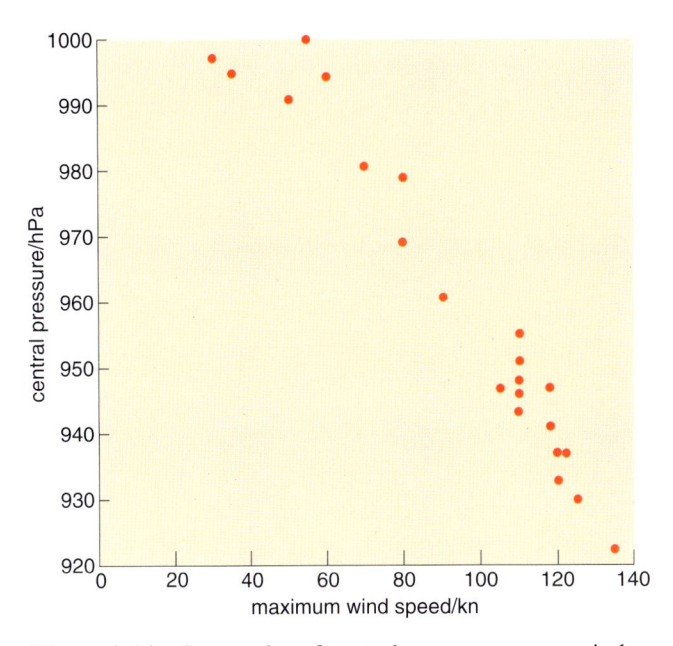

Figure 3.14 Scatterplot of central pressure versus wind speed data for Hurricane Andrew (from Table 3.8).

Activity 3.4

The duration of the torrential rain (and severe winds) of the eyewall in Homestead was 20×10^3 m/(8 m s^{-1}) = 2.5×10^3 s = 42 minutes.

Activity 4.1

(a) The duration of the tornado's track across Kansas was 73 minutes (from 17.57 until 19.10, local time).

(b) The approximate proportion of the tornado's lifetime spent producing severe damage or worse was the number of F values of 3 or greater recorded along the tornado's track divided by the total number of recordings: 14/25 = 0.56 = 56%.

(c) The length of the tornado path was approximately 80 km, so its average speed was
80×10^3 m/(73 × 60 s) = 20 m s^{-1}.

(d) The width of the tornado's damage swathe when it hit the Mobile Home Park was approximately 600 m.

Answers to questions

Question 3.1

The warmth of the sea surface is a very important part of the recipe for the formation and persistence of hurricanes, typhoons and cyclones (the same phenomenon with a different name, depending on the ocean).

We know that a surface value of 26 °C or warmer, to a depth of some 60 m, is critical in that water that is warm is very likely to spark off deep convective clouds. Also, it means that vast quantities of water vapour are evaporated from the warm ocean to provide significant energy from latent heat to the growing disturbance, when the vapour condenses into cloud droplets. If the sea surface is cooler than this, convection is much less likely and the supply of water vapour is correspondingly smaller.

Sea-surface temperature varies markedly within the tropical oceans (Figure 3.2a). The South Atlantic and the South-East Pacific are both partly supplied by marked cool currents that flow towards the Equator along the western coasts of Southern Africa and South America. These are the Benguela and Humboldt Current respectively (Figure 3.2b). The transport of cool water by these currents is such that both ocean areas are anomalously cold for their latitude, so much so that it's too cool to support the development of hurricanes and the like.

Question 3.2

Anticyclonic outflow (clockwise in the Northern Hemisphere) is a critical component in the uppermost few kilometres of hurricanes, typhoons and cyclones. This is so because it is the way in which mass is exported from the extensive column of air that represents the body of the hurricane. The prime source of mass for the system is the powerful cyclonic (anticlockwise in the Northern Hemisphere) inflow that occurs in the lowest couple of kilometres above the sea surface.

If, then, the mass flowing out aloft is greater than the mass flowing in at low levels, within the column that is the hurricane, then the surface pressure will fall, and the surface winds are likely to strengthen. It is therefore very important to be able to quantify the extent and strength of the inflow and outflow.

The hurricane's winds are sensed in the main by the Hercules aircraft of the Hurricane Hunters. At the highest of altitudes, however, high-flying jet aircraft are occasionally used to monitor the pattern and intensity of the outflow. This knowledge is crucial in the process of aiding the prediction of whether the surface central pressure will change, and therefore if the winds will change too.

Weather satellites image clouds. The geostationary ones image their patterns every 30 minutes, which means that it's possible to study the motion of the cirriform clouds that mark the upper outflow of air. The gradual outward-spiralling pattern of these icy, filament-like clouds acts as a tracer of the flow and thus permits an estimate to be made of the mass flow too.

Acknowledgements for Topic 1
Extreme Weather

Grateful acknowledgement is made to the following sources for permission to reproduce material in this book:

Figure 1.1: © 2002 Freefoto.com Inc.; *Figures 1.2 and 2.4*: © 2002 EUMETSAT; *Figure 2.2*: © NERC Satellite Station, University of Dundee; *Figure 3.2*: B. W. Meeson *et al.* (1995) *ISLSCP Initiative 1 — Global Data Sets for Land–Atmosphere Models, 1987–1988*, vols 1–5, published on CD by NASA; *Figure 3.7*: © 1998 Honduras.com; *Figure 3.9*: Official US Air Force photo, © hurricanehunters.com; *Figures 3.10, 3.11*: © NOAA; *Figure 4.1*: © NSSL/NOAA; *Figure 4.8*: Adapted from a NOAA figure; *Figures 4.9, 4.10*: © National Weather Service/NOAA.

Every effort has been made to trace all the copyright owners, but if any has been inadvertently overlooked, the publishers will be pleased to make the necessary arrangements at the first opportunity.

TOPIC 2
ATMOSPHERIC CHEMISTRY and POLLUTION

Kiki Warr

1	Introduction	51
2	**The natural components of the air: sources and sinks**	**53**
2.1	Introduction	53
2.2	Surface sources	55
2.3	Atmospheric sources and sinks	59
2.4	Surface sinks	60
2.5	Taking stock and looking forward	61
2.6	Summary of Section 2	63
3	**Chemistry driven by the Sun**	**65**
3.1	Photons and photochemistry	67
3.2	Photochemistry in the stratosphere: the ozone layer	71
3.3	Photochemistry in the troposphere	74
3.4	Summary of Section 3	80
4	**Atmospheric budgets: local and global**	**83**
4.1	Atmospheric lifetimes	84
4.2	Lifetimes and atmospheric transport	86
4.3	The influence of human activities	89
4.4	Summary of Section 4	95
5	**Ozone air pollution**	**96**
5.1	Introduction: evidence of a perturbed atmosphere	96
5.2	Ozone and photochemical smog	98
5.3	The role of thermal inversions	106
5.4	A regional perspective	109
5.5	Ozone exposure: air quality standards	111

5.6	**Controlling ozone pollution**	**116**
5.7	**Summary of Section 5**	**124**

Learning outcomes for Topic 2   125

Comments on activities   126

Answers to questions   129

Acknowledgements   133

Introduction

Figure 1.1 A photograph taken from the Space Shuttle showing the Earth's atmosphere, with clouds silhouetted by the Sun against the pearly-white, aerosol-rich lower atmosphere (the troposphere). Molecular scattering higher up gives the blue layer.

The Earth supports a rich variety of life, protected and nourished by a thin film of gas (Figure 1.1) with a composition that is unique amongst the planets of the solar system. Oxygen, one of the two main constituents, clearly plays a pivotal role. We suffocate if the air we breathe contains less than about 5–10% of oxygen. But the atmosphere also contains smaller amounts of a host of other gases and many of these are also crucially important. Familiar examples include water vapour and carbon dioxide, CO_2 — the two greenhouse gases that are largely responsible for maintaining surface temperatures suitable for life.

From the perspective of the total 'Earth system' developed earlier in the course, these and other so-called 'trace gases' are part of a continual cycling of material that links the atmosphere with the Earth's surface — its soils, rocks and sediments, oceans and freshwaters, and all the plants, animals and microbes that inhabit these various environments. Thus, evaporation supplies water vapour to the atmosphere; winds carry moisture around; cloud formation (evident in Figure 1.1) and rainfall return it to the ground. Likewise, atmospheric CO_2 is central to the functioning of the global carbon cycle. Trace gases containing nitrogen and sulfur contribute to the biogeochemical cycling of these two elements.

Here we shift the emphasis, and focus attention on the life cycles of *individual* trace gases. What processes control their atmospheric concentrations? Some of the answers are already familiar. Like water vapour and CO_2, most trace gases have sources at the Earth's surface. But the vast majority of trace species differ from these two constituents in one important respect: they are removed from the air by *chemical* interactions *within* the atmosphere. And some atmospheric gases, notably ozone, are also generated in this way. This unseen ferment of chemical activity is the stuff of **atmospheric chemistry**.

Topic 2 Atmospheric Chemistry and Pollution

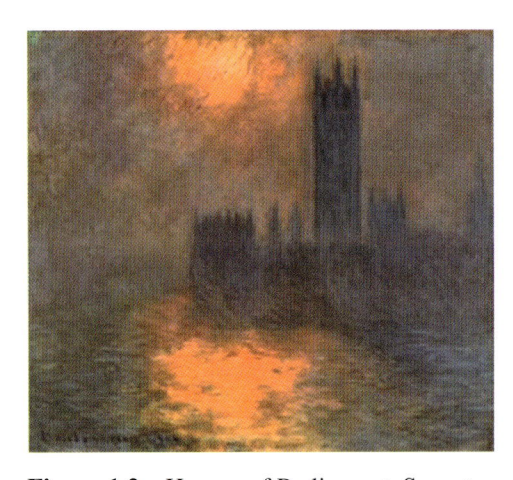

Figure 1.2 Houses of Parliament, Sunset 1904 by Claude Monet, one of a series of paintings inspired by his many visits to London during the foggy winter months. To Monet: 'Without the fog London would not be a beautiful city … its regular and massive blocks become grandiose in that mantle …'

The main aim of this topic is to provide an introduction to atmospheric chemistry, and the part it plays in maintaining the complex chemical composition of the 'natural' atmosphere. That will take us through to the end of Section 4. We then turn our attention to the influence of human activities. It is now evident that the latter can modify the atmosphere substantially. By the end of the twentieth century, *global* environmental issues associated with such changes were well established on the world's agendas. Almost daily, we read about the potential consequences for climate of the on-going accumulation of atmospheric CO_2.

On a more local or regional scale, however, air pollution has been a problem for centuries. To this day, we can relate to the views of the twelfth century Hebrew philosopher and scientist Moses Maimonides (1135–1204):

> Comparing the air of cities to the air of deserts and arid lands is like comparing waters that are befouled and turbid to waters that are fine and pure. In the city, because of the height of its buildings, the narrowness of its streets, and all that pours forth from its inhabitants and their superfluities … the air becomes stagnant, turbid, thick, misty and foggy …
>
> Quoted in V. Goodhill, 'Maimonides — Modern Medical Relevance', *Transactions of the American Academy of Ophthalmology and Otolaryngology*, p. 463, May–June, 1971.

More recent manifestations of air pollution range from the smokes and fogs of industrial cities (Figure 1.2), through the summer smogs that began to afflict southern California during the 1940s (Figure 1.3), to acid deposition (discussed elsewhere in the course).

Atmospheric chemistry plays a part in all these issues, and we shall touch on several of them. However, the one we consider in some detail in Section 5 is the generation of ozone in photochemical smog — a problem that afflicts cities and their environs in many parts of the world. As we shall see, ozone turns out to be a key player in the chemistry of the *unpolluted* troposphere. That chemistry, in turn, provides vital clues to the processes that are responsible for enhanced ozone concentrations in many *polluted* environments. A sound scientific understanding of these processes is essential to the formulation of effective strategies against this pervasive and damaging form of air pollution.

Figure 1.3 Los Angeles after formation of 'photochemical smog' — a noxious brew of chemicals (including ozone) and particles. Such episodes result from the action of sunlight on the mix of gaseous pollutants in vehicle exhaust fumes.

The natural components of the air: sources and sinks

2.1 Introduction

The bulk composition of Earth's atmosphere can be summarized succinctly. Leaving aside highly variable amounts of water vapour, more than 99.9% of the gases that make up the atmosphere are of just three kinds: molecular nitrogen (N_2) and oxygen (O_2), and the chemically inert noble gas argon (Ar, present as individual atoms). These major components are listed in Table 2.1, along with a selection of the minor constituents — or **trace gases** — that together make up the remaining 0.1% of the atmospheric gas. The entry for each gas also records a value for its **mixing ratio**, here taken to mean the 'mixing ratio *by volume*'. (Bear in mind, however, that mixing ratios *by mass* are often used in other areas of environmental science.) Consult Box 2.1 if you want to refresh your memory about the significance of the term mixing ratio in an atmospheric context or about the units — parts per million (ppmv), for example — that are used to report values for gases present in trace amounts.

Table 2.1 A selection of the gases naturally present in the Earth's atmosphere, and the current (mid-1990s) average mixing ratio of each in dry air at ground level in remote regions.

Gas	Formula	Mixing ratio
Major constituents		
nitrogen	N_2	0.781
oxygen	O_2	0.209
argon	Ar	0.0093
Trace gases		
ozone	O_3	5–30 ppbv
carbon compounds		
carbon dioxide	CO_2	365 ppmv
carbon monoxide	CO	30–100 ppbv
methane	CH_4	1.72 ppmv
nitrogen compounds		
nitrous oxide	N_2O	312 ppbv
nitric oxide	NO	5–20 pptv
nitrogen dioxide	NO_2	5–20 pptv
ammonia	NH_3	1 ppbv
sulfur compounds		
sulfur dioxide	SO_2	0.2 ppbv
hydrogen sulfide	H_2S	0.05 ppbv

Box 2.1 Measures of atmospheric composition

Atmospheric scientists commonly refer to the mixing ratio (by volume) as a measure of the atmospheric concentration of a given constituent, even though the mixing ratio is not, strictly, a measure of concentration at all! Taking O_2 as an example, the formal definition goes as follows:

$$\text{mixing ratio} = \frac{N(O_2)/V}{N(\text{total})/V} = \frac{N(O_2)}{N(\text{total})} \tag{2.1}$$

Here, $N(\text{total})/V$ is the **number density** of air (the total number of molecules per unit volume V of air, commonly taken to be $1\,\text{cm}^3$), and $N(O_2)/V$ is the number density of oxygen (the number of molecules of O_2 per unit volume).

Thus, the mixing ratio is actually a *ratio* of concentrations: strictly, it tells us about the 'fractional abundance' or proportion of a given species. Expressing the value as a number (necessarily less than 1) or as a percentage (by multiplying by 100) is fine for the major atmospheric constituents (see the entries in Table 2.1), but it becomes unwieldy for trace gases. The mixing ratios of such species are usually recorded as ppmv (**parts per million**, 10^6) or as ppbv (**parts per billion**, 10^9) — or even as pptv (**parts per trillion**, 10^{12}) for the least abundant species. (Note that the abbreviation ppt sometimes means parts per *thousand* in other areas of environmental science.)

○ In Table 2.1, the mixing ratio of nitrous oxide (N_2O) is given as 312 ppbv. Express this value as a fractional abundance (in scientific notation), then in ppmv and in pptv.

● A value of 312 ppbv means that in every billion molecules of air, 312 will, on average, be molecules of N_2O. So 312 ppbv is equivalent to:

$$\frac{312}{10^9} = 3.12 \times 10^{-7} \text{ or } 0.312 \times 10^{-6} \text{ (0.312 ppmv)}$$

or $312\,000 \times 10^{-12}$ (312 000 pptv).

Table 2.1 refers to *dry* air. Of course, the atmosphere also contains water vapour, present in amounts which typically range from 0.5 to 3% at ground level, depending on temperature and relative humidity. Analysis of air samples reveals trace amounts of hundreds of other substances. Yet more are found in atmospheric aerosols (fine solid particles and liquid droplets suspended in the air), or dissolved in rain and cloud water. We shall meet a few of these many compounds later on.

The concentrations of trace gases in the atmosphere can be very variable in space and over time. All those included in Table 2.1 are now commonly regarded as air 'pollutants', in one sense or another. On a local or global scale, their atmospheric concentrations are increased by a wide range of human activities, especially (but not exclusively) by the burning of fossil fuels — coal, oil and natural gas. Indeed, this was one reason for making this particular selection of trace gases. However, they are all *natural* components of the air, present long before human population growth and the coming of the industrial age began to make their mark. For example, the list of natural greenhouse gases also includes methane, nitrous oxide and ozone.

The natural components of the air: sources and sinks

To begin our study of atmospheric chemistry, we put the influence of human activities to one side, and concentrate on the processes that govern the composition of the natural atmosphere. This is still a bit of a moveable feast as far as the trace-constituent species are concerned, but we shall take the concentrations recorded in Table 2.1 to be representative of what we shall refer to as the **background atmosphere**: the values are typical of those found in air sampled in the more remote, and relatively unpolluted, corners of the world (Figure 2.1).

Our immediate interest is to follow the general path of trace constituents from their **sources** to their **sinks**. You have met this kind of language already, in the context of the global carbon cycle for example, where a crucial ingredient is the cycling of atmospheric CO_2 to and from the Earth's surface.

Figure 2.1 The Cape Grim Baseline Air Pollution Station in Tasmania, established by the Australian government in 1976. Tasmania is an ideal location for monitoring background levels of trace gases in the Southern Oceans maritime air mass. For much of the year, the island is battered by the 'roaring forties' — strong westerly winds that carry pollution-free air thousands of kilometres across the southern oceans.

Question 2.1

Think back to your study of the carbon cycle. List the main natural sources and sinks of atmospheric CO_2.

Carbon dioxide is chemically unreactive in the atmosphere, and so there is little to add here about the processes that govern its atmospheric concentration. But to a greater or lesser extent, all the other trace gases in Table 2.1 do take part in chemical reactions *within* the atmosphere. As a result, their life cycles are fundamentally different from that of CO_2.

2.2 Surface sources

The nitrogen and sulfur-containing gases in Table 2.1 all play a part in the biogeochemical cycling of these two elements between the atmosphere and the Earth's surface. Their main *surface* sources (identified earlier in the course) are summarized in Table 2.2, along with comparable information for methane (CH_4) and carbon monoxide (CO).

Table 2.2 Major natural sources at the Earth's surface of selected trace gases in the atmosphere.

Gas	Natural sources
Carbon compounds	
CH_4	enteric fermentation in grazing animals; termites; wetlands (bogs, swamps, etc.); oceans
CO	forest fires; oceans
Nitrogen compounds	
N_2O	soils; oceans
NO, NO_2	forest fires; soils
NH_3	soils; animal wastes
Sulfur compounds	
SO_2	volcanic activity
H_2S	soils and sediments; volcanoes and fumaroles

○ If you had to collect the sources given in Table 2.2 into three broad categories, which ones would you choose?

● At first sight, there is a bewildering variety of sources, some of which (e.g. soils, oceans, wetlands, etc.) are not that easy to categorize. However, it would be a good (and correct) guess that emissions from these environments originate with the activities of living things. So the three broad categories are 'biological', 'volcanic' and 'forest fires'.

From our perspective, volcanic activity can be viewed as a form of *natural* air pollution. Explosive eruptions spew vast clouds of dust and gas high into the atmosphere. The most prevalent gaseous components are water vapour and CO_2 (though in amounts that are dwarfed by their other natural sources). However, volcanic emissions are *the* major natural source of sulfur dioxide (SO_2) at the Earth's surface (Figure 2.2), together with smaller quantities of other sulfur gases such as hydrogen sulfide (H_2S, notorious for its 'rotten eggs' smell).

Figure 2.2. Condensing steam and fumes rise from Mount Etna (Sicily), Europe's largest active volcano. It may release as much as 4×10^6 kg of SO_2 *per day*. Averaged over many years, the annual release from all the world's volcanoes amounts to some 10^{10} kg of sulfur (largely as SO_2). About 70% of this leaks passively; the rest comes from periodic explosive events.

Forest fires are another form of natural air pollution. Every year bolts of lightning start innumerable fires all over the world. They are a normal part of the healthy functioning of many forest ecosystems, and of savannah grasslands during the dry season in the subtropical belts of southern Africa (Figure 2.3) and South America. It is easy to lose sight of this today. Compelling images of tropical forests being deliberately hacked down and burnt, especially in Amazonia and parts of Southeast Asia, have (rightly) fuelled widespread concern about the environmental impacts of such wholesale destruction.

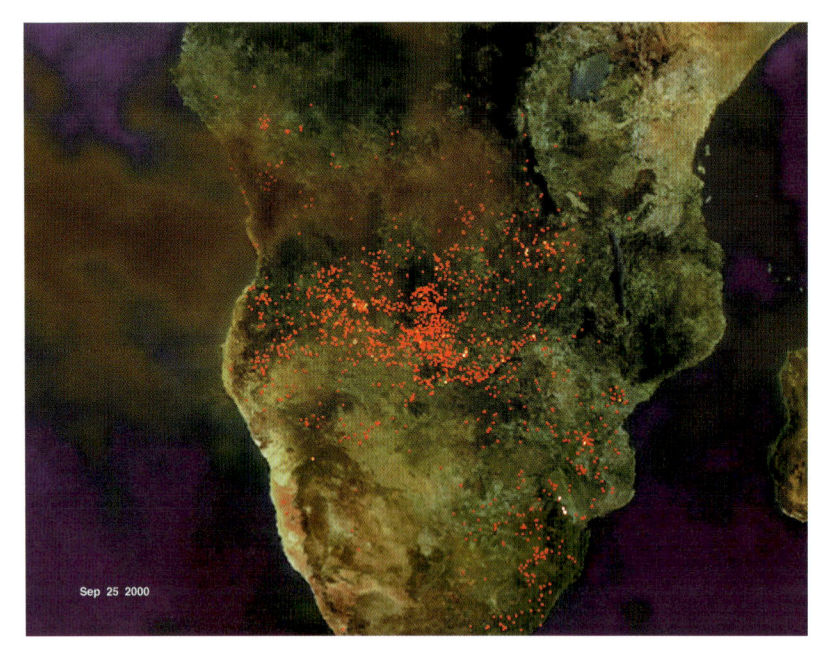

Figure 2.3 Fires routinely scorch the southern African landscape during the dry season, blackening an area larger than the UK, Germany, France, Belgium and the Netherlands combined. Throughout the region, natural wild fires are augmented by fires set deliberately for agriculture (to stimulate a flush of new grass for livestock), hunting, reducing pests, etc. August to October are the peak months, the region's heaviest burning being concentrated in the subtropical belt. Constructed from data collected by satellite-borne instruments, the image shows multiple fires burning across the region in September 2000, together with the large amount of particulate matter (overlaid on top) they generate.

Box 2.2 Biomass and respiration

As you saw earlier in the course, the elemental composition of all living things can be approximated by a single chemical formula, $(CH_2O)_n$, where n is a whole number. Strictly, the term **biomass** means the total quantity of living organisms in a given area, but the word is also used to describe dead organic matter (e.g. food).

There is little actual CH_2O in the real world (although there is some in the atmosphere; Section 2.3): CH_2O is a smelly and carcinogenic compound called formaldehyde. However, there is plenty of $(CH_2O)_6$, which is a sugar called glucose (usually written $C_6H_{12}O_6$). More generally we can regard biomass, $(CH_2O)_n$, as being synonymous with carbohydrates.

When animals digest biomass, they oxidize it, liberating CO_2 and the energy needed to synthesize other vital molecules. This controlled 'burning' of biomass is called **aerobic** (oxygen-using) **respiration**, expressed at its simplest as:

$$(CH_2O)_n + nO_2 = nCO_2 + nH_2O$$

Trees, grasses and scrub are 'biomass'(Box 2.2) — and so are the creatures unfortunate enough to be consumed by fires, however ignited. In principle, one might expect that burning biomass would result in the carbon it contains being converted to CO_2. It does, but not completely. High-temperature combustion of carbon-based fuels (petrol in a car engine is another example) is a complex process. In the heat of a forest fire, organic compounds in biomass are vaporized: some escape unchanged; others are literally torn apart, and may be stripped down to bare carbon atoms, or end up combined with varying amounts of oxygen. The upshot is that fires are a large source of soot and other particulate matter (detectable from space, Figure 2.3) as well as a variety of *partially* oxidized species, especially carbon monoxide (CO). On a global scale, fires contribute about half of the naturally emitted CO — and also some 25% of the nitric oxide (NO), mainly derived from the nitrogen-containing organic compounds in biomass. Burning biomass can also produce *reduced* species (such as CH_4), depending on the type of vegetation and the way it burns. (If you want to refresh your memory about oxidation and the converse reaction, reduction, consult Box 2.3).

Box 2.3 Oxidation and reduction

In the present context, the most useful definition of oxidation is exemplified by the combustion reaction that will be familiar if you use natural gas to keep your home warm or for cooking. The chief constituent of natural gas is methane, and (given a plentiful supply of oxygen) it burns to give carbon dioxide and water vapour:

$$CH_4 + 2O_2 = CO_2 + 2H_2O$$

We say that methane has been oxidized to CO_2 because *the proportion of oxygen atoms in the molecule has increased* — from zero to two.

○ What has happened to the proportion of hydrogen atoms in the methane molecule during this reaction?

● It has decreased — from four atoms per molecule (in CH_4) to zero per molecule (in CO_2).

This is another way of identifying an oxidation process — *the loss of hydrogen from a compound.*

Reduction is the converse of oxidation. In a reduction, the proportion of oxygen in a compound is reduced and/or the proportion of hydrogen is increased. In the combustion of methane, for example, oxygen in the air is reduced: some of it ends up combined with hydrogen to form water.

These definitions of oxidation and reduction can also be used to *compare* simple compounds of a given element. For example, carbon is said to be in a reduced state in CH_4, and in a highly oxidized state in CO_2. Carbon monoxide (CO) clearly lies somewhere in between: the proportion of oxygen is higher than that in CH_4, but lower than that in CO_2. CO can be described as a *partially oxidized* species. The terminology is not that precise, but it does reflect what happens if natural gas is burned in a gas fire or water heater with a blocked flue. Then a proportion of the CH_4 is partially oxidized to carbon monoxide, CO (rather than being fully oxidized to CO_2), and there is danger of a build-up of this deadly gas. Those who die in fires are usually killed by CO before their burns become fatal.

Question 2.2

Consider the nitrogen-containing species: NO, N_2O, N_2, NH_3, NO_2 and NO_3^- (the nitrate ion). Place these in order, starting with the least oxidized (most reduced) species and finishing with the most oxidized.

We come now to the influence of the biosphere. On land, and in the seas, the major biogenic sources of trace gases originate with the activities of countless microbes, which form an abundant living component in soils, leaf moulds and sediments. For example, ammonia (NH_3) is generated by the microbes that degrade the nitrogen-containing compounds in urine and other less savoury animal wastes. Like animals, many microbes derive their energy by oxidizing organic material (dead biomass) via aerobic respiration (Box 2.2). However, in **anoxic** (oxygen-poor) environments, which can develop in soils and sediments, there are communities of microbes that can oxidize biomass using the oxygen in nitrate ions, NO_3^-, (e.g. denitrifying bacteria) or sulfate ions, SO_4^{2-}, in place of free O_2. As a result, the nitrate or sulfate ions are reduced to N_2 or N_2O (often accompanied by NO) or to H_2S, respectively. Similar activities in marine environments result in emissions of these gases from the oceans.

Finally, the major natural sources of methane (CH_4) can all be traced back to the breakdown of organic matter by bacteria under **anaerobic** (oxygen-free) conditions (see Question 2.3 in Section 2.6) — principally in waterlogged soils (bogs, swamps and marshes; Figure 2.4) and in the intestines of many grazing animals. In that respect, digestion (known as enteric fermentation) in these animals effectively provides a mobile wetland source of CH_4; currently, their flatulence contributes an astonishing 16% of all methane emissions! Anaerobic decomposition in the hindgut of some termites and other insects also makes a small, but significant, contribution.

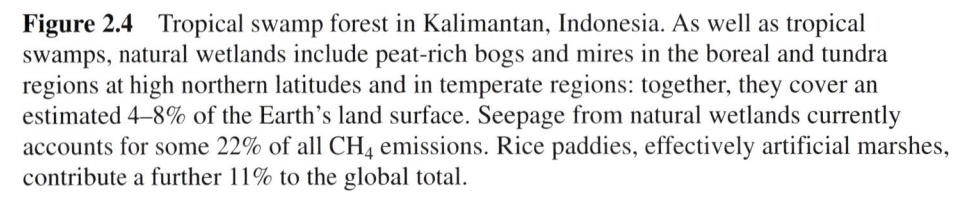

Figure 2.4 Tropical swamp forest in Kalimantan, Indonesia. As well as tropical swamps, natural wetlands include peat-rich bogs and mires in the boreal and tundra regions at high northern latitudes and in temperate regions: together, they cover an estimated 4–8% of the Earth's land surface. Seepage from natural wetlands currently accounts for some 22% of all CH_4 emissions. Rice paddies, effectively artificial marshes, contribute a further 11% to the global total.

To conclude this brief review of surface emissions, note that the list of trace gases with biogenic sources includes CH_4, H_2S, NH_3, N_2O and NO.

○ In view of the material on oxidation and reduction summarized in Box 2.3, how would you characterize these compounds?

● They are all *reduced* (CH_4, H_2S and NH_3), or only *partially* oxidized (N_2O and NO) species. Certainly, for each element there exists a familiar molecule or ion that is more highly oxidized, namely CO_2, SO_4^{2-} and NO_3^-.

In this sense, the other two gases in Table 2.2 (CO and SO_2) are also only partially oxidized species. It is not hard to imagine what the *chemical* fate of all these compounds might be once released into the atmosphere.

2.3 Atmospheric sources and sinks

The Earth's atmosphere contains 21% molecular oxygen, O_2; it provides an *oxidizing* environment. Not surprisingly, the reaction pathways in the atmosphere often involve the oxidation of reduced gases through to more oxidized forms. For our collection of carbon, nitrogen and sulfur trace gases, these pathways can be summarized in a schematic way, as shown in Figure 2.5.

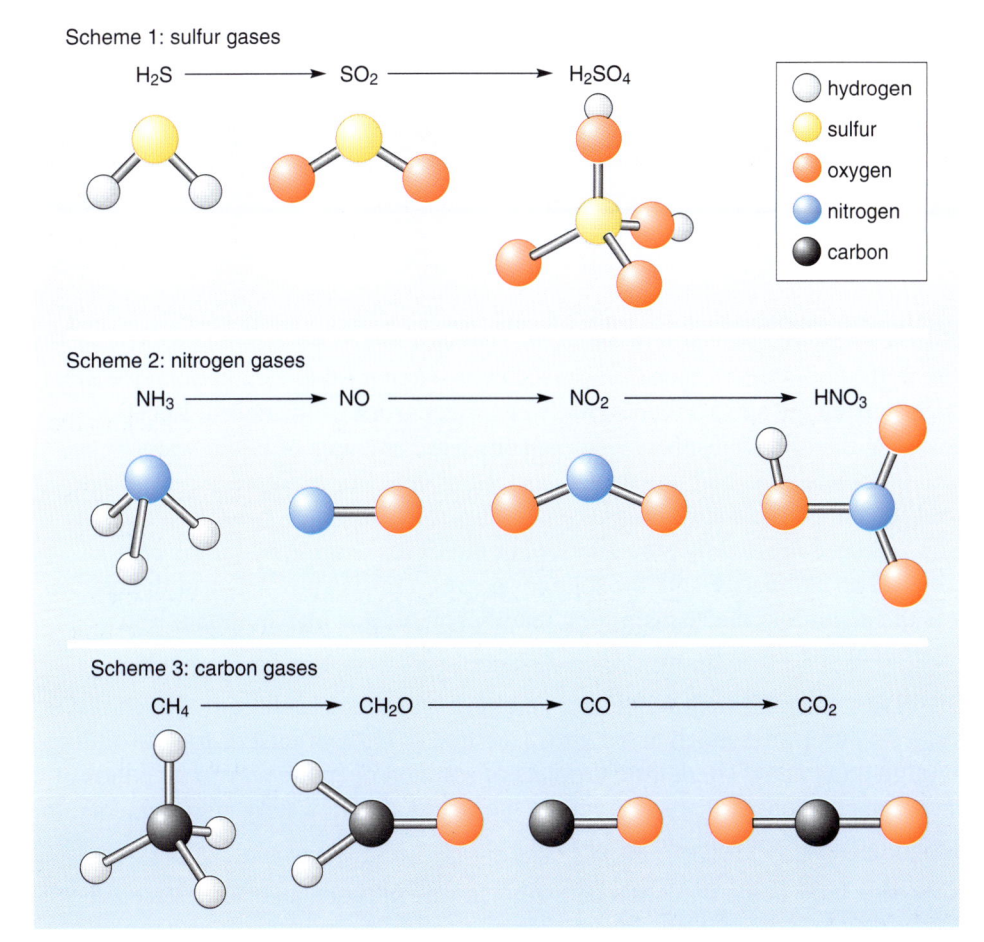

Figure 2.5 A schematic overview of the pathways whereby sulfur, nitrogen and carbon containing trace gases are converted into more oxidized species within the atmosphere. The figure includes a ball-and-stick representation of the molecule of each species; different colours have been used to distinguish the atoms of the various elements.

The point to note about these schemes is that *each* stage represents the destruction of one trace gas and the generation of another. Thus in Scheme 1, for example, the oxidation of H$_2$S to SO$_2$ acts as an *in situ* sink for the reactant (H$_2$S) and an *in situ* source of the product (SO$_2$). In practice, this is a relatively minor natural source of SO$_2$, compared with the amount pumped into the atmosphere by volcanic activity. But such atmospheric sources are important for some trace gases. One case in point is provided by the oxidation pathway of methane. As shown in Scheme 3, this proceeds via formaldehyde (CH$_2$O; Box 2.2) and carbon monoxide, and serves as a very significant natural source for both of these gases.

○ Think back to your study of the nitrogen cycle earlier in the course. Do you recall a further atmospheric source for one of the trace gases in Figure 2.5?

● Lightning can 'fix' atmospheric nitrogen as nitric oxide. At the high temperatures (they can be well over 3000 °C) generated by the shock waves from a lightning stroke, N$_2$ and O$_2$ in the air react rapidly to form NO.

2.4 Surface sinks

Chemical transformations within the atmosphere act as sinks for the reactants. But like CO$_2$, some are also removed by processes that deposit them on the Earth's surface. These processes are divided into 'dry deposition' and 'wet deposition'.

The term **dry deposition** refers to the transfer of a gaseous species directly from the air to the leaves of a plant, say, or a soil surface or body of water, and so on. Obviously, for this to occur the air containing the trace constituent has to be reasonably close to the surface in question. Even then, this direct transfer acts an effective sink only where some interaction with the surface ensures that the gas is removed permanently from the air. One might, for example, view the uptake of CO$_2$ by plant photosynthesis as a specific biological interaction of this kind. Likewise, microbial sinks are known for soil removal of carbon monoxide (CO). Another more general factor is the amount of moisture on a surface, since this will enhance the uptake of moderately soluble gases, such as SO$_2$.

The description 'dry' deposition is used even when a soluble compound transfers directly from the air to a thoroughly wet surface, such as the ocean. Recall, however, that the atmosphere also contains large amounts of liquid water — as cloud droplets, mist and fog, as well as falling rain. For example, a typical fair-weather cumulus cloud contains over 10^3 kg of liquid water. Incorporation into cloud water, followed by precipitation, or direct scavenging by falling rain provides an alternative sink for soluble species. This is known as **wet deposition**. It is the major loss mechanism for a readily soluble gas like ammonia, and a significant one for SO$_2$.

Dissolve a soluble gas in water, and you have an aqueous solution — even if it is just a tiny drop suspended in the air (in a cloud or bank of mist or fog), or falling through it (as rain). These little droplets of solution open the way to the kinds of chemical reactions that are common in the world's more watery environments. In particular, *ionic* species can be formed.

Now look back at the oxidation pathways for the nitrogen and sulfur gases in Figure 2.5. In each case, the most oxidized species is a substance (either nitric acid or sulfuric acid) that is probably more familiar as an aqueous solution, containing hydrogen ions (H$^+$) and either nitrate (NO$_3^-$) or sulfate (SO$_4^{2-}$) ions. In the atmosphere, these acids may be formed as molecular species (H$_2$SO$_4$ or HNO$_3$) that are readily soluble in water, or as ions in aqueous droplets (following the take up of a more reduced species like SO$_2$). Either way, the acids are rapidly removed by wet deposition. This, then, marks the end of the oxidation pathways for the nitrogen and sulfur gases. It also explains how emissions of these gases can lead eventually to **acid rain**. This idea is illustrated for the sulfur gases in Figure 2.6.

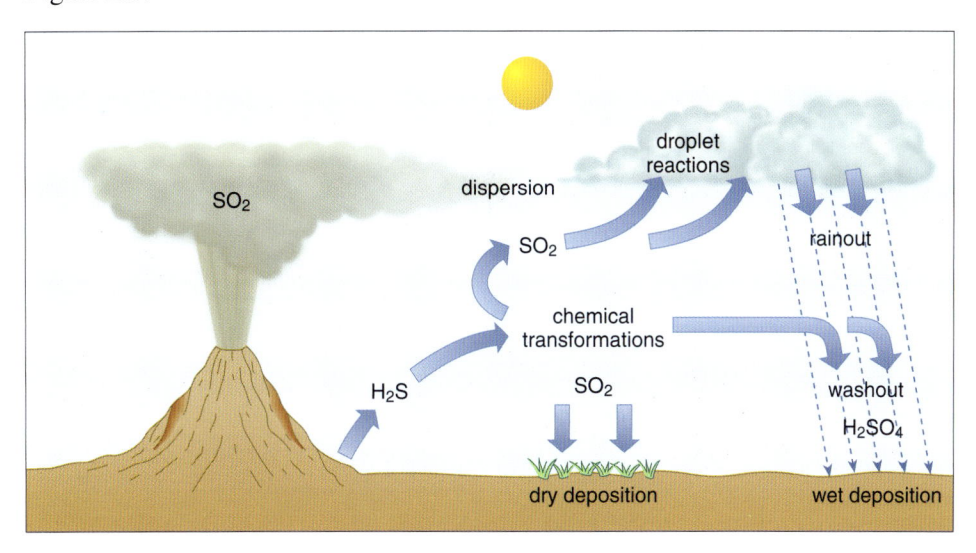

Figure 2.6 Processes which may be involved between the emission of sulfur gases (H$_2$S and SO$_2$) and their ultimate deposition to the ground as acid rain.

The environmental impacts of acid rain are explored elsewhere in the course, and the atmospheric processes involved in its formation are discussed in more detail in that context. We shall not consider them further.

2.5 Taking stock and looking forward

These preliminary skirmishings with the sources and sinks of selected trace gases have identified three kinds of process that may be involved:
- emissions at the Earth's surface;
- chemical transformations within the atmosphere, which can act as both sources and sinks;
- wet or dry deposition to the surface.

For a particular gas, some sub-set of these possible processes will be dominant. It will have its own 'budget' of sources and sinks, and its atmospheric concentration will depend on the balance struck between these various inputs and outputs. To gain insight into the part that atmospheric chemistry plays in these individual budgets, we need to look more closely at the oxidation pathways that were summarized in Figure 2.5. For example, the most abundant of the reactive trace gases is methane (Table 2.1). Just how is a molecule of methane (CH$_4$) converted to one of formaldehyde (CH$_2$O), and then on to CO and finally to CO$_2$? What other species are involved in these transformations?

The implications of these questions can be exposed by pausing to consider how the atmospheric oxidation of methane compares with what happens when the gas burns in air (Figure 2.7). The *overall* transformation is the same: in both cases CH$_4$ is eventually oxidized to CO$_2$ (Box 2.3). But there the immediate

Topic 2 Atmospheric Chemistry and Pollution

resemblance ends. No flames accompany the atmospheric process, and the journey through to CO_2 takes a very long time indeed. On average, a molecule of methane can be moved around in the atmosphere for about a decade before it even embarks on this journey. Rather a long time to wait for your dinner to cook!

To explore this comparison further, read the information in Box 2.4 and then attempt Activity 2.1.

Figure 2.7 A natural-gas hob in use. Natural gas is typically about 92% methane by volume.

Box 2.4 Burning methane

Natural gas is mainly methane. Premixed with air (as it would be in a well-adjusted natural gas burner) and ignited, methane burns smoothly with a familiar blue flame (Figure 2.7). The balanced chemical equation for this combustion process is the one we gave in Box 2.3:

$$CH_4 + 2O_2 = CO_2 + 2H_2O$$

Just thinking about the ingredients in the reaction suggests that it is unlikely to proceed in a single step. For that to happen, all of the chemical bonds in the methane and oxygen molecules would have to break — *at the same time* — and then the individual atoms would have to recombine in such a way as to form just CO_2 and H_2O molecules. If we could view what was happening at the molecular level, then we would see that the combustion reaction actually proceeds in a series of discrete steps, with each step involving a relatively minor rearrangement of chemical bonds. The full set of steps that taken together account for an overall chemical transformation is called the **reaction mechanism**.

As you know from experience, if a gas tap is turned on, the escaping gas doesn't immediately burst into flames. To start, or initiate, the reaction, a spark or flame is required. This supplies the energy needed for one of the preliminary discrete steps in the combustion process — the rupture of a carbon–hydrogen bond, which we represent as follows:

$$CH_4 \longrightarrow CH_3 + H$$

Notice two things. First, the arrow sign ($\longrightarrow$), rather than the equality sign (=), is the agreed sign for an individual step in a reaction mechanism. Second, the fragments that are formed (Figure 2.8) each contain an atom that is not fully bonded. This means that they will be highly reactive species. Indeed, they will react fairly indiscriminately with any other fragment or molecule they come across (including O_2 in the air), with the result that the number of highly reactive species increases. In the ensuing 'chemical turmoil', the hydrogen atoms of methane are stripped off one by one, oxygen atoms are added, and we end up eventually with carbon dioxide and water vapour as the major products.

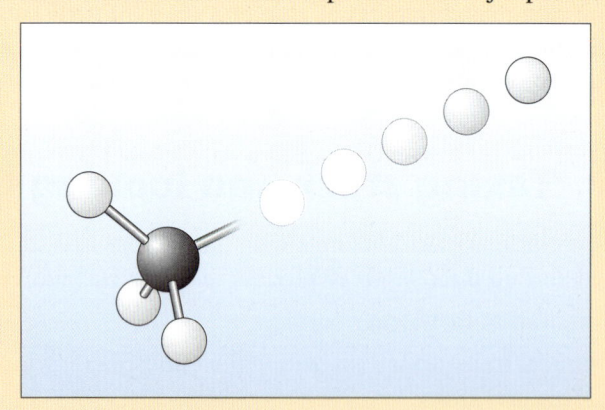

Figure 2.8 A ball-and-stick model of methane being fragmented by breaking a carbon–hydrogen bond. One fragment is a carbon atom which is involved in only three chemical bonds. The other is a hydrogen atom, which is involved in none. (The colour code is as in Figure 2.5)

The species (intact molecules or fragments) that are involved along the way are called *intermediates*, and they include both formaldehyde and CO. In a hot natural-gas flame (temperatures can be over 1200 °C) formaldehyde and other more reactive intermediates are consumed as fast as they are formed: they do not survive long enough to escape the reaction zone. Some carbon monoxide can escape, however, especially if the oxygen supply is compromised for any reason (Box 2.3).

Activity 2.1

Having studied the material in Box 2.4, draw up a table with the column headings: 'combustion process' and 'atmospheric process'. In the first column list what you take to be the key features of the combustion of methane. One entry might be 'occurs at high temperature', for example. Then, alongside each entry, note down any ways in which the atmospheric process appears to be similar or different. If you can't make a judgement either way just enter 'no information'.

To crystallize the points raised in Activity 2.1, and answer some of them: the oxidation of methane in the atmosphere proceeds at ambient temperature, which is rarely more than 30 °C or so at ground level and well below zero higher up. We can regard it as a *low-temperature* combustion process. Somewhat surprisingly, we shall find that the detailed mechanism of this atmospheric process — the individual discrete steps whereby it proceeds — also involves an array of highly reactive atoms and molecular fragments. The latter are called free radicals, or **radicals** for short. Only in the atmosphere, the formation of these radical species is initiated, not by a spark or a flame, but by sunlight.

2.6 Summary of Section 2

1. The bulk composition of the atmosphere is essentially a 4:1 mixture of nitrogen (N_2) and oxygen (O_2). Trace gases make up less than 0.1% by volume, the most abundant reactive species being methane (CH_4).
2. The budget of an individual chemical species in the atmosphere can have inputs (sources) and outputs (sinks) of three kinds:
   - *Emissions* Chemical species are emitted at the Earth's surface by various natural sources. Gases of biogenic origin tend to be reduced species (CH_4, H_2S, NH_3) or only partially oxidized (N_2O, NO). Other sources of partially oxidized species (SO_2, CO, NO) include volcanoes and biomass burning.
   - *Chemistry* The atmosphere is an oxidizing medium. Oxidation pathways lead to the removal of some species and the formation of others.
   - *Deposition* This takes two forms: 'dry deposition', involving direct uptake at the surface, and 'wet deposition' involving incorporation into falling rain or cloud droplets. Chemical transformations may be important precursors of wet deposition.
3. The oxidation pathway of atmospheric methane can be regarded as a low-temperature combustion process. It proceeds via intact molecules (CH_2O and CO) that survive long enough to have an independent existence in the atmosphere, but the detailed mechanism also involves highly reactive atoms and molecular fragments known as radicals.

Question 2.3

The decomposition of biomass under anaerobic conditions that results in emissions of methane can be represented at its simplest by the following net reaction:

$$2(CH_2O)_n = nCH_4 + nCO_2$$

Is the carbon in biomass oxidized or reduced during this process? Explain your answer.

Question 2.4

One of the molecular fragments involved in the atmospheric oxidation of methane is derived from water vapour. Suggest what this species is likely to be.

Question 2.5

The natural sources and sinks of CH_4 and CO identified in Sections 2.2 to 2.4 are collected in Figure 2.9, along with those for atmospheric CO_2 (Question 2.1).

(a) The treatment of the global carbon cycle earlier in the course emphasized the cycling of CO_2 shown on the right-hand side of Figure 2.9. It also equated the atmospheric carbon 'reservoir' with the mass of carbon in atmospheric CO_2. Why is this a simplification, and why is it justified?

(b) Earlier on, we said that CO_2 is chemically unreactive in the atmosphere, and 'so there is little to add here about the processes that govern its atmospheric concentration'. What information would you need in order to justify this assertion?

(c) What human activities are likely to have augmented the natural sources of CH_4 and CO?

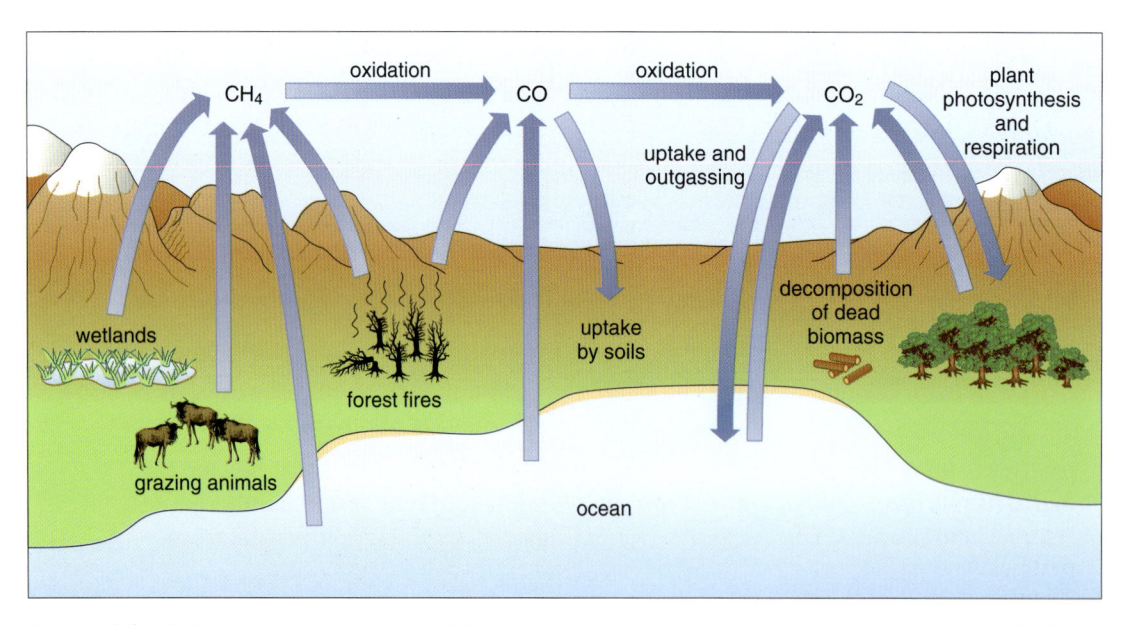

Figure 2.9 Schematic representation of the main natural sources and sinks of atmospheric CH_4, CO and CO_2.

Chemistry driven by the Sun

3

Energy from the Sun powers the Earth's heating and air-conditioning systems, and is ultimately responsible for the weather phenomena — clouds, rain, winds, etc. — that are familiar manifestations of the atmosphere's *physical* behaviour. But sunlight is potent stuff. Its influence is central to an understanding of atmospheric *chemistry* as well. To unpick this idea, Figure 3.1 serves as a reminder that the incoming radiation from the Sun peaks — is strongest, or most intense — in the visible region of the electromagnetic spectrum. Nevertheless, there are significant amounts of radiation to either side of this region, both at longer wavelengths (in the infrared) and at shorter wavelengths (in the ultraviolet).

Earlier in the course, you saw that certain atmospheric gases absorb infrared radiation — the origin of the greenhouse effect. But molecules in the atmosphere also absorb radiation of shorter wavelengths, notably in the ultraviolet (uv) region of the spectrum. And that can have a dramatic effect, because there is a link between the wavelength of radiation and the energy associated with it (see Box 3.1): *the shorter the wavelength of radiation, the higher the energy of the associated photons.*

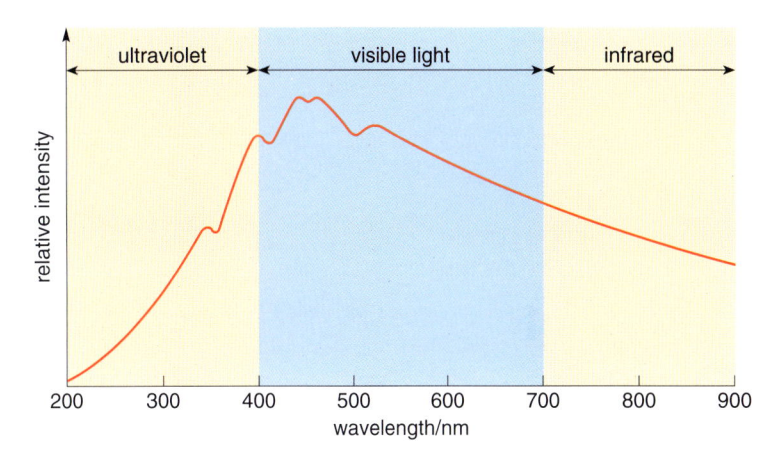

Figure 3.1 The variation of the intensity of incoming solar radiation with wavelength at the top of the atmosphere. Note that here wavelength is measured in nanometres, where 1 nm = 10^{-9} m.

Box 3.1 Electromagnetic radiation: waves and 'packets' of energy

Visible light, and ultraviolet and infrared radiation are all forms of **electromagnetic radiation** (Figure 3.2). All electromagnetic radiation propagates through space as though it is a wave, but each type is characterized by a particular range of wavelengths. The **wavelength** is just the distance between successive crests of a wave; it is given the symbol λ (Greek letter lambda). Visible light corresponds to the wavelength range from about 400 nm (violet light) to 700 nm (red light).

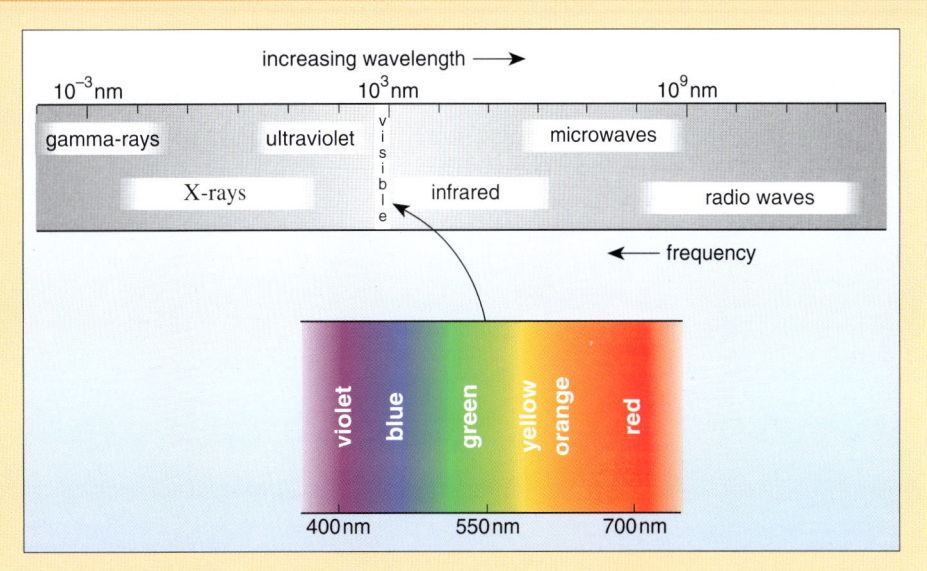

Figure 3.2 A portion of the electromagnetic spectrum. The wavelengths indicated in the expanded visible part of the spectrum are typical, but in approximate positions. Note that the wavelength changes by a *factor of ten* for each division along the top scale.

Another characteristic property of a wave is its **frequency**, usually given the symbol f. Think of this as simply the number of crests (and hence wavelengths) that pass a fixed point in one second. The unit of frequency can then be thought of as 'cycles per second' or just s^{-1}. For all waves, there is a simple relationship between wavelength and frequency. For electromagnetic radiation, it is:

$$\lambda = \frac{c}{f} \qquad (3.1)$$

where the symbol c represents the speed of light. All electromagnetic radiation passes through space at a *constant* speed, which is $c = 3.00 \times 10^8 \text{ m s}^{-1}$ in a vacuum.

○ According to Equation 3.1, what happens to the wavelength λ if the frequency f is increased?

● Remember that the speed of light c is a constant. Thus, if f increases, the term on the right of the equation must become smaller; i.e. the wavelength gets shorter.

Put the other way about: *the shorter is the wavelength of the radiation, the higher is the frequency.*

Everyday experience tells us another important feature of electromagnetic radiation: it 'carries' energy — an idea we visited earlier in the course. We can feel the warming effect of radiation from a desk lamp or a physiotherapist's infrared lamp, for example. But using a solarium requires care: ultraviolet radiation carries enough energy to damage unprotected skin.

Just how does radiation carry energy? The answer lies at the heart of quantum physics, one of the most important scientific revolutions of the twentieth century. Expressed in modern terms, light (or any other electromagnetic radiation) interacts with matter as though it consists of a stream of particles, called photons. Each **photon** can be regarded as a discrete 'packet of energy' (a quantum), the specific amount of energy (E) in a given photon being determined by the frequency (f) of the associated radiation. The formal link is given by a very simple equation:

$$E = hf \qquad (3.2)$$

where the symbol h is used to represent the Planck constant (named in celebration of Max Planck (1858–1947), the father of the quantum revolution): $h = 6.63 \times 10^{-34}$ J s (joule seconds).

○ According to Equations 3.1 and 3.2, which has the higher energy, a photon of ultraviolet radiation or a photon of infrared radiation?

● A photon of ultraviolet radiation. This has a shorter wavelength than infrared radiation (Figure 3.2). It therefore has a higher frequency (Equation 3.1), and so the photons associated with it will have a higher energy (Equation 3.2).

The link between photon energy (E) and wavelength (λ) can be formalized by combining Equations 3.1 and 3.2 to give:

$$E = hc/\lambda \qquad (3.3)$$

Don't worry about the mathematical manipulations required to get to Equation 3.3. The important point for our purposes is to get a feel for the size of these photon energies.

○ Calculate the energy of a photon of violet light ($\lambda = 400$ nm).

●
$$E = \frac{(6.63 \times 10^{-34} \text{ J s}) \times (3.00 \times 10^8 \text{ m s}^{-1})}{400 \times 10^{-9} \text{ m}}$$

$$= 4.97 \times 10^{-19} \text{ J}$$

Clearly, an individual photon represents a very tiny 'packet' of energy — appropriate to interactions at the level of a single atom or molecule. When dealing with the world around us, however, it is more convenient to consider energies for *one mole* of material. This is achieved by multiplying the photon energy by the Avogadro constant ($N_A = 6.02 \times 10^{23} \text{ mol}^{-1}$), which is just the number of units (atoms, molecules — or in this context, photons) in one mole.

'Scaled' in this way, the *energy per mole of photons* becomes:

$$E = N_A hc/\lambda \qquad (3.4)$$

Substituting values for N_A, h and c, we find that with E and λ expressed in kilojoules per mole (kJ mol^{-1}) and nanometres (nm), respectively:

$$\underset{\text{photon energy in kJ mol}^{-1}}{E} = \frac{1.20 \times 10^5}{\underset{\text{wavelength in nm}}{\lambda}} \qquad (3.5)$$

Absorbing uv radiation imparts more energy to a molecule than does the absorption of infrared wavelengths — and it can be sufficient to break chemical bonds, thus splitting the molecule into fragments. This process is known as **photodissociation** or **photolysis** (literally 'splitting by light'). It has one all important *chemical* consequence: it generates highly reactive atomic and radical species — an idea foreshadowed in Section 2.5. It is through this kind of **photochemical** process that the Sun can be said to drive much of the chemical change within the atmosphere.

This is where **ozone** (O_3, Figure 3.3) comes centre stage, because *all* ozone in the atmosphere is formed photochemically. The gas occurs in the troposphere and in the stratosphere, but its formation and significance in the two zones could not be more different.

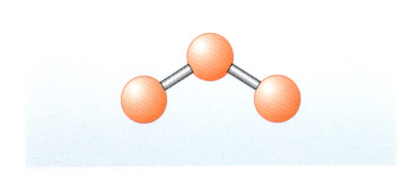

Figure 3.3 Ozone is a form of oxygen, first recognized as a new substance by Christian Friedrich Schönbein in 1839. In the laboratory, the gas is made by passing an electric discharge through ordinary oxygen. It has a distinctive pungent smell (*ozein* is Greek for 'to smell'), sometimes perceptible near sparking electrical equipment (you may have noticed this smell when travelling on the London Underground).

We begin this section by looking a bit more closely at the process of photolysis. Armed with that background, we then examine (in Section 3.2) the processes that maintain the ozone layer in the stratosphere — an appropriate place to start when considering the influence of *incoming* solar radiation. As we shall see, events up in this more rarefied region also have an important influence over the chemical pathways in the lower atmosphere, a subject we return to in Section 3.3.

3.1 Photons and photochemistry

What does the photolysis of an atmospheric constituent — an O_2 molecule, say — actually entail? To answer this question, we need to take a step back, and ask a more fundamental one: why, and how, is uv radiation absorbed? One might, for example, expect that a given molecule could pick up energy from any passing photon, but this is not so. *Only photons with strictly defined energies can be absorbed*.

In this respect, molecules are just like atoms (see Box 3.2). Each type of molecule has its own distinctive 'ladder' of electronic energy levels. To make a molecule jump from its ground state to an excited state, photons of the correct energy must be supplied — and it so happens that these often correspond to radiation at ultraviolet wavelengths.

Topic 2 Atmospheric Chemistry and Pollution

Box 3.2 Exciting atoms

The electrons of an atom of oxygen (or any other chemical element) usually adopt a distribution about the nucleus which has the lowest possible energy; when this is so, the atom is said to be in its electronic **ground state**.

There are also states of higher energy in which the electrons are distributed about the nucleus in other ways. The distinctive feature of the atomic world is that there are only a *limited* number of these higher energy or **electronically excited states** — and so only a limited set of higher energy levels are open to the atom (Figure 3.4a). Consequently, an atom can absorb radiation *only if the photon energy (and hence wavelength) is exactly the amount needed to raise (or excite) the atom to one of its limited number of excited states* — as shown schematically in Figure 3.4b.

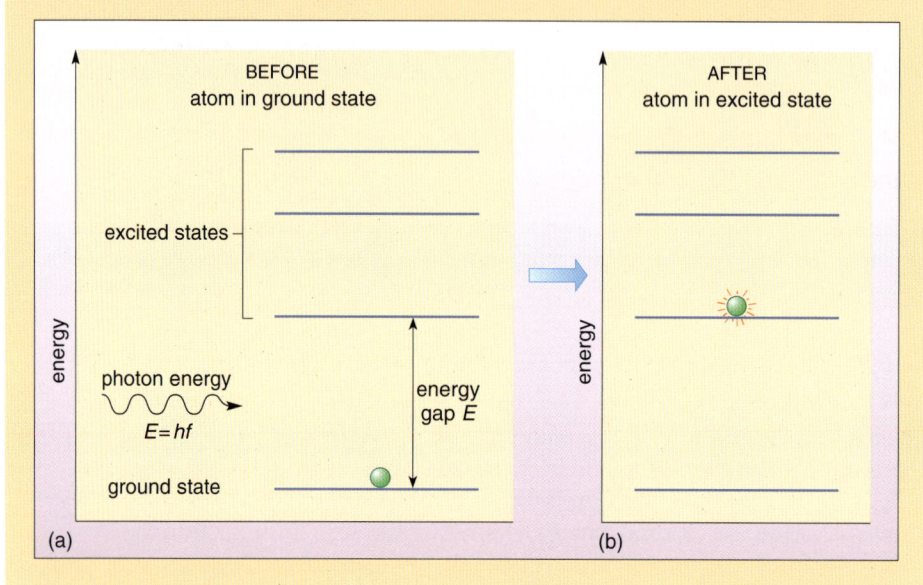

Figure 3.4 (a) The set of allowed energy levels for an atom is like a ladder; in its ground state the atom occupies the lowest rung, the higher rungs (excited states) being empty. (b) Absorbing a photon with an energy that exactly matches the energy gap between the two lowest rungs makes the atom jump from its ground state to one of its excited states.

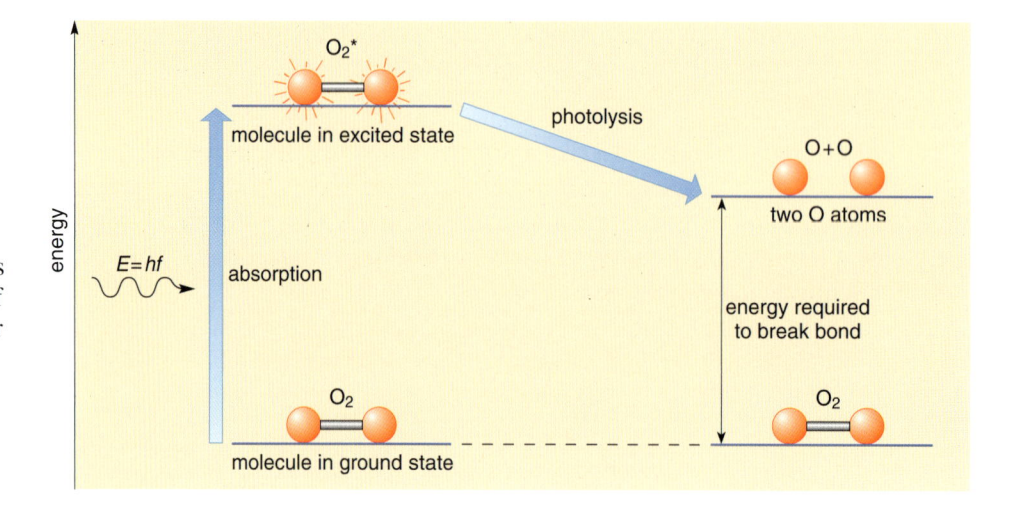

Figure 3.5 A schematic representation of the photolysis of O_2 initiated by absorption of a photon of energy hf. (In other texts, you may see photons represented as $h\nu$, where ν (Greek letter nu) is an alternative symbol for frequency.)

With this in mind, we can think of the photolysis of a molecule of O_2 as happening in two stages, as shown schematically in Figure 3.5. On the left of the figure, absorption of a photon of the correct energy — written in shorthand as *hf* — promotes the molecule to a higher-energy excited state, indicated by attaching an asterisk (as in O_2*). The energy required to break the oxygen–oxygen bond in O_2, known as the **bond energy**, is shown on the right of the figure. In the situation depicted in Figure 3.5, the absorbed photon provides more than enough energy to fragment the molecule, and so leads to photolysis, which can be represented as follows:

$$O_2 + hf \longrightarrow O + O \tag{R1}$$

Including the photon as a 'reactant' emphasizes that this is a *photochemical* reaction.

From this brief discussion, we can draw an important general conclusion. The photolysis of a given atmospheric constituent depends on two factors that are *intrinsic* to the molecule in question:

- the wavelengths it absorbs;
- the bond energies of the bonds that hold the molecule together.

In general, small molecules show intense electronic absorption at shorter uv wavelengths than more complex molecules. For example, O_2 absorbs strongly for wavelengths below about 240 nm (written in symbols as $\lambda < 240$ nm), H_2O for $\lambda < 180$ nm, CO_2 for $\lambda < 165$ nm[†] while N_2 absorbs significantly only for $\lambda < 100$ nm. A feel for the significance of the second factor above can be gained from the information collected in Table 3.1. This lists a few of our atmospheric constituents, together with the chemical effect of breaking one of the bonds in the molecule and the corresponding bond energy. Clearly, some bonds are a lot stronger than others! For example, it takes well over four times as much energy to break the oxygen–oxygen bond in O_2 as it does to break one of these bonds in ozone. But how do these bond energies compare with the energy carried by solar radiation?

Table 3.1 A selection of atmospheric constituents, together with the chemical effect of breaking a bond in the molecule and the corresponding bond energy, in kilojoules *per mole*: each value represents the energy required to break up a mole of molecules.

Gas	Result of bond breaking	Bond energy/kJ mol^{-1}
O_2	$O_2 \longrightarrow O + O$	498
N_2	$N_2 \longrightarrow N + N$	945
CO_2	$CO_2 \longrightarrow CO + O$	532
H_2O*	$H_2O \longrightarrow H + OH$	499
CH_4*	$CH_4 \longrightarrow CH_3 + H$	430
NO_2	$NO_2 \longrightarrow NO + O$	306
O_3	$O_3 \longrightarrow O_2 + O$	107

* The molecular fragments formed by breaking an O–H bond in H_2O and a C–H bond in CH_4 were introduced earlier: see Question 2.4 and Box 2.4, respectively.

[†] Don't get confused here. Absorption of these short uv wavelengths by water vapour and carbon dioxide leads to *electronic* excitation. The absorption of longer wavelength *infrared* radiation by these molecules (the origin of their role as greenhouse gases) makes the molecules vibrate.

Question 3.1

Use Equation 3.5 in Box 3.1 to calculate the photon energies (in kJ mol^{-1}) for light at the extreme ends of the visible band, i.e. violet light ($\lambda = 400$ nm) and red light ($\lambda = 700$ nm).

Comparable calculations for a range of the wavelengths spanned by incoming solar radiation give the smooth curve shown in Figure 3.6. The labelled points superimposed on this curve provide the comparison with bond energies that we are after. Each point is located on the horizontal energy scale according to the bond energy of the molecule (as shown explicitly for N$_2$). Reading across to the vertical scale then allows us to estimate the wavelength of radiation that could, in principle, lead to photolysis of the molecule — about 130 nm for N$_2$, for example.

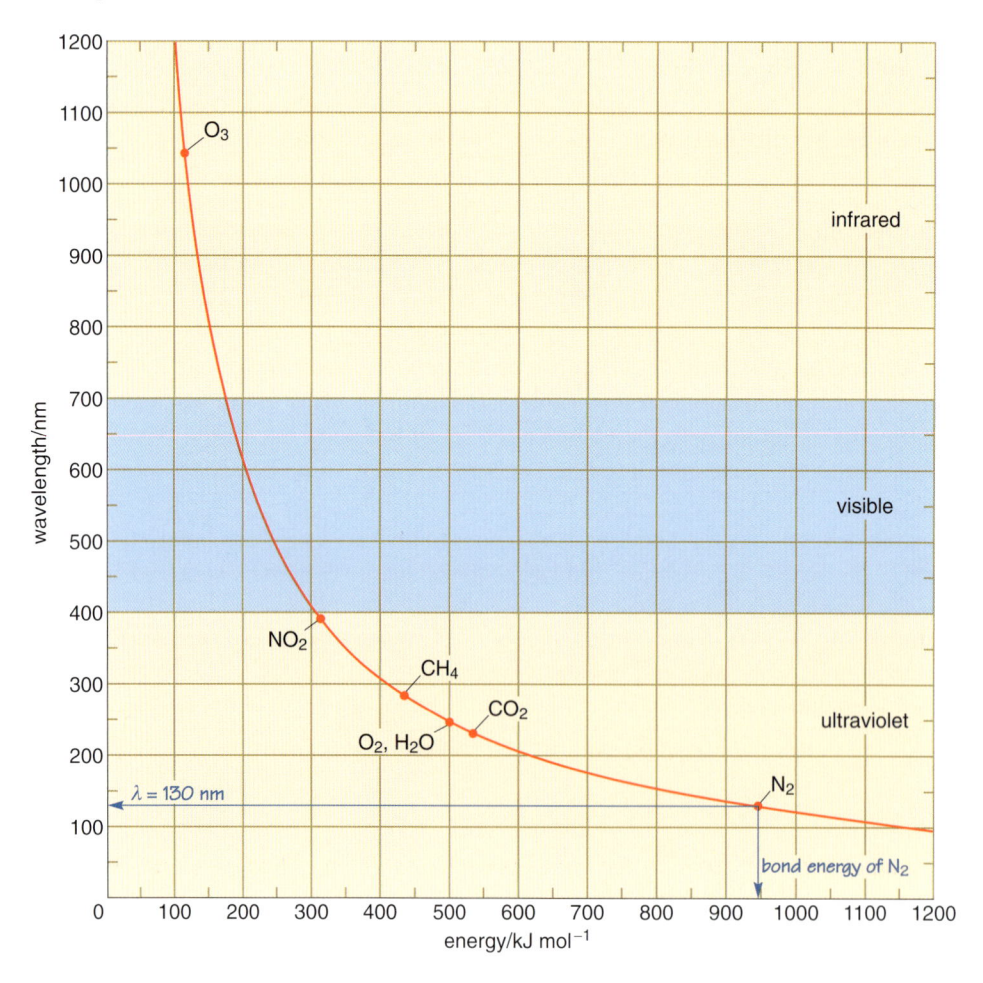

Figure 3.6 How the bond energies of atmospheric constituents compare with the energy carried by solar radiation. See text for further details.

○ What is the most striking general feature of the comparison in Figure 3.6?

● With the sole exception of ozone, the energies required for photolysis of these atmospheric constituents all correspond to wavelengths in the uv region.

Just as clearly, however, the actual uv wavelength needed to split up a given molecule varies markedly — from about 130 nm for N_2 to a value close to that of violet light (400 nm) for NO_2.

Question 3.2

Use Figure 3.6 to estimate the wavelength of radiation required for the photolysis of O_2. Can the wavelengths absorbed by O_2 ($\lambda < 240$ nm) lead to photolysis?

3.2 Photochemistry in the stratosphere: the ozone layer

Unlike the trace gases discussed in Section 2, only about 10% of the Earth's ozone is in the **troposphere** (Figure 3.7); the bulk of it (some 90%) is in the **stratosphere**. The profile sketched in Figure 3.7b is typical of the way in which the concentration (strictly, mixing ratio) of ozone varies with altitude. It is the distinctive bulge in this profile in the stratosphere that has led to the term **ozone layer**.

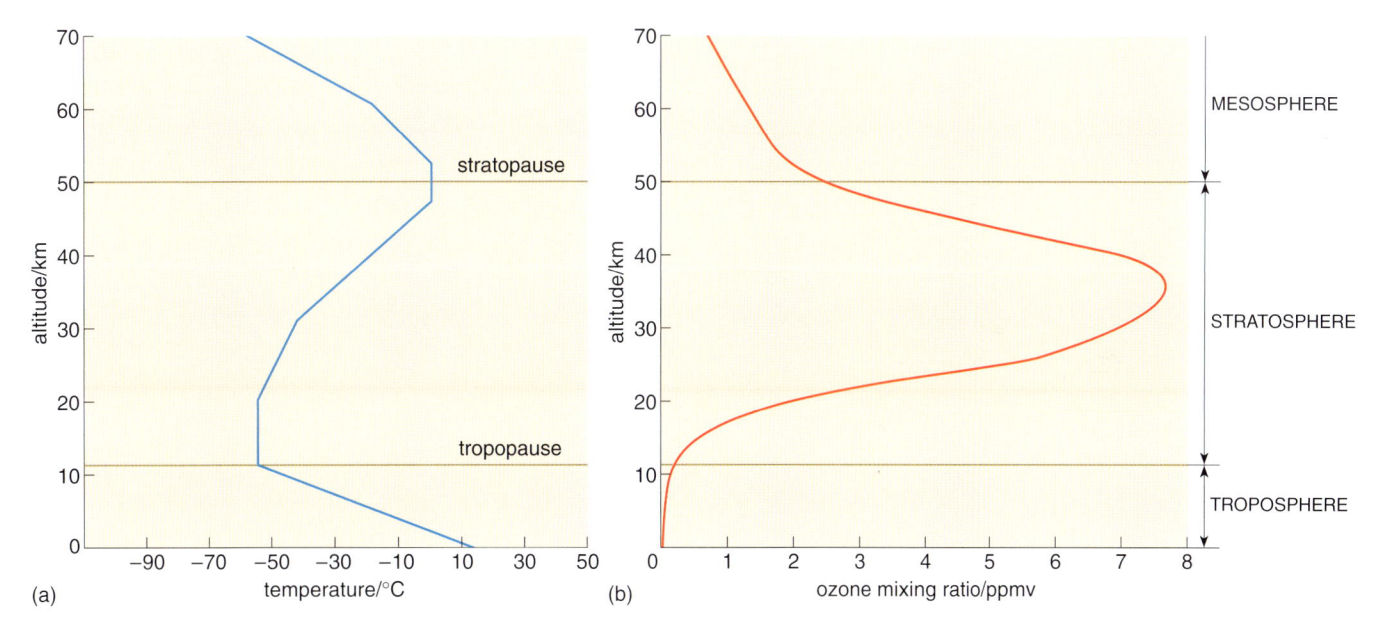

Figure 3.7 (a) A reminder that the vertical structure of the atmosphere is defined by the way in which temperature varies with altitude. Typically, the temperature falls with increasing altitude throughout the troposphere, pauses at the tropopause, and then increases again on moving up through the stratosphere. (b) The altitude profile of the ozone mixing ratio, expressed in parts per million (ppmv). The temperature and ozone profiles sketched here are typical for mid-latitudes.

But where does the ozone in the stratosphere come from? The raw materials are ordinary oxygen molecules and solar uv radiation, but the processes involved depend on how *both* oxygen and ozone interact with that radiation.

Molecular oxygen absorbs solar uv radiation with wavelengths below about 240 nm (Section 3.1): this provides enough energy to split up the molecule (Question 3.2):

$$O_2 + hf \longrightarrow O + O \quad \text{(R1)}$$

The free oxygen atoms generated in this way can then combine with oxygen molecules to form ozone:

$$O + O_2 \longrightarrow O_3 \quad \text{(R2)}$$

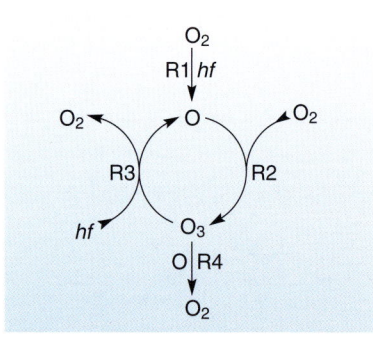

Figure 3.8 Ozone is created when short-wavelength uv radiation breaks up an O_2 molecule (R1), freeing its atoms to combine with other O_2 molecules (R2). The ozone so formed is repeatedly broken up (R3) and reformed (R2), until it is destroyed by reaction with an oxygen atom (R4).

But ozone itself absorbs strongly in the ultraviolet — mainly in a band between 200 and 300 nm, with weaker absorption of wavelengths up to 320 nm. Since the bonds in O_3 are much weaker than the one in O_2 (Table 3.1), absorption of these longer wavelengths again leads to photolysis (Figure 3.6):

$$O_3 + hf \longrightarrow O_2 + O \tag{R3}$$

It would seem that the odds are stacked against ozone! Concentrations build up provided there is enough molecular *and atomic* oxygen around to offer numerous opportunities for the interaction in reaction R2 to take place. In effect, ozone cycles through the round of reactions depicted in the centre of Figure 3.8: it absorbs uv, breaking into its constituent parts, only to be formed again. Trips around this cycle come to an end when an O_3 molecule encounters a free oxygen atom, and is converted back into ordinary oxygen:

$$O_3 + O \longrightarrow O_2 + O_2 \tag{R4}$$

The reactions collected in Figure 3.8 comprise our first example of a reaction mechanism (in the sense defined in Box 2.4) — in this case, a description of the individual discrete steps, both chemical and photochemical, believed to contribute to the maintenance of the ozone layer. It was first proposed as such in 1930 by the British geophysicist Sydney Chapman. It is now clear that the detailed mechanism is actually a good deal more complicated than this. Nevertheless, this simple scheme does capture the essence of the ozone budget in the stratosphere.

One key point is that ozone is constantly being created and destroyed in the stratosphere. But given constant conditions, a dynamic steady state is set up: the concentration of ozone at a given altitude stays more or less the same because the *rate* at which it is formed (driven by O_2 photolysis) is balanced by the rate of its removal (through reaction R4). A useful analogy here is with a bucket of water: think of it being filled from a tap at a steady rate, but also drained through a hole in the bottom of the bucket at the *same* rate. The level of water in the bucket (analogous to the steady-state ozone concentration) does not change.

As ozone cycles through its round of creation and destruction, the solar energy absorbed by both O_2 and O_3 is ultimately released as heat. It is this *in situ* heating that produces the switch from decreasing to increasing temperatures at the tropopause (Figure 3.7a).

○ Can you see how the scheme in Figure 3.8 could also account for the characteristic bulge in the steady-state ozone profile (Figure 3.7b)?

● Making O_3 is driven by interactions between O_2 molecules and incoming uv radiation (with $\lambda < 240$ nm). So you might (rightly) expect the rate of formation to depend on both the number of O_2 molecules around and the intensity of this radiation. But these change with altitude in exactly opposite ways. Too high in the atmosphere, there are insufficient O_2 molecules to absorb much uv radiation and to associate with O atoms to make O_3; too low, most of the radiation with $\lambda < 240$ nm has already been 'filtered' out. This trade-off means that the steady-state ozone concentration peaks somewhere in between.

The way that absorption by O_2 and O_3 in the upper atmosphere acts as a filter of the Sun's uv radiation is shown schematically in Figure 3.9. Here the red line marks the altitude below which solar radiation of a given wavelength cannot penetrate further into the atmosphere. Only at wavelengths longer than about 320 nm does sunlight reach the Earth's surface at *full* intensity.

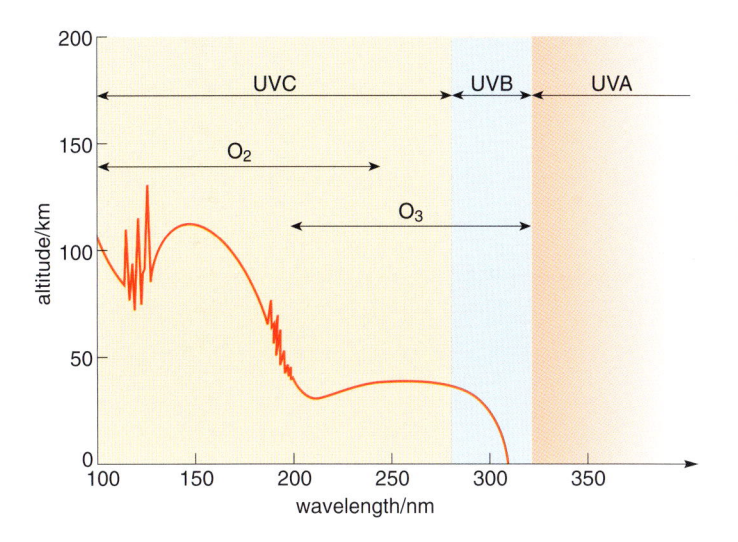

Figure 3.9 The depth of penetration of solar radiation into the atmosphere as a function of wavelength, and the principal absorbing gases. Shorter uv wavelengths ($\lambda < 100$ nm) are absorbed high up in the atmosphere by other gases, notably N_2. By convention, incoming solar uv is split into three bands: UVC (100–280 nm), UVB (280–320 nm) and UVA (320–400 nm).

No living organism can safely absorb wavelengths in the band labelled UVC in Figure 3.9: such radiation is lethal to microbes and carries enough energy to damage biologically important molecules such as DNA. Up to about 240 nm, absorption by O_2 provides an effective filter. But in the range 240–280 nm, protection from UVC is due entirely to the ozone layer. Weaker absorption by O_3 in the band labelled UVB greatly attenuates the solar input at these wavelengths, but does not block it completely. Even the relatively low levels of UVB that normally get through to ground level can have harmful effects, causing sunburn, eye damage and skin cancer in humans, for example.

Figure 3.10 Two publicity posters produced by the New Zealand Cancer Society. The posters are part of a public education programme prompted by concern about increased UVB reaching the country as a result of reductions in stratospheric ozone.

Topic 2 Atmospheric Chemistry and Pollution

The prospect of more UVB reaching the ground (Figure 3.10) is the reason most often cited for concern about the possible consequences of damage to the ozone layer — manifested most dramatically as the so-called 'ozone hole' over Antarctica (Figure 3.11), although there is evidence of reductions in stratospheric ozone elsewhere as well. These changes provide a sobering lesson in the impact that human activities can have. Put simply, they result from a disturbance to the natural ozone balance in the stratosphere — caused by the presence of chlorine atoms derived from the breakdown of synthetic chemicals such as cholorofluorocarbons (CFCs). We shall not pursue the details, because here our main concern is with the *lower* atmosphere. Our immediate interest is with the implications of the screening out of incoming uv radiation for the *chemistry* of this region: photochemically active radiation that gets through to the troposphere is at longer wavelengths — essentially all above 300 nm — than that absorbed higher up.

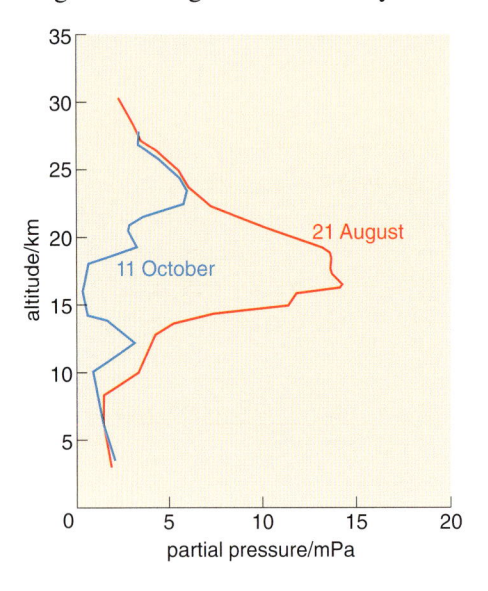

Figure 3.11 The development of the ozone hole over the South Pole between late August (red line) and early October (blue line) 1992: notice that at some altitudes ozone was *completely* removed. Changes like those shown here now recur each southern spring, and regularly extend over the entire Antarctic continent (and beyond), with ozone concentrations recovering again during the summer. (The horizontal axis records the partial pressure of ozone; this is directly proportional to its mixing ratio.)

○ What does this imply as far as the formation of ozone is concerned?

● Photolysis of molecular oxygen is impossible in the troposphere, so ozone cannot now be generated in this way.

So where does the ozone in the troposphere come from?

Question 3.3

Look back at Figure 3.6. Could the absorption of wavelengths longer than 300 nm lead to the photolysis of any of the atmospheric constituents shown there? If so, which ones?

3.3 Photochemistry in the troposphere

Only some 10% of all atmospheric ozone is located in the troposphere, with volume mixing ratios in *unpolluted* environments ranging from just a few ppbv near the surface in the tropical Pacific, to about 100 ppbv (0.1 ppmv) in the upper troposphere. Until the beginning of the 1970s, it was generally thought that this ozone was largely of stratospheric origin.

○ It is often said that the tropopause acts like a lid, separating the turbulent zone in which we live from the calm, stable stratosphere above. In what sense would you expect the stratosphere to be 'calm and stable'?

● With warmer air lying above cooler air, the stratosphere is inherently stable; that is, conditions are *not* conducive to convection — the process that produces large-scale, and often rapid, vertical mixing in the lower atmosphere. (Slow vertical transport does occur in the stratosphere, but by different mechanisms.)

In reality, this tropospheric lid is decidedly leaky. Rapidly rising air can, and does, overshoot the tropopause, carrying its constituents into the lower stratosphere. Mostly this happens over the tropics, often in the updraught of violent storms. And there are return routes as well: tongues of stratospheric air descend into the upper troposphere, mainly at middle latitudes. In short, there is a constant, if relatively slow, exchange of air across the tropopause.

Transport down does indeed provide an important source of ozone to the upper troposphere. But it is now clear that O_3 can also be produced *within* the troposphere, via mechanisms that are interwoven with the oxidation pathways of CH_4 and CO, and dependent on the availability of two nitrogen oxides — NO and NO_2, commonly known collectively as **NO_x**. Along with the unravelling of these mechanisms has come the recognition that the relatively small amount of ozone in the troposphere is of critical importance for the composition of Earth's atmosphere.

Things can get a little complicated at this point. So it is helpful to proceed in stages, by addressing each of the following questions in turn:

• Why is tropospheric ozone so important?
• How is ozone generated within the troposphere?

3.3.1 The role of tropospheric ozone

The key role of the ozone in the lower atmosphere again goes back to its interaction with solar uv radiation. As a quick glance back at Figure 3.6 will confirm, the threshold energy required for the photolysis of ozone corresponds to radiation in the infrared region. When ozone absorbs *ultraviolet* radiation (strictly, of wavelengths shorter than about 320 nm), this imparts so much extra energy that photolysis then generates an atomic fragment that is itself in an electronically excited state. We write it as O*. *Tropospheric* production of O* is due entirely to the presence of ozone; it takes place in a narrow wavelength band between 300 and 320 nm:

$$O_3 + hf \longrightarrow O^* + O_2 \quad (\lambda < 320 \text{ nm}) \tag{R5}$$

Most of the O* atoms generated in this way lose energy (and return to the ground state) when they bump into other molecules in the air. They can then re-form ozone, via reaction R2. However, the excess energy possessed by an *excited* oxygen atom (Box 3.2) has one very important consequence: if an O* atom happens to collide with a water molecule, it has enough energy — *unlike a ground-state oxygen atom* — to strip off a hydrogen atom:

$$O^* + H_2O \longrightarrow HO + HO \tag{R6}$$

The molecular fragments formed in reaction R6 (recall Question 2.4) are **called hydroxyl radicals**. From now on, we shall identify these, and other, radical species by attaching a dot to the atom in the molecular fragment that is not fully bonded, as in HO•. This will help you to distinguish radicals from 'intact' molecules (see Box 3.3).

> ### Box 3.3 An important distinction
>
> It is important to be clear about the distinction between the hydroxyl radical, HO•, and the *hydroxide ion*, usually written OH⁻. Both species can be *thought* of as derived from the H_2O molecule by breaking one of the O–H bonds — but in different ways:
>
> - In forming the HO• radical, one hydrogen atom is stripped away *complete with its single electron*. Neither fragment carries an electric charge.
> - By contrast, in the more familiar dissociation reaction
>
> $H_2O = H^+ + OH^-$
>
>   one hydrogen departs as a bare proton (H^+), its electron being retained by the other fragment (OH^-).

Reactions R5 and R6 act as the primary source of hydroxyl radicals within the troposphere. Note that together, they also lead to the destruction of ozone.

Like others of their kind, hydroxyl radicals are highly reactive species — so reactive, in fact, that on a globally averaged basis, their mixing ratio in the troposphere is only about 4×10^{-14} (0.04×10^{-12} or just 0.04 parts per trillion; Box 2.1). Nevertheless, it is attack by this ultra-minor constituent that triggers the oxidation pathways we considered briefly in Section 2.3 (Figure 2.5) — and not direct reaction with O_2, which is some 10^{13} times more abundant. Truly remarkable!

Figure 3.12 summarizes the immediate effect of reaction with HO• for some of the reduced or partially oxidized trace gases we discussed in Section 2. Notice that such attack sometimes produces intact molecules that are already easily removed from the atmosphere, either by wet deposition (e.g. nitric acid, HNO_3) or by direct uptake at the Earth's surface (e.g. CO_2). More often, however, reaction with HO• is just the first step. It generates free atoms (e.g. hydrogen) and/or further radicals — such as the **methyl radical**, •CH_3 (recall Figure 2.8) — which are themselves highly reactive. A sequence of reactions then leads on to increased oxidation, increased aqueous solubility, and eventual scavenging from the atmosphere.

SULFUR GASES
$H_2S + HO• \longrightarrow HS• + H_2O$
$SO_2 + HO• \longrightarrow HSO_3•$

NITROGEN GASES
$NH_3 + HO• \longrightarrow •NH_2 + H_2O$
$NO_2 + HO• \longrightarrow HNO_3$

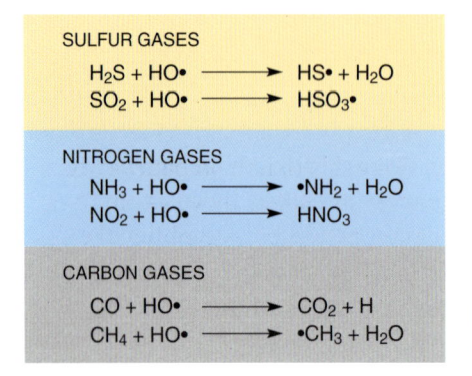

CARBON GASES
$CO + HO• \longrightarrow CO_2 + H$
$CH_4 + HO• \longrightarrow •CH_3 + H_2O$

Figure 3.12 The effect of attack by hydroxyl radicals on sulfur, nitrogen and carbon containing trace gases.

It is clear from Figure 3.12 that the hydroxyl radical is pretty cavalier about its molecular victims. But compounds that contain hydrogen atoms are particularly prone to attack. Figure 3.12 includes three examples: in each case, the hydroxyl radical reverts to the more stable water molecule by abstracting an H atom from its target. Such **hydrogen-abstraction reactions** initiate the removal from the atmosphere not only of natural trace gases, but also of a vast array of organic compounds released by human activities (more on which in Section 5). For this reason, the hydroxyl radical is often called 'the atmosphere's detergent'. *Without it, the composition of the atmosphere would be very different.*

In the background troposphere, remote from the direct effects of human activities, some 60% of the HO• radicals react with CO and most of the rest with CH_4 and its oxidation products (which include CO, remember). But attack on these molecules does not necessarily lead to the removal of the hydroxyl radical. It merely triggers a sequence of reactions that may, in turn, compensate for the initial HO• loss. The simplest example is provided by one oxidation pathway for CO:

$$CO + HO\bullet \longrightarrow CO_2 + H \quad (R7)$$
$$H + O_2 \longrightarrow HO_2\bullet \quad (R8)$$
$$HO_2\bullet + O_3 \longrightarrow HO\bullet + 2O_2 \quad (R9)$$
$$\text{Net: } CO + O_3 = CO_2 + O_2 \quad (R10)$$

The species formed in reaction R8 is called the *hydroperoxy* radical, $HO_2\bullet$ (Figure 3.13). This, and other, **peroxy radicals** also play a central role in tropospheric chemistry, as we shall see.

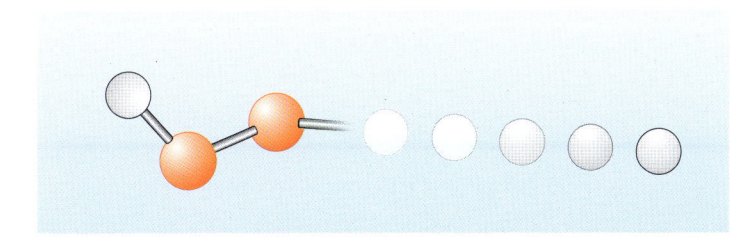

Figure 3.13 The $HO_2\bullet$ radical can be *thought* of as being derived from hydrogen peroxide (H_2O_2) by the rupture of an O–H bond (whence its name). But $HO_2\bullet$ is *not* formed in this way in the atmosphere (just as HO• and $CH_3\bullet$ are not formed by direct photolysis of their 'parent' molecules, H_2O and CH_4). Like $HO_2\bullet$, other peroxy radicals contain the group –O–O•, where the terminal oxygen atom is involved in only one bond.

○ The net effect of the individual discrete steps in this mechanism was obtained by adding together the reactants and products in reactions R7 to R9. Have a go at that, and hence check that the overall reaction given (R10) is correct.

● Adding up gives:

$$CO + \cancel{HO\bullet} + \cancel{H} + \cancel{O_2} + \cancel{HO_2\bullet} + O_3 = CO_2 + \cancel{H} + \cancel{HO_2\bullet} + \cancel{HO\bullet} + \cancel{2}O_2$$

Cancelling out the species that appear on both sides (i.e. HO•, H, $HO_2\bullet$ and *one* O_2), as shown, then leaves reaction R10.

Reactions R7 to R9 spell out the various steps along this oxidation pathway for CO, but they do little to emphasize the essential feature noted above — the regeneration of the hydroxyl radical. That is best achieved by assembling the steps as a cycle, as shown in Figure 3.14. *You do not need to remember the details of this oxidation pathway — nor of others we shall meet shortly. But you should be able to interpret a diagram like the one in Figure 3.14: see Box 3.4.*

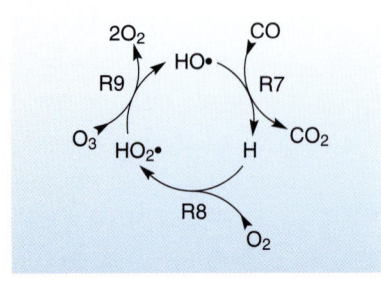

Figure 3.14 The essential steps in one pathway for the tropospheric oxidation of carbon monoxide (reaction R10) initiated by reaction with HO•.

> **Box 3.4 Interpreting reaction cycles**
>
> Taking the cycle shown in Figure 3.14 as an example, there are two points to note:
>
> - The HO• radical appears at the top of the cycle, such that one clockwise rotation from this point gives the sequence of steps in the order in which they would appear in the conventional representation of the mechanism — the one shown in reactions R7 to R9 in this case.
> - The reactive intermediates (here HO•, H and HO$_2$•) remain *within* the cyclic pathway. By contrast, the reactants (CO and O$_3$) and products (CO$_2$ and O$_2$) in the *overall* transformation (reaction R10) enter or leave the cycle at various points.
>
> So there is a simple way of identifying the *net* effect of a reaction cycle. Just note the species that enter it, and those that leave: these will be the reactants and products, respectively.

You may be struck by one puzzling feature of the CO-oxidation pathway in Figure 3.14. Each trip around the cycle — each molecule of CO that is oxidized to CO$_2$ — results in the *destruction* of a molecule of ozone. Yet earlier we implied that the atmospheric oxidation of CO (and CH$_4$) is linked to ozone *production*. It is, but only in the presence of NO$_x$.

3.3.2 Tropospheric ozone production: the critical role of NO$_x$

Making ozone requires a source of atomic oxygen. The only known candidate in the troposphere is the photolysis of NO$_2$ by the longer uv wavelengths that penetrate below the stratosphere (recall Question 3.3):

$$NO_2 + hf \longrightarrow NO + O \quad (\lambda < 400\,nm) \tag{R11}$$

$$O + O_2 \longrightarrow O_3 \tag{R2}$$

In the absence of other gases, reactions R11 and R2 are two parts of a chemical triad that links NO$_x$ and O$_3$. The third reaction of the group is:

$$NO + O_3 \longrightarrow NO_2 + O_2 \tag{R12}$$

Since ozone is needed to effect this conversion, and NO$_2$ is necessary for ozone formation, it is apparent that this triad of reactions does not, *by itself*, have any effect on ozone levels in the troposphere. Rather it represents a so-called null cycle, which has no net effect, evident by inspecting Figure 3.15.

Net ozone production is possible only if NO is converted into NO$_2$ by reacting with something *other than* O$_3$ itself. This is where the oxidation pathways of CO and CH$_4$ come into the picture. The crucial reaction partners for NO are various peroxy radicals, such as HO$_2$• (Figure 3.13), formed in the course of these pathways (e.g. via reaction R8). Thus, for example, the following reaction:

$$HO_2\bullet + NO \longrightarrow HO\bullet + NO_2 \tag{R13}$$

effectively bypasses the ozone consumption reaction R12. At the same time, it also provides an alternative means of regenerating hydroxyl radicals — one that can bypass yet another reaction (R9) that leads to ozone loss:

$$HO_2\bullet + O_3 \longrightarrow HO\bullet + 2O_2 \tag{R9}$$

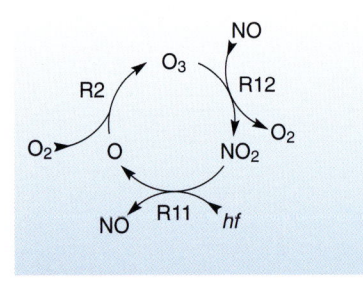

Figure 3.15 Null cycle showing the destruction/regeneration of O$_3$ and NO$_2$ via reactions R12, R11 and R2. The species entering and leaving the cycle are the same: it has no net effect.

If reaction R13 closes the loop in Figure 3.14 — rather than reaction R9 — we get the oxidation cycle shown in Figure 3.16. Now the oxidation of CO has very different consequences, because the net effect of *this* cycle is as follows:

$$CO + O_2 + NO = CO_2 + NO_2 \tag{R14}$$

Each trip around the cycle now produces an NO_2 molecule — and hence, potentially at least, a molecule of ozone (via reactions R11 and R2).

Question 3.4

Check that the net effect of the cycle in Figure 3.16 is indeed as shown in reaction R14. By then including the photolysis of NO_2 (R11) and the formation of O_3 (R2), show that the overall effect of this CO-oxidation pathway can be represented as:

$$CO + 2O_2 = CO_2 + O_3 \tag{R15}$$

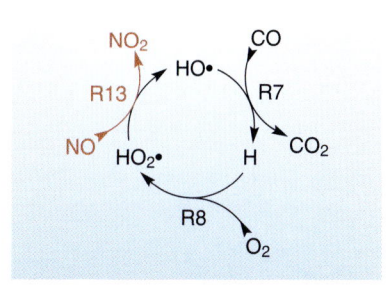

Figure 3.16 The essential steps in the tropospheric oxidation of CO in the presence of NO (reaction R14).

In conclusion: *depending on the availability of NO*, the oxidation of CO can follow two different pathways — one leading to the loss of ozone (Figure 3.14), and the other to its formation (Figure 3.16).

Similar conclusions emerge from a study of the atmospheric oxidation of methane — first to formaldehyde (CH_2O) and then on to CO and CO_2 (as shown schematically in Figure 2.5). As you might expect, the reaction pathways are both more complex than those for CO, and more numerous. We shall not dwell on the details. But again, the availability of NO determines the results, and the key player is a peroxy radical — formed this time from the methyl radical by a reaction analogous to R8:

$$•CH_3 + O_2 \longrightarrow CH_3O_2• \tag{R16}$$

Like $HO_2•$, this organic peroxy radical can, in turn, convert NO into NO_2 (by a reaction analogous to R13) — and this opens the way to oxidation pathways that generate ozone.

In an NO-rich environment, the net effect of one such pathway (explored in Activity 3.1) can be represented as follows:

$$CH_4 + 8O_2 = CO_2 + 2H_2O + 4O_3 \tag{R17}$$

Under these circumstances, other possible fates for various intermediate species (notably formaldehyde) can result in a net gain of not only O_3, but also $HO•$ radicals. By contrast, *in an NO-poor environment*, methane is again oxidized to CO_2 via formaldehyde and CO — but now *without ozone formation*. Such pathways can also result in a net loss of $HO•$ radicals.

The results summarized above are of central importance for the chemistry of the *background* troposphere. On a global scale, the oxidation of CH_4 and CO together play a dominant role in the ozone and hydroxyl budgets. CH_4 and CO are the principal reaction partners of $HO•$. The formation of this molecular scavenger acts as a sink for ozone (via reactions R5 + R6). And the oxidation pathways of CH_4 and CO can act as either sources or sinks for both ozone and $HO•$ — *the critical factor here being the concentration of NO_x*. In the oxidation of CO, for example, the crossover from the 'high-NO_x' (ozone generating) regime to the 'low-NO_x' (ozone loss) regime occurs at NO_x concentrations below about 10 pptv (10×10^{-12}).

- Notice that we have reverted to the collective term NO_x. Can you now see why it is appropriate to lump NO and NO_2 together as NO_x?

- Reactions within the atmosphere constantly act to interconvert the two oxides — via (R13 + R11), for example, or via the null cycle in Figure 3.15.

In practice, NO_x is emitted at the Earth's surface mainly as NO. During the day, however, this is rapidly converted to NO_2, and back again. Now this gives pause for thought, because it would seem to imply that NO_x should linger indefinitely within the atmosphere and just go on accumulating there.

○ Look back at Figure 3.12. Can you spot a process that acts as a sink for NO_x, removing it from the atmosphere?

● Attack by HO• (again!) converts NO_2 — and hence NO_x — to nitric acid, which is rapidly rained out in the troposphere.

For reasons we return to in Section 4, the influence of this sink means that the NO_x threshold noted above — low as it may seem — is nevertheless not always exceeded. As a result, Earth's atmosphere is host to both 'low-NO_x' and 'high-NO_x' regimes at different times and places.

3.4 Summary of Section 3

1. The shorter the wavelength (λ) of electromagnetic radiation, the higher the energy of the associated photons (packets of energy). The photon energies of ultraviolet radiation are comparable with both the energies required to promote molecules to electronically exited states, and the energies needed to break chemical bonds.

2. Different atmospheric constituents absorb different wavelengths in the incoming solar uv radiation. Absorption has two important effects:
   - It leads to photolysis, splitting molecules into highly reactive fragments (free atoms and radicals), and hence initiating chemical change.
   - It acts as a filter: the uv radiation that penetrates to the troposphere is of longer wavelength ($\lambda > 300$ nm), and hence lower energy, than that absorbed higher up (in and above the stratosphere).

3. Ozone (O_3) occurs in trace amounts throughout the atmosphere, but its concentration varies markedly with altitude. Concentrations peak in the ozone layer in the stratosphere. The presence of ozone causes the reversal in the Earth's temperature profile at the tropopause.

4. All ozone in the atmosphere is formed photochemically — from the photolysis of O_2 (at $\lambda < 240$ nm) in the stratosphere, and from the photolysis of NO_2 (at $\lambda < 400$ nm) in the troposphere. Some of the ozone formed in the stratosphere is transferred to the troposphere.

5. An important sink for tropospheric ozone is photolysis (at $\lambda < 320$ nm), followed by reaction of the excited O* atoms so formed with water vapour (reactions R5 + R6 in Section 3.3.1). This also provides the primary source of hydroxyl radicals, HO•.

6   The hydroxyl radical acts as the atmosphere's detergent. Attack by HO• is the first step in the oxidation pathways of most natural trace gases. Subsequent reactions lead to more oxidized compounds that are readily removed from the troposphere either by wet deposition (e.g. SO_2, H_2SO_4, HNO_3) or by dry deposition (e.g. CO_2). Substances released by human activities are also scrubbed from the atmosphere in this way.

7   In the background atmosphere, the principal reaction partners for HO• are CO and CH_4. Hydroxyl attack initiates the oxidation of CO and CH_4 (via CH_2O and CO) to CO_2. Alternative oxidation pathways can act as either a source or a sink for both O_3 and HO•. In this respect, the outcome depends on the availability of NO_x.

8   NO_x is a collective term for NO and NO_2. Mostly NO_x is emitted at the Earth's surface as NO. In the absence of other gases, rapid interconversion of NO and NO_2 occurs via a null cycle that links NO_x and O_3 (Figure 3.15): this has no *net* effect on ozone concentrations. Formation of HNO_3 acts as a sink for NO_x.

9   Peroxy radicals, such as HO_2• and CH_3O_2•, are formed in the course of oxidation of CO and CH_4. These radicals can convert NO into NO_2 without ozone consumption. As a result, in NO_x-rich environments, the oxidation of CO and CH_4 leads to O_3 formation. Ozone destruction dominates in NO_x-poor environments, mainly via reaction with HO_2• (reaction R9 in Section 3.3.1).

The following activity gives you an opportunity to practise the interpretation of reaction cycles (an important learning outcome for this section), and to check that you have grasped key features of the chemistry that controls ozone levels in the unpolluted troposphere.

Activity 3.1

This question is concerned with the atmospheric oxidation of CH_4 to CH_2O, and then on to CO, and finally CO_2 in a NO_x-rich environment.

(a) Under these circumstances, attack by HO• initiates the oxidation of CH_4 to CH_2O via the pathway shown in Figure 3.17a. Explain why this cycle also leads to ozone formation, according to the following overall reaction:

$$CH_4 + 4O_2 = CH_2O + H_2O + 2O_3 \quad \text{(R18)}$$

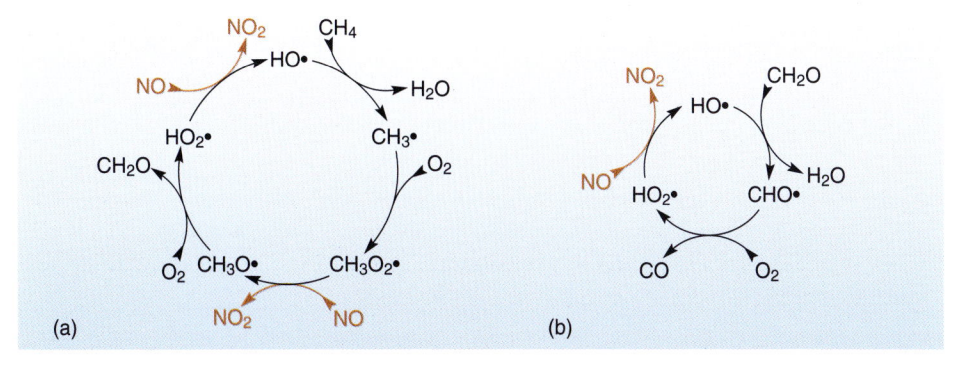

Figure 3.17 Tropospheric oxidation in a NO_x-rich environment of (a) CH_4 to CH_2O, and (b) CH_2O to CO.

(b) Reaction with HO• can also be the first step in the further oxidation of CH_2O to CO, according to the mechanism given in Figure 3.17b. (i) Write an equation for the second step in this mechanism; and (ii) show that the net effect of the cycle can be represented as:

$$CH_2O + 2O_2 = CO + H_2O + O_3 \qquad \text{(R19)}$$

(c) Under NO_x-rich conditions, CO is oxidized to CO_2 via the cycle shown earlier in Figure 3.16. How many O_3 molecules are produced per molecule of CO oxidized?

(d) In the oxidation of CH_4 to CO_2 by the mechanism outlined in parts (a)–(c), how many O_3 molecules are produced per molecule of CH_4 oxidized?

(e) Present-day global emission estimates are about 3×10^{13} moles per year for CH_4, and 4×10^{13} moles per year for CO. Using your results from parts (c) and (d), estimate the global production rate of ozone in the troposphere. Explain briefly why this approach might be expected to produce an overestimate. *(A short paragraph)*

Atmospheric budgets: local and global

Take a sample of air anywhere in the troposphere or stratosphere, and its bulk composition will be the same: essentially a 4:1 mixture of N_2 and O_2. Large-scale air movements throughout these regions, and up and down across the tropopause, ensure that they are 'well mixed' as far as the major atmospheric constituents are concerned.

The same cannot be said for many trace gases. For example, we have seen that ozone concentrations vary markedly with altitude. Within the troposphere, there are strong geographical and seasonal variations as well. By contrast, one of the main natural precursors of tropospheric ozone — methane — *is*, in fact, reasonably well mixed throughout the lower atmosphere. Yet this is not so for the other one, carbon monoxide. Concentrations are 30–100 ppbv in remote parts of the world (Table 2.1), but can be as high as several ppmv in heavily polluted urban areas (such as Mexico City), where prevailing CO concentrations are considered a hazard to human health. The abundance of NO_x can show even more striking variations: values of several hundred ppbv have been recorded in city air, compared with the 5–20 pptv typical of remote regions.

There are two factors at work here. Winds transport chemical species away from their point of origin (at the surface or within the atmosphere). But the influence this has on the overall distribution of a given species depends on its atmospheric 'lifetime' — its **residence time** in the atmosphere. The second factor is the influence of human activities.

Try the following question to refresh your memory about the meaning of the term 'residence time'.

Question 4.1

Figure 4.1 is a much simplified representation of the global water cycle discussed earlier in the course. Here the total amount of water on Earth has been carved up into just two 'boxes' (or reservoirs) — that in the atmosphere (mainly water vapour), and that everywhere else (whether liquid or frozen). The arrows represent the movement of water to and from the atmosphere by virtue of its sources (evaporation) and sinks (precipitation) at the surface, all across the globe.

(a) According to the estimates in Figure 4.1, the amounts of water entering and leaving the atmosphere each year balance. What would happen were this not the case?

(b) Use the information in Figure 4.1 to estimate the average residence time of water in the atmosphere.

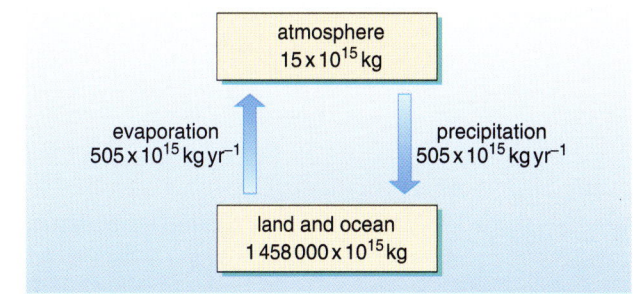

Figure 4.1 A simple box diagram to show the annual balance between water entering the atmosphere by evaporation and leaving it by precipitation.

atmosphere 15×10^{15} kg

evaporation 505×10^{15} kg yr^{-1}

precipitation 505×10^{15} kg yr^{-1}

land and ocean $1\,458\,000 \times 10^{15}$ kg

4.1 Atmospheric lifetimes

To estimate the mean residence time of a trace gas, we need information about its total atmospheric stock, and about the rate of input or output. For water this is relatively straightforward. True, water vapour is an important ingredient in tropospheric chemistry (Section 3.3), but the reactions that create and destroy H_2O molecules within the atmosphere have little impact on the overall residence time. That is determined by the much larger fluxes to and from the atmosphere through the *physical* processes of evaporation and precipitation — a message revisited in Question 4.1.

The rate at which water is lost from the atmosphere can be estimated from records of the mean annual precipitation from sites around the world. But how do we estimate the rate of loss for a gas like methane, for which the major removal mechanism is the following reaction *within* the atmosphere?

$$CH_4 + HO\bullet \longrightarrow \bullet CH_3 + H_2O \qquad (R20)$$

The necessary information comes from laboratory studies of the *rate* of this reaction — and how this depends on the concentrations of CH_4 and $HO\bullet$, and on other factors, notably temperature. The end result is an equation (see Box 4.1) that effectively summarizes all this information. For methane loss via reaction R20, this 'rate equation' is as follows:

$$\text{rate of loss} = k \times [CH_4] \times [HO\bullet] \qquad (4.1)$$

where k, the *rate constant* for reaction R20, has a fixed value (determined experimentally) at a given temperature, and enclosing the molecular formula in square brackets is the standard shorthand for 'the concentration of' a substance.

Making the crude assumption of *a constant concentration* of hydroxyl radicals in the atmosphere (a point we return to later on), and introducing a modified rate constant, $k' = k \times [HO\bullet]$, Equation 4.1 takes the simplified form:

$$\text{rate of loss} = k' \times [CH_4], \text{ where } k' = k \times [HO\bullet] \qquad (4.2)$$

We now have a *quantitative* link between the rate of loss of methane and its atmospheric concentration. Because the link is with concentration, rather than total stock (expressed in mass units, say, as in Question 4.1), it is convenient to define a slightly modified measure of the residence time. Thus, for reactive gases like methane the **atmospheric lifetime** is defined as:

$$\text{lifetime} = \frac{\text{atmospheric concentration}}{\text{rate of loss}} \qquad (4.3)$$

Applied to methane, this definition leads to a very simple result:

$$\text{lifetime} = \frac{[CH_4]}{k' \times [CH_4]} = \frac{1}{k'} \qquad (4.4)$$

So the lifetime of methane is independent of its atmospheric stock. All we need is information about the rate constant for reaction R20, and the global mean concentration of $HO\bullet$ radicals. Estimates of the latter rely on ingenious indirect assessments, the value commonly used in this context being $[HO\bullet] = 1 \times 10^6$ molecules cm^{-3}. Note that this is a *true* concentration (a number density), *not* a mixing ratio (recall Box 2.1).

Box 4.1 What influences the rate of a reaction?

Experiments show that many reactions can be made to go faster in two general ways:

- by increasing the reactant concentrations; and
- by raising the temperature.

The reaction that effectively removes methane from the atmosphere shows this behaviour:

$$CH_4 + HO\bullet \longrightarrow \bullet CH_3 + H_2O \quad \quad (R20)$$

This is an example of a **bimolecular reaction** — a single step in a mechanism that involves the *direct* interaction of just two chemical species (atoms, molecules or radicals).

Why should the rate of a reaction like this depend on the reactant concentrations? Intuitively, it seems fairly obvious that reactant molecules must meet or 'collide' before they can react. So the reaction rate should be related to how often this happens in a given time, i.e. to the collision rate.

○ What happens on a molecular level when the *concentration* of a substance is increased?

● The number of molecules in a given volume is increased.

Put another way, the 'given volume' becomes more crowded, so we might expect the collision rate to increase as well. Think of the difference between walking along a relatively deserted street, and struggling to make progress through a crowd.

This intuitive conclusion is confirmed by detailed experimental study of reaction R20. Like other bimolecular reactions, there is a simple proportionality relationship between the reaction rate *at a given temperature* and the reactant concentrations. For reaction R20, the reaction rate is a measure of the 'rate of loss' of methane, so we can write:

reaction rate = rate of loss of methane $\propto [CH_4] \times [HO\bullet]$

○ According to this relationship, what happens to the reaction rate if the concentration of HO• radicals is kept constant, but that of methane is doubled?

● If the value of [HO•] does not change, but the value of [CH_4] doubles, then the reaction rate doubles — the reaction goes twice as fast.

The relationship above can be turned into an equation by introducing a constant of proportionality — known in this context as the **rate constant, k**; that is:

$$\text{rate of loss} = k \times [CH_4] \times [HO\bullet] \quad \quad (4.1)$$

Equation 4.1 is called the **rate equation** for reaction R20. It holds at a *fixed* temperature (25 °C, say). At any other temperature, the rate equation has the same form as equation 4.1, but usually the value of k is different. In this case laboratory measurements show that the rate constant increases by a factor of 3 between −25 °C and +25 °C, temperatures typical of different altitudes in the troposphere. This explains why the reaction goes faster at the higher temperature (assuming the same reactant concentrations).

Rate equations analogous to Equation 4.1 can be written for other simple bimolecular reactions — a category that includes all of the reactions with the hydroxyl radical collected in Figure 3.12. For each reaction, the value of k at a given temperature is intrinsic to that reaction, and so is the way in which k changes with temperature.

Question 4.2

Expressed in the same concentration unit, $k = 6.2 \times 10^{-15}$ (molecules cm^{-3})$^{-1}$ s^{-1} for reaction R20 at 25 °C. Use this information, together with the value of [HO•] given above, to estimate the atmospheric lifetime of methane.

Comparable information for some other trace gases is collected in Table 4.1. Note that the estimated lifetimes recorded there are based on data at 25 °C, and relate to loss via reaction with the hydroxyl radical. Both provisos can be important. For methane, for example, estimates that incorporate the influence of tropospheric temperatures below 25 °C (recall Figure 3.7a) produce a mean lifetime of around 12 years — more than double the figure in Table 4.1. Conversely, the influence of additional sinks — notably dry and wet deposition — can significantly reduce the mean lifetime of some species; it comes down to 3–4 days for SO_2, for example.

Topic 2 Atmospheric Chemistry and Pollution

Table 4.1 Estimated lifetimes for loss via reaction with the hydroxyl radical, with [HO•] = 1 × 10⁶ molecules cm⁻³; $k' = k \times$ [HO•], where k is the rate constant for the reaction with HO• at 25 °C

Gas	k'/s^{-1}	Lifetime
CH_4	6.2×10^{-9}	5 years
CO	2.0×10^{-7}	2 months
NO_2	1.4×10^{-5}	1 day
SO_2	1.5×10^{-6}	8 days

○ What other assumption is embedded in the estimated lifetimes in Table 4.1?

● The assumption of a constant HO• concentration.

In reality, the average HO• radical persists for only a few seconds in the atmosphere, so hydroxyl concentrations are actually quite variable. Clearly, the assumption of a constant concentration is more justifiable for the long-lived methane, for which fluctuations in HO• will average out, than for a short-lived species like NO_2.

4.2 Lifetimes and atmospheric transport

Air motions play a key role in the transport and mixing of chemical species: they are the means whereby substances released at the surface (or formed within the atmosphere) are distributed about. Close to the ground, winds gust and change direction frequently — and vertical movement is strongly influenced by local factors. The way that conditions in this region, commonly called the boundary or mixed layer, can result in severe air pollution episodes is a subject we return to in Section 5.

Above about 2–3 km, however — in the so-called free troposphere— there is a recognizable large-scale pattern of horizontal air movements. This *general circulation*, discussed earlier in the course, is the province of meteorology: it is manifest in the prevailing winds and weather systems experienced by different regions around the world. We shall not revisit the details. It is sufficient to recall that the general circulation is driven by the temperature contrast between low (warm) and high (cold) latitudes, but strongly influenced by the Earth's rotation about its axis. As a result, the regular wind regimes at all latitudes have a strong *zonal* (that is, east–west) component — and horizontal transport is fastest in this direction. As shown in Figure 4.2, it takes only a few weeks for air to circumnavigate the Earth at a given latitude. Within each hemisphere, there is a slower exchange of air between latitude bands — in the *meridional* (that is, north–south) direction. Transport *between* the hemispheres is slower still.

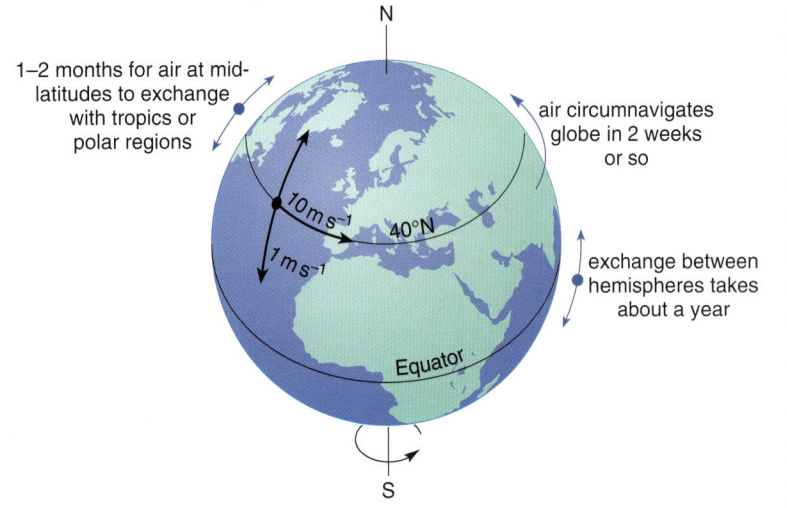

Figure 4.2 Typical timescales for global horizontal transport in the troposphere. Wind speeds around a given latitude band are about 10 m s⁻¹. Meridional transport is slower, with wind speeds of around 1 m s⁻¹.

○ What hinders the mixing of air across the Equator?

● As you saw earlier in the course, each hemisphere has it own Hadley circulation (or Hadley cell), with warm air rising near the Equator and sinking at higher subtropical latitudes (about 30° north or south).

The easterly Trade Winds in each hemisphere are the surface branches of these cells. The winds converge near the Equator, at the Intertropical Convergence Zone (ITCZ) — a region marked by strong convection, towering cumulonimbus clouds and heavy rainfall. The interhemispheric exchange of air takes place in part by the mixing of storm outflows at the ITCZ, and in part by the seasonal migration of the ITCZ north and south of the Equator (Figure 4.3) — causing, as one author has put it, 'tropical air to slosh between hemispheres'.

(a)

(b)

Figure 4.3 The band of cloud cover across Africa is linked with the ITCZ and shifts with the seasons. (a) In July, the ITCZ lies just north of the Equator and clouds barely extend to the thirsty Sahel (10–15° N) to drop a load of very unpredictable rainfall. (b) In January, the ITCZ moves south of the Equator, leaving the tropical forests free of cloud cover.

A comparison can now be made between the mean lifetimes of various species and the timescales for different transport processes. Figure 4.4 provides a simple way of visualizing that comparison. The point of interest is the distance or geographical scale (indicated along the bottom of the figure) over which a species can be redistributed during its lifetime.

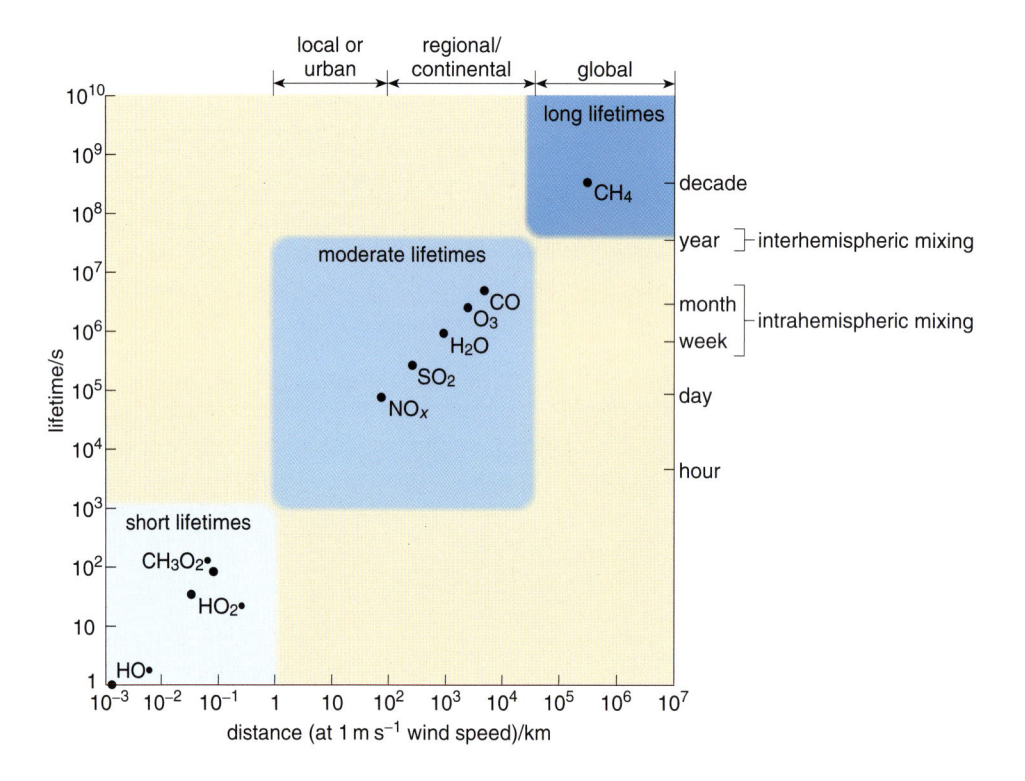

Figure 4.4 Comparison of mean lifetimes for some important trace species, and the spatial scale for their redistribution by atmospheric transport (assuming a mean wind speed of 1 m s^{-1}).

Clearly, radicals like HO• essentially live and die where they are born. At the other extreme, methane survives long enough to become reasonably well-mixed throughout the lower atmosphere. There is little *spatial* variability in its tropospheric concentration, despite the fact that surface emissions of methane are not evenly distributed around the world. Wetland sources, for example, are mainly concentrated in the tropics and in the boreal and tundra regions at high northern latitudes (Figure 2.4).

The situation is very different for compounds with moderate lifetimes — a day to a month or so — like those collected in the centre of Figure 4.4. Such species are removed from the air more quickly than the timescales required for large-scale atmospheric mixing, but can still be transported downwind on local or regional scales. As a result, their concentrations show high spatial variability even in the background atmosphere — reflecting more closely the geographical distribution of their sources.

Such considerations also have a direct bearing on the geographical extent — local, regional or global — of the **air pollution** resulting from human activities.

4.3 The influence of human activities

As we said at the outset, many natural trace gases are now classified as pollutants, by virtue of the enhanced atmospheric concentrations resulting from human activities. We concentrate here on three species whose life-cycles are interwoven with that of tropospheric ozone: CH_4, CO and NO_x. As well as providing a backdrop for the discussion of ozone air pollution in Section 5, the range of lifetimes displayed by these species also serves to illustrate the final point in Section 4.2.

For the long-lived (and well-mixed) CH_4, measurements from monitoring stations scattered around the world are sufficient to discern any trends in its concentration throughout the troposphere. Some recent values are included in Figure 4.5b, alongside evidence (Figure 4.5a) that the current situation continues a rising trend that began towards the end of the eighteenth century. Methane levels today (at 1.72 ppmv) are more than double the so-called 'pre-industrial' value (around 0.7 ppmv from Figure 4.5a). This increase coincided with a period of explosive population growth (from under 1 billion worldwide in 1800 to over 6 billion by the year 2000), thereby implicating human activities as the cause.

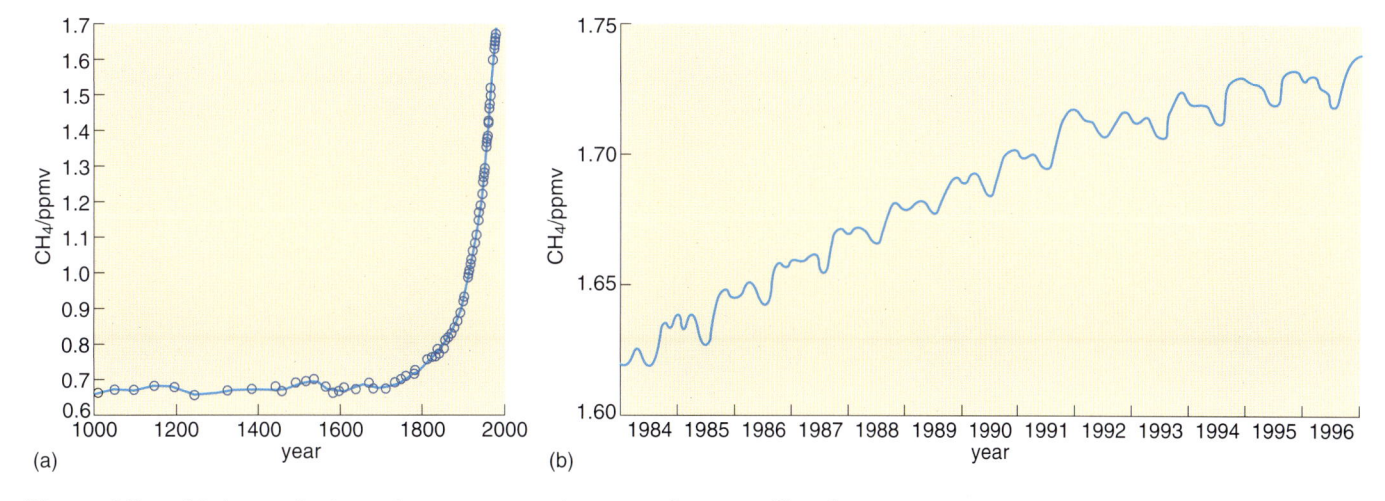

Figure 4.5 (a) Atmospheric methane concentrations over the past millennium, determined from ice cores drilled in Antarctica and Greenland, and from direct monitoring after 1970. As glacier ice is formed by compaction of successive layers of snow, small bubbles of air become trapped. When a sample of ice is drilled out, these air bubbles can be dated quite accurately, and when analysed, form an archive of past atmospheres. (b) Global-mean CH_4 concentrations from 1984 to 1996.

From the perspective of a simple box model (analogous to the one depicted in Figure 4.1 for H_2O), the stock of methane in the global atmospheric box is clearly *not* in a steady state: there is an imbalance between its sources and sinks. Table 4.2 gives a global budget for atmospheric methane in the mid-1990s, based on a recent assessment by the Intergovernmental Panel on Climate Change (IPCC). The entries under sources are current best estimates of the annual input of CH_4 from various identified sources, both natural and **anthropogenic** (human-related): a more detailed breakdown of the contribution each makes is given in Figure 4.6.

Table 4.2 Estimated sources and sinks of atmospheric methane.

	Best estimate/ 10^9 kg yr^{-1}
sources (total)	535
natural	160
anthropogenic (biogenic)	275
anthropogenic (fossil fuels)	100
sinks (total)	515
reaction with HO•	485
soils	30

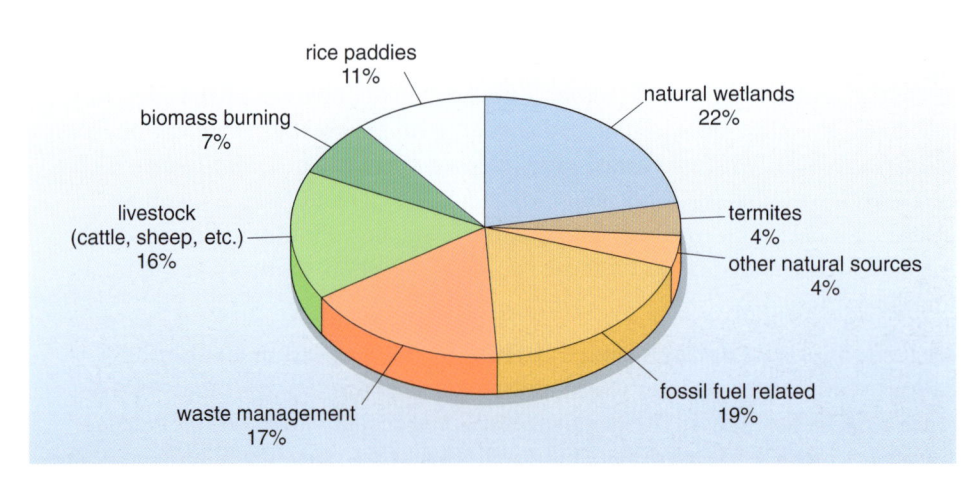

Figure 4.6 Contributions from identified natural and anthropogenic sources as a percentage of the total annual emissions (535×10^9 kg) of CH_4 worldwide.

○ Why should the IPCC be interested in the atmospheric burden of CH_4?

● Like CO_2, methane is a greenhouse gas. The build-up of CH_4 will enhance any global warming induced by the ongoing accumulation of atmospheric CO_2.

One striking feature highlighted by Table 4.2 is the dominant contribution (over 50%) from biogenic sources that are nevertheless associated with human activities. As shown in Figure 4.6, this category includes the emissions linked to food production mentioned earlier (Question 2.5), along with CH_4 released during waste management (by rotting organic matter in landfill sites, for example; Figure 4.7) and biomass burning. Under the 'fossil fuels' heading come emissions through leakage from natural gas pipelines, and by venting to the atmosphere at oil and gas production sites (Figure 4.8) and from coal mines.

Figure 4.7 Methane produced by the anaerobic decomposition of organic matter in landfill sites can build up underground and cause explosions. Chimneys like this one at a site in the UK allow release and burn-off of the gas.

Figure 4.8 An early evening view of the Arabian Desert and Gulf area, photographed by the crew of the Space Shuttle *Columbia* during a mission in 1990. The white areas are cities and the orange lights seen throughout the region are gas flares from oil exploration and production facilities both on- and offshore.

○ Are the estimates in Table 4.2 consistent with an ongoing increase in atmospheric CH_4?

● Yes, in the sense that there is a shortfall (of 20×10^9 kg per year) between the estimated sinks and sources.

This shortfall is broadly consistent with the observed rate at which CH_4 is accumulating in the atmosphere, estimated at 37×10^9 kg per year in the mid-1990s. Even this measure of agreement may be illusory, however. In reality, there are large uncertainties associated with the 'best estimates' recorded in Table 4.2 — as there are in attempts to track the way the various factors have contributed to the build up of CH_4 since pre-industrial times.

Is there any evidence of a *global* trend in CO levels? This is not an easy question to answer for this relatively short-lived species. Records from several monitoring stations in the northern hemisphere indicate an upward trend of around 1% a year for several decades up to the 1990s. Since then, CO concentrations appear to have steadied, or even declined somewhat, although there is no agreed explanation for this altered behaviour.

However that turns out, the telling point is that the budget for CO is now dominated by anthropogenic sources — as is that for the even shorter-lived NO_x. A breakdown of the various identified sources of these two species is given in Figure 4.9, where the global totals are estimated to be 2250×10^9 kg per year for CO and 52×10^9 kg per year for NO_x (expressed as the mass of nitrogen).

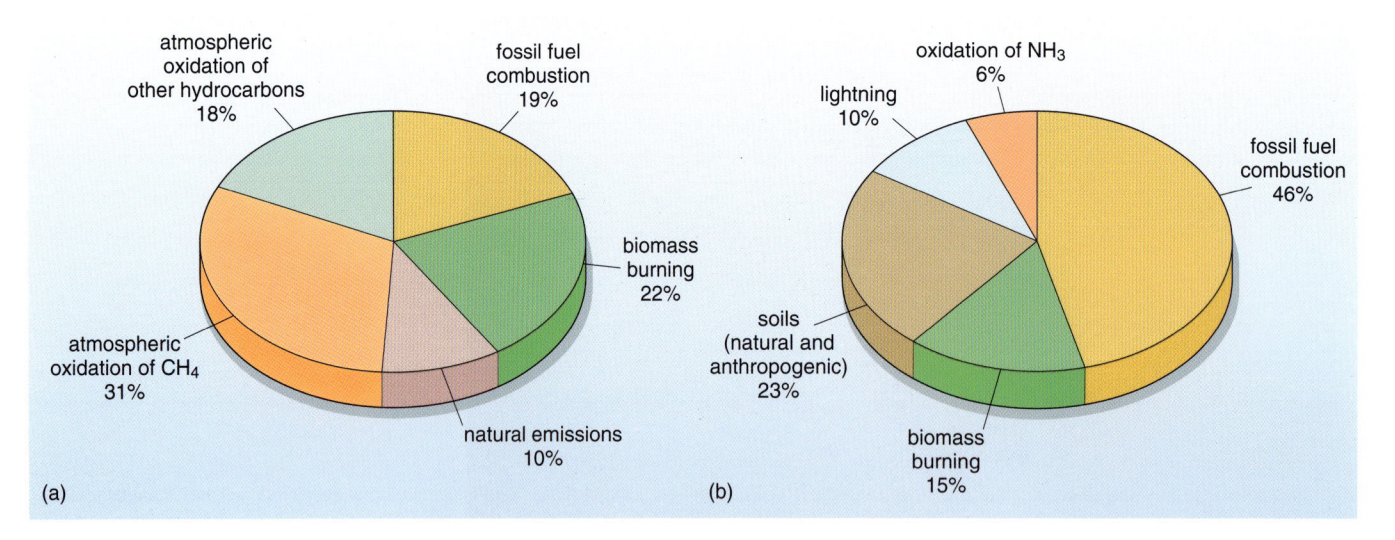

Figure 4.9 Contributions from identified natural and anthropogenic sources as a percentage of the total annual input worldwide for (a) CO and (b) NO_x.

Concentrating first on Figure 4.9a, note that the breakdown includes the CO generated *within* the atmosphere, as well as surface emissions. Given the discussion above, there is clearly a significant anthropogenic element to the CO derived from the oxidation of methane — as there is to that released by the deliberate burning of grasslands and tropical forests. As we shall see in Section 5, there are industrial and combustion sources of larger hydrocarbons than CH_4: again CO is one product of their atmospheric oxidation, as indicated in the figure. However, over half of the input from the oxidation of these *non-methane* hydrocarbons is believed to be related to *natural* emissions of such compounds from trees and other types of vegetation: more on this in Section 5.

Turning now to NO_x (Figure 4.9b), we find a familiar mix of biogenic and fossil fuel related sources linked to human activities. Emissions of NO from soils are known to be stimulated by agricultural practices, such as ploughing and the application of nitrogenous fertilizers and manures, etc. But clearly the main anthropogenic sources stem from the high-temperature combustion of living (i.e. biomass) or 'fossil' carbon. The bulk of the NO_x released by burning fossil fuels comes from the 'fixing' of atmospheric nitrogen. At the high temperatures involved in a car engine (typically in excess of 1500 °C), for example, nitrogen and oxygen in the air drawn in with the fuel may combine to form nitric oxide — much as they do during a lightning stroke:

$$N_2 + O_2 = 2NO$$

On leaving the exhaust system, NO can be oxidized to NO_2 (as discussed in Section 3.3.1).

The key point about relatively short-lived gases like CO and NO_x is that anthropogenic sources can have a major impact on atmospheric concentrations at the local or regional level. This is evident in the pattern of NO_2 concentrations for the UK shown in Figure 4.10. Note the 'hot spots' associated with large conurbations, especially London — and the rapid fall off in NO_2 levels away from urban areas.

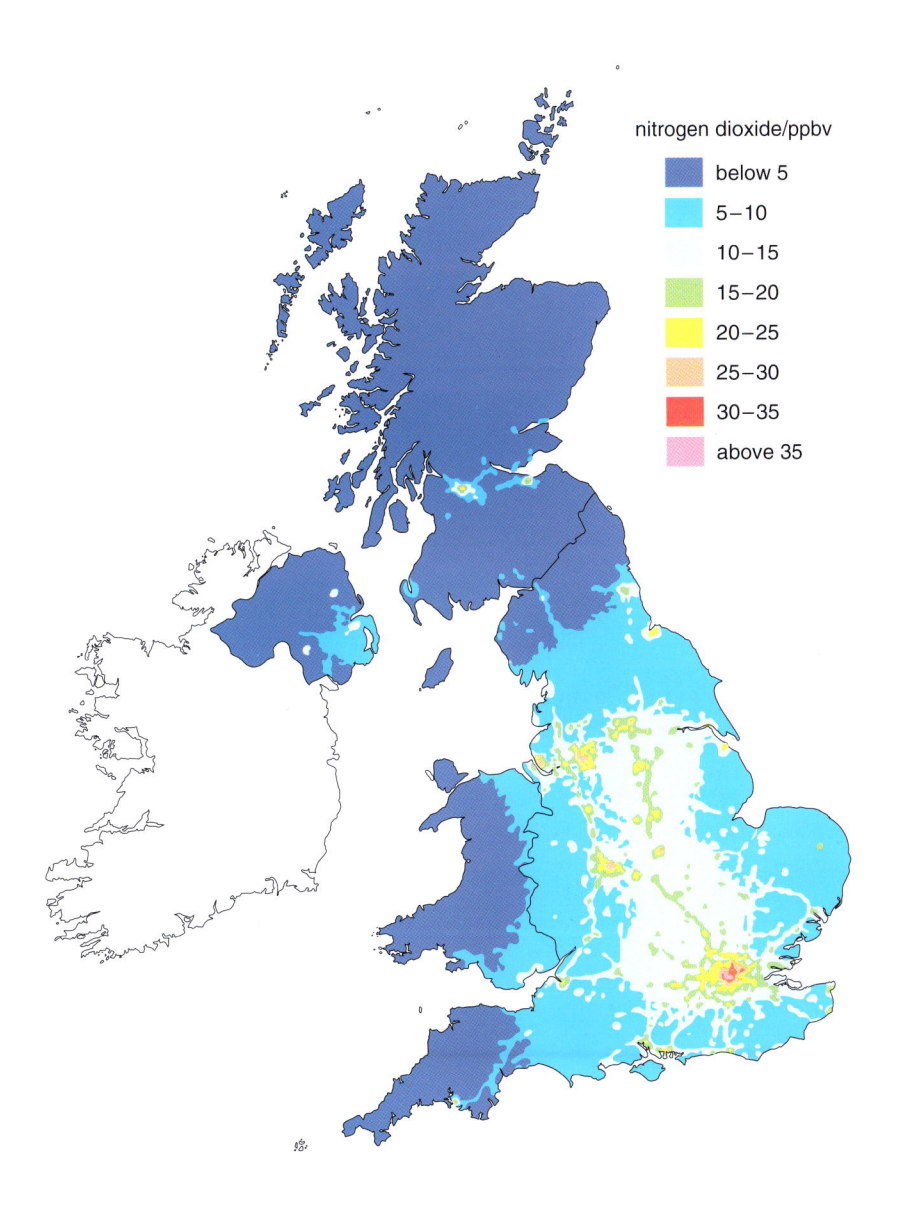

Figure 4.10 Annual mean NO$_2$ concentrations at ground level for the UK in 1996. The figures were assembled from a network of monitoring sites (in both urban and rural locations) spread through the UK.

○ What source of NO$_x$ is likely to be responsible for these urban hot spots?

● There are other combustion sources of NO$_x$, as we shall see in Section 5, but the main culprit in urban areas is undoubtedly road transport. Enhanced concentrations of NO$_2$ associated with the motorway network in England are also evident in Figure 4.10.

For a more global view, techniques now exist for the 'remote sensing' of certain minor atmospheric constituents using instruments aboard satellites and space shuttles. In the present context, 2001 saw release of the first dramatic images of the geographical distribution of CO around the world — and the way it changes from day to day throughout the year. Two examples are included in Figure 4.11.

30 April 2000

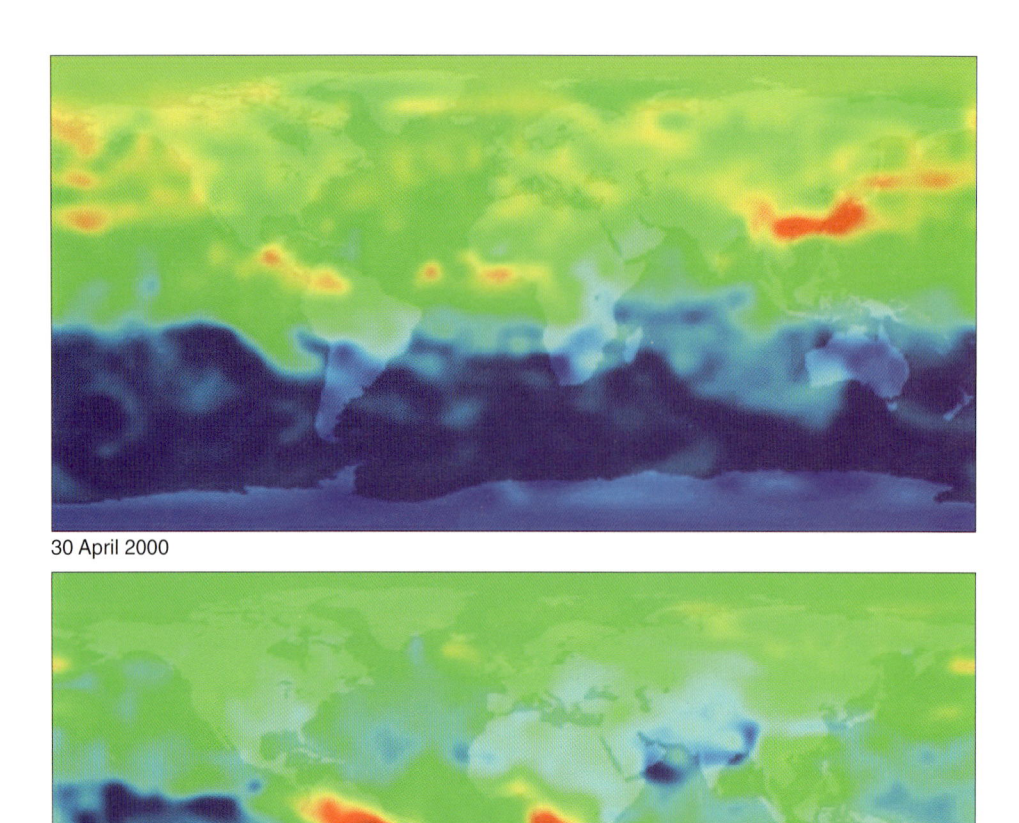

30 October 2000

carbon monoxide concentration/ppbv 50 — 220 — 390

Figure 4.11 Geographical distribution of CO mixing ratios in the free troposphere (4–5 km above the surface) on two days in 2000. The images come from data collected by an innovative global air pollution monitor, know as MOPITT (Measurements of Pollution in the Troposphere), onboard NASA's Terra spacecraft. MOPITT is making the first long-term global observations of CO as Terra circles the Earth from pole to pole, 16 times every day. (The Equator runs across the middle of each image; the locations of the continents are more evident in the lower one.)

We close this section with an activity that invites you to study Figure 4.11, and then attempt to interpret the observed variations in CO concentrations.

Activity 4.1

Describe briefly the *main* features of the spatial and seasonal variations in CO mixing ratios evident in Figure 4.11. Include a plausible explanation for the variations you describe, in terms of the main surface sources of CO. *(A couple of short paragraphs)*

4.4 Summary of Section 4

1. The atmospheric lifetime of a trace gas (A, say) can be defined as lifetime = [A]/L, where [A] is the atmospheric concentration of A, and L is the rate of loss. If the main removal mechanism is reaction with the hydroxyl radical, then $L = k \times [HO\bullet] \times [A]$, where the rate constant k is intrinsic to the reaction, usually dependent on temperature and determined by laboratory measurements. Additional sinks (e.g. wet and dry deposition) shorten the mean lifetimes of some species.

2. A long-lived species like CH_4 is well-mixed in the troposphere. By contrast, the concentrations of short-lived species (e.g. CO and NO_x) are very variable in space and time, and dependent on the strength of local or regional sources.

3. Today, the global budgets of CH_4, CO and NO_x are dominated by anthropogenic sources — both enhanced biogenic emissions (linked to food production, waste management and biomass burning) and emissions related to the extraction and combustion of fossil fuels.

4. Human activities have increased the total atmospheric burden of CH_4 since pre-industrial times, with possible consequences for global climate. Anthropogenic sources of CO and NO_x perturb background levels on local or regional scales.

Question 4.3

Table 4.2 (Section 4.3) includes an estimate of the mass of methane removed from the atmosphere each year by reaction with the HO• radical. How would you go about making an estimate like this? What information would you need?

(*Hint*: As a starting point, note that Equation 4.4 in Section 4.1 implies that $k' = 1/\text{lifetime}$; then try substituting this expression for k' into Equation 4.2.)

Question 4.4

Uptake by vegetation is often an important sink for ozone close to the ground (in the boundary layer) in rural areas. The rate of loss by this dry deposition is proportional to the prevailing ozone concentration, and (by analogy with Equation 4.2) can be expressed as:

$$\text{rate of loss} = k\,[O_3]$$

A typical value of the 'rate constant' for dry deposition of ozone is $k = 1 \times 10^{-5}\,s^{-1}$. What is the lifetime of ozone with respect to this process?

Question 4.5

The rate constant (k) for the reaction between CO and HO• does *not* depend on temperature. Yet CO actually has a very variable atmospheric lifetime, increasing from about a month in the tropics to close to a year at high latitudes in winter. Try to suggest a plausible explanation for this behaviour.

(*Hint*: Think about the primary source of HO•, discussed in Section 3.3.1)

5 Ozone air pollution

5.1 Introduction: evidence of a perturbed atmosphere

Tropospheric ozone is the ultimate precursor of HO• (Section 3.3.1), and so plays a key role in maintaining the oxidizing power of the lower atmosphere — and hence in controlling the composition of this region. But one can have too much of a 'good thing'. In the lower atmosphere, O_3 acts as a greenhouse gas (like CH_4) and is also extremely hazardous to life. Breathed in by people, or taken up by the leaves of plants, ozone's vigorous oxidizing powers become all too apparent — with damaging effects for both human health and vegetation. This is why enhanced ozone concentrations at ground level are viewed as a form of air pollution.

The *potential* for O_3 formation in the lower atmosphere is large. Indeed, there is evidence from old ozone records to suggest that O_3 levels were much lower a century ago. During the late nineteenth century several sets of ozone data were obtained using a technique devised by Schönbein (the discoverer of the gas; see Box 5.1)

Figure 5.1 compares recent re-evaluations of some of these data sets with more modern measurements for three remote sites (Arkona, off the Baltic coast of Germany, Ispra in Italy and Cape Grim, Tasmania). Instead of the average O_3 concentrations of 20–40 ppbv typically observed in the 1980s, the levels a century earlier were closer to 10 ppbv throughout the year. Notice too the much more marked seasonal variations in the modern measurements, with a local summer maximum (June–July in the north; December–January in the south).

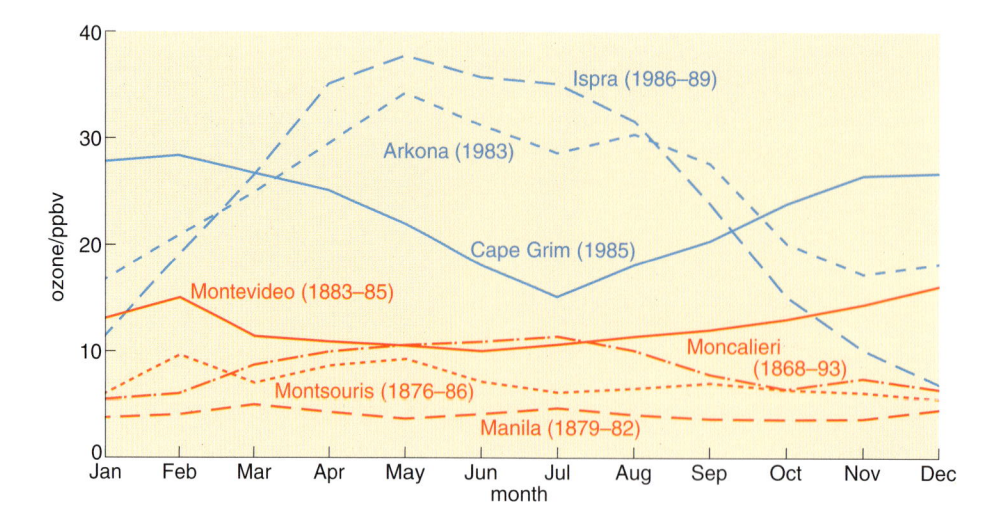

Figure 5.1 Average monthly concentrations of ozone at several sites in the nineteenth century (Montsouris is southwest of Paris; Moncalieri is in Italy) (in red) compared with others during the twentieth century (in blue).

Box 5.1 Schönbein's legacy

As Richard P. Wayne has written (*Chemistry of Atmospheres*, third edition, Oxford University Press, 2000, pp. 390–391):

> Christian Friedrich Schönbein [Figure 5.2] is often regarded as the 'father of air chemistry' because, soon after his discovery of ozone in the laboratory in 1839, he demonstrated the presence of ozone in the atmosphere of several European cities. For this purpose, he employed test papers coated with potassium iodide. As Schönbein had found out already, ozone liberates iodine from the iodide ions, and the iodine can interact with the starch in the paper (or deliberately added) to form a dark-blue complex. By comparing the depth of colour with tints on a calibration chart after the test papers had been exposed to air for a fixed period, it was possible to reach an estimate of the atmospheric ozone concentration. *(Our parenthesis)*
>
> Problems beset the use of these Schönbein papers, but it has proved possible to allow for what are thought to be the major sources of error, and hence re-evaluate early measurements made using this technique.
>
> A variant of the Schönbein method (based on liberating iodine from iodide) is still used today in 'ozone sondes'. Here a balloon-borne device released at the surface can track ozone concentrations up to heights of about 30 km (where the balloon bursts), providing information about the ozone altitude profile in a given location (and *changes* therein, of the kind shown in Figure 3.11, for example). Other techniques for measuring atmospheric ozone rely mainly on its absorption of solar uv radiation, using instruments on the ground, or deployed on aircraft or satellites.

Figure 5.2 Born in Swabia in southern Germany, Schönbein (1799–1868) was a fascinating character, full of contradictions. He was a passionate experimentalist, yet there was a speculative, anti-materialistic cast to his thought that ultimately blocked his efforts to uncover the true nature of ozone. As he wrote to Liebig in 1853: 'The popular French–English fiction, which wants to view the entire chemical world of phenomena as an atomistic game and tries to explain everything by the 'arrangement des molecules', stands very low in my estimation, as you may know'. It is a cruel irony that ozone was later seen as one of the most emphatic arguments in favour of the ideas Schönbein so distrusted.

It is important to emphasize that the modern records in Figure 5.1 are typical of the pattern found today in relatively *remote* areas: the past century saw an increase in the *background* concentration of tropospheric ozone. Thus, O_3 generated in the kind of pollution episodes to be examined later in this section are effectively superimposed on a baseline level that is already higher than it used to be.

As a starting point for the discussion to come, Figure 5.3 is a reminder of the sources and sinks of tropospheric ozone that were identified earlier (Section 3.3.2) — with one addition. Dry deposition of ozone at the surface (the origin of its damaging effects) also acts as a sink for the gas (as noted in Question 4.4).

Figure 5.3 Schematic summary of the main sources and sinks of tropospheric ozone.

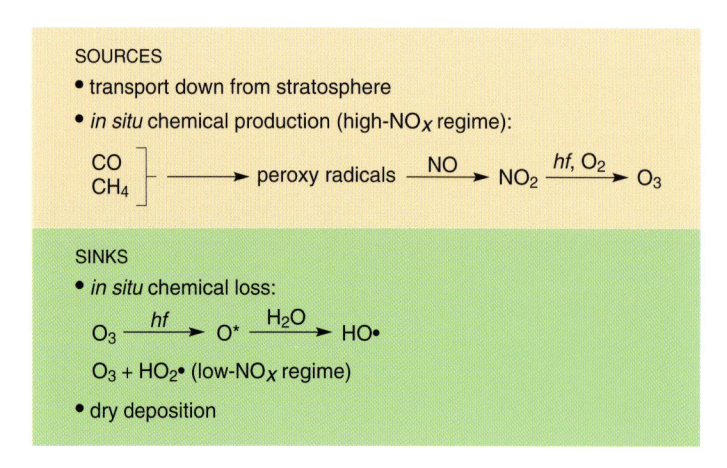

SOURCES
- transport down from stratosphere
- *in situ* chemical production (high-NO_x regime):

$$\left.\begin{array}{c}CO\\CH_4\end{array}\right\} \longrightarrow \text{peroxy radicals} \xrightarrow{NO} NO_2 \xrightarrow{hf, O_2} O_3$$

SINKS
- *in situ* chemical loss:

$$O_3 \xrightarrow{hf} O^* \xrightarrow{H_2O} HO\bullet$$

$O_3 + HO_2\bullet$ (low-NO_x regime)

- dry deposition

It is now fairly well established that, on a global scale, the ozone gained by stratospheric injection roughly balances that lost via dry deposition — and that both are anyway relatively minor terms in the overall O_3 budget in the lower atmosphere. In other words, the abundance of tropospheric ozone is largely controlled by *in situ* production and loss. *Now try the following Activity.*

The following activity gives you an opportunity to use your understanding of the material in Section 3.3 to think through the likely impacts on tropospheric ozone of the kinds of human activities reviewed in Section 4.3.

Activity 5.1

Given the final point above (together with the information in Figure 5.3), explain briefly, *but carefully*, why the key to enhanced levels of tropospheric ozone in the background atmosphere is likely to be larger and more widespread sources of NO_x. Include in your answer the link with the levels of atmospheric CO and CH_4, and how these have been influenced by human activities.

Structure your answer under the headings 'Introduction', 'Discussion' and 'Conclusion', using bullet points to identify the meat of the argument.

5.2 Ozone and photochemical smog

Book 3 of *Our Mutual Friend* by Charles Dickens opens with the following graphic description:

> It was a foggy day in London, and the fog was heavy and dark. Animate London, with smarting eyes and irritated lungs, was blinking, wheezing and choking; inanimate London was a sooty spectre … Even in the surrounding country it was a foggy day, but there the fog was grey, whereas in London it was, at about the boundary line, dark yellow, and a little within it brown, and then browner, and then browner, until at the heart of the City — which call Saint Mary Axe — it was rusty-black.

The word did not exist at the time, but Dickens is clearly describing 'smog' — the devastating combination of *sm*oke (from burning coal) and *f*og that has afflicted many industrialized cities. Reports of such conditions in London date back to the fourteenth century, but the most tragic episode occurred in December 1952 (Figure 5.4). Such events were all but eliminated in the UK by a combination of measures — the most important being the Clean Air Acts of 1956 and 1968, and the introduction of smokeless zones.

Figure 5.4 The smog of 1952 brought London to a halt for a week and left 4000 people dead. Those who lived through it endorse Dickens' description, and recall its foul taste and smell. London's unhappy place in smog history owes much to the size and extensive industrialization of the city. In addition, British coal has a high tar and sulfur content, producing large amounts of smoke and SO_2 when it burns. This combination is very effective in fog nucleation under stagnant, humid conditions during the winter months.

Meanwhile, the late 1940s saw reports of a new type of **smog** (now taken to mean any sort of hazy pollution) in the Los Angeles area (Figure 1.3). In marked contrast to the London version, this was characterized by air containing high levels of ozone and other eye-watering and plant-damaging pollutants, and occurred on *hot* days with *bright* sunshine: hence the term **photochemical smog**. Much research has been devoted to studying the smogs of the Los Angeles region. Pollutants and reactions first characterized there have since been detected in the air of cities all over the world — in Greece, Mexico (Figure 5.5), Brazil, Japan and Australia, and even the UK. In short, photochemical pollution is now recognized to be a worldwide problem.

Figure 5.5 View over Mexico City, showing the murky haze of photochemical smog trapped below an inversion layer (see Section 5.3), with clear skies above.

Figure 5.6 Satellite image and map of the region of southern California around Los Angeles. Forest-covered mountains lie to the north and east of the city.

Clues to the basic requirements for photochemical smog formation are provided by the factors that made the problem so severe in Los Angeles and southern California. The Los Angeles Basin is open to the Pacific to the south-west, but is otherwise nearly encircled by mountains (Figure 5.6). This geographical setting, together with certain meteorological conditions, often results in pollutants being trapped in the basin. Intensely sunny days are common. And there is a very high density of motor vehicles — a major source of pollutants in urban areas. Here, we look first at the action of sunlight on the mix of **primary pollutants** in vehicle exhaust fumes, which produces various **secondary pollutants**, including ozone. The influence of geographical and meteorological factors is taken up in Section 5.3.

5.2.1 What is observed on a smoggy day?

Several reproducible features of a classic smog episode in southern California are evident in Figure 5.7. Concentrations of nitric oxide (NO) increase during the early morning rush hour. Subsequently, NO is replaced by NO_2 and ozone levels start to build-up — peaking shortly after noon, when the NO concentration drops to a low value. By this stage, there is a mucky haze (evident in Figure 5.5) caused by fine airborne particles, and the eyes water because substances that act as powerful lachrymators (eye irritants) are also present.

What of the chemistry that gives rise to this unpleasant sequence of events? Clearly, one key ingredient is NO — produced by fixing atmospheric nitrogen in the internal combustion engine (Section 4.3), and released in the exhaust gases. As we emphasized earlier, however, the vital precursor for O_3 generation in the lower atmosphere is NO_2 (an idea revisited in Activity 5.1), not NO. So we need a means of converting the primary pollutant NO into NO_2, before ozone can be formed.

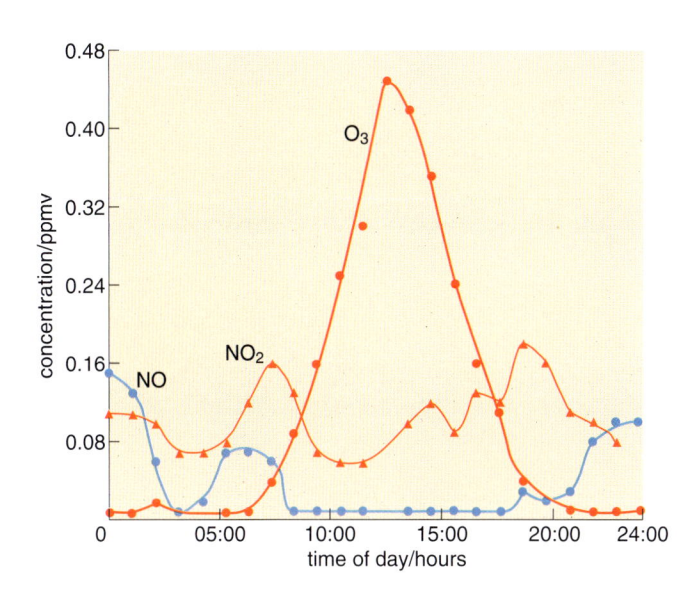

Figure 5.7 Variations in the mixing ratios of some primary (blue) and secondary (red) pollutants during the course of a smoggy day in southern California in the 1970s.

The situation mirrors that in the background troposphere discussed in Section 3.3.2. Thus the reaction of NO with O_3:

$$NO + O_3 \longrightarrow NO_2 + O_2 \tag{R12}$$

would do the trick, but requires O_3 to be available already. In the polluted urban atmosphere, one effect of reaction R12 seems to be to prevent a build up of O_3 until most of the free NO has been converted to NO_2, as suggested by the observed behaviour in Figure 5.7.

○ What species effect the conversion of NO to NO_2 without ozone consumption in the unpolluted troposphere?

● Peroxy radicals, generated from CO and CH_4.

This is where the other primary pollutants in the exhaust gases come into the picture.

5.2.2 The role of road transport

The most important reaction in a petrol engine — the one that provides the energy to drive the vehicle — is the combustion of fuel in air. Petrol can contain well over 300 different **hydrocarbons** (HCs; see Box 5.2), with carbon numbers ranging from C_4 to C_{12}. The mix includes saturated and unsaturated compounds and aromatics, but we can take octane (C_8H_{18}) as a typical constituent. Then in an ideal engine, combustion would be complete, with CO_2 and steam as the only products:

$$2C_8H_{18} + 25O_2 = 16CO_2 + 18H_2O \tag{R21}$$

In practice, the internal combustion engine falls short of this ideal. *Incomplete* combustion of the fuel leads to emissions of CO and a wide range of unburnt hydrocarbons and other partial oxidation products. These emissions are particularly high during both idling (traffic jams come to mind) and deceleration, when insufficient air is taken in for complete combustion to occur.

Box 5.2 Hydrocarbons

Crude oil, petrols and diesels are all mixtures containing many different hydrocarbons, each with its own characteristic molecule. But in every case, the molecule contains just carbon and hydrogen atoms, linked together according to two simple rules:

1. Each H atom forms just one bond, which must link it to a C atom.
2. Each C atom forms four bonds, any one of which may be linked to either an H atom or another C atom.

We can get a handle on the bewildering variety of hydrocarbons by grouping them into classes or families, in terms of their *structural formulae*. Some typical examples (structures **1** to **8**) are collected under three headings in Figure 5.8.

Alkanes are hydrocarbons in which the only type of bond between two carbon atoms is the single C–C bond seen in structure **2**: this is ethane (C_2H_6), a minor constituent of natural gas. In the normal, or *straight-chain alkanes*, all the C atoms, and the single bonds between them, can be written down as a single chain. One then links H atoms to the chain of C atoms until rule 2 is satisfied, as shown in structure **3** — which is octane (C_8H_{18}).

However, this is not the only structural formula with the molecular formula C_8H_{18}. Structure **4** shows a modification of the octane structure which contains C atoms linked to more than two other C atoms — an example of a *branched-chain alkane*. It used to be known as *iso*-octane (as shown in Figure 5.8), and is important because it is the octane of the well-known octane number of petrol.

In structure **4**, methyl groups (CH_3–) — derived from the simplest hydrocarbon, methane (structure **1**) — branch off the main chain of C atoms. Chemists commonly use the symbol R to represent such **hydrocarbon groups**, leading to the simplified notation RH for *any* alkane.

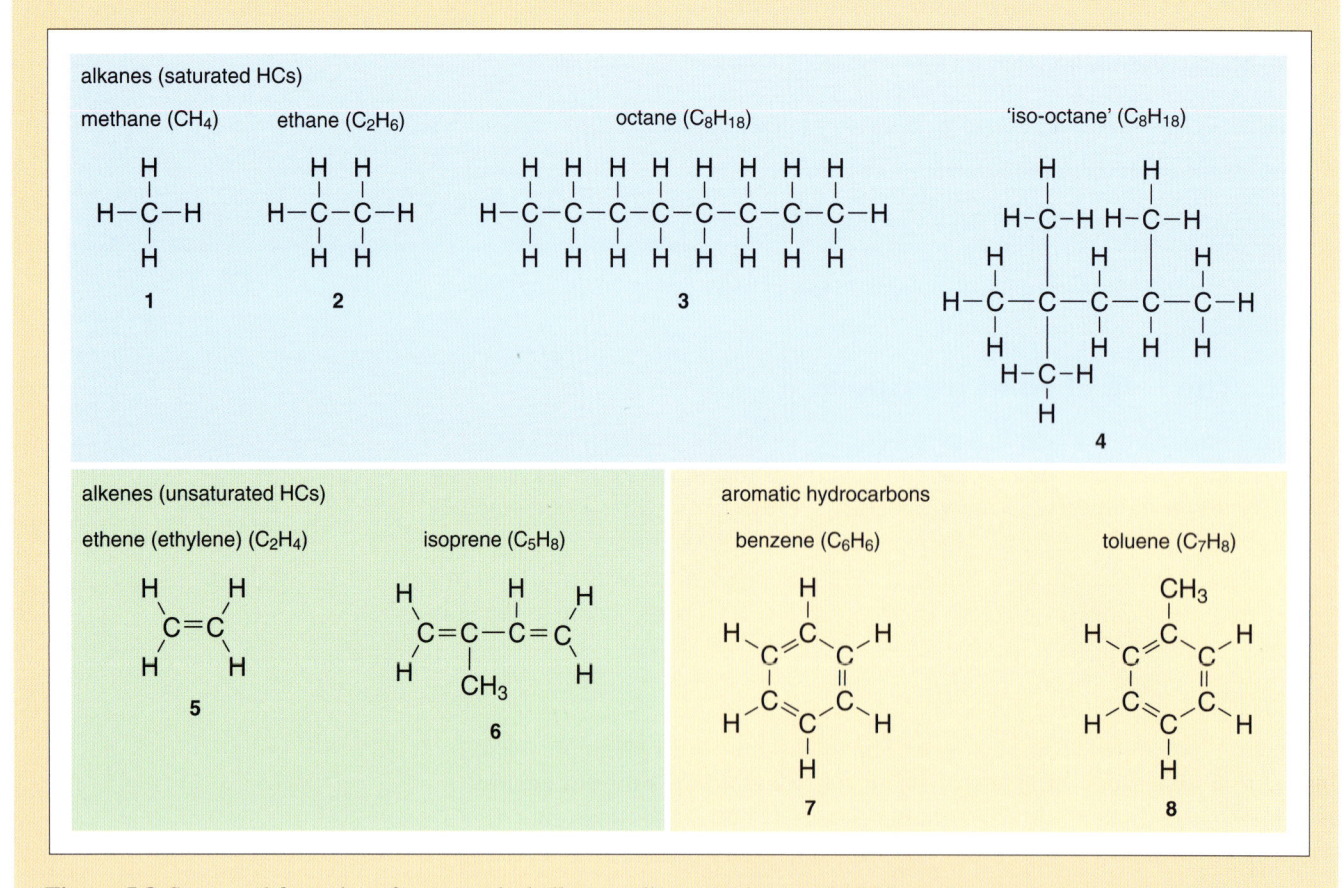

Figure 5.8 Structural formulae of some typical alkanes, alkenes and aromatic hydrocarbons.

○ If RH is used to represent ethane (structure **2**), what does the R stand for in this case?

● Removing one H atom from structure **2** leaves a CH_3 group followed by a CH_2 group; that is, CH_3–CH_2–. This is known as the ethyl group, which can be abbreviated further to C_2H_5–.

In all alkanes, any two neighbouring C atoms are linked by just a single C–C bond. These compounds are also known as **saturated hydrocarbons** because they contain the highest possible ratio of hydrogen to carbon atoms. But there exist other hydrocarbons in which two adjacent C atoms use two of their four bonds to bind themselves to one another — by means of a carbon–carbon double bond, C=C. These compounds are called **alkenes**, the simplest example being ethene (structure **5**). Alkenes are examples of **unsaturated hydrocarbons**, because (in accordance with rule 2) they necessarily contain less hydrogen than the corresponding alkane: compare the structural formulae of ethane (**2**) and ethene (**5**), for example.

Plants generate and emit a large variety of unsaturated hydrocarbons, including both ethene (ethylene in old parlance) and isoprene (structure **6**). Ethene is known to play a key role in controlling plant growth and development: seed germination, flowering, fruit ripening, senescence of flowers and leaves, and sex determination are all affected by exposure to ethene. Isoprene probably accounts for about half of the global total of HC emissions from living plants, but its function (or functions) is uncertain.

Notice that the main carbon chain in isoprene (structure **6**) consists of alternate single and double carbon–carbon bonds. A chain of six C atoms linked in this way, but connected up nose to tail to form a ring, is the characteristic feature of **aromatic hydrocarbons** — the final category in Figure 5.8. Benzene (structure **7**) is the simplest example. A family of aromatic hydrocarbons can be built up by replacing one or more of the H atoms in benzene with a hydrocarbon group. Toluene (structure **8**) is the simplest compound of this type.

The information collected in Table 5.1 emphasizes just how significant road transport is as a source of these carbon-based pollutants — and of NO_x. The data are for the UK, but the breakdown for other industrialized nations shows a broadly similar pattern.

Table 5.1 Contributions from different anthropogenic sources of some of the principle primary air pollutants in the UK in 1994.

Source	% of total emissions		
	NO_x	CO	HCs
road transport (including fuel production and evaporation)	56	91	41
other transport	7	1	1
electricity supply industry	24	<1	<1
other industry	8	1	20
solvent use	–	–	31
domestic/commercial	5	6	2
other	<1	1	5
total/10^6 kg	2218	4833	2117

Just add sunlight to the brew of chemicals in the exhaust gases from a combustion engine, and the scene is set for ozone generation.

Like methane, other hydrocarbons are subject to hydrogen abstraction by the HO• radical. Using the simplified notation RH for any alkane (Box 5.2), this initial attack yields a hydrocarbon radical, R• (analogous to •CH$_3$), which can then go on to form an organic peroxy radical, RO$_2$•, via a reaction analogous to reaction R16:

$$RH + HO• \longrightarrow R• + H_2O \tag{R22}$$

$$R• + O_2 \longrightarrow RO_2• \tag{R23}$$

Like HO$_2$• and CH$_3$O$_2$•, these larger peroxy radicals rapidly convert NO into NO$_2$:

$$RO_2• + NO \longrightarrow RO• + NO_2 \tag{R24}$$

Photolysis of NO$_2$ then leads to ozone production, as well as regenerating NO for re-oxidation to NO$_2$.

In effect then, smog ozone chemistry is just a grotesquely exaggerated version of the ozone chemistry in the unperturbed troposphere discussed in Section 3.3. Higher concentrations of the key species (NO$_x$, CO and HCs) are present in polluted environments, together with a wider variety of saturated, unsaturated and aromatic HCs (Box 5.2). For example, small alkenes (e.g. ethene) are typical products of engine combustion, and a proportion of the aromatics (e.g. benzene, toluene) present in the fuel often escapes combustion altogether. *All these compounds are more potent ozone precursors than CH$_4$ and CO.*

One simple measure of a hydrocarbon's potency in this respect is how fast it reacts with HO•, the process that drives formation of the all important peroxy radicals (R23). Table 5.2 lists values of the rate constant, k, for reaction with HO• for a variety of hydrocarbons, including some of the compounds typically found in vehicle exhausts: comparison with the values for methane and CO underlines the higher *intrinsic* reactivity of the other compounds. Notice in particular the enhanced reactivity of the unsaturated compounds (e.g. ethene and propene) compared with their saturated counterparts (ethane and propane). The presence of a double bond (C=C) provides an alternative site for rapid attack by HO• (alternative to hydrogen abstraction via reaction R22, that is). This again leads on to the formation of peroxy radicals, but we shall not pursue the details. The important general point is that reactive HCs, rather than CH$_4$ and CO, are the dominant source of peroxy radicals in polluted environments.

Table 5.2 Values of the rate constant, k, at 25 °C for the reaction between HO• and selected hydrocarbons (and CO).

Compound	$\dfrac{k}{10^{-12} \text{ (molecules cm}^{-3})^{-1}\text{ s}^{-1}}$
carbon monoxide	0.20
alkanes	
methane	0.0062
ethane (C$_2$)	0.27
propane (C$_3$)	1.1
alkenes	
ethene (C$_2$)	8.5
propene (C$_3$)	26
isoprene	101
aromatics	
benzene	1.2
toluene	6.0

Question 5.1

What do you conclude about the atmospheric lifetimes of the species in Table 5.2, compared to those of methane and CO?

5.2.3 Other aspects of photochemical smog

Reactions R22–R24 above are only part of the smog story. To explain the haze, and the unpleasant eye-watering effects, we must follow the fate of the RO• radicals formed in reaction R24. Again, there are parallels with the atmospheric oxidation of CH$_4$, where R = CH$_3$.

○ Look back at Figure 3.17a in Activity 3.1. Write an equation to represent what happens to CH$_3$O• in this oxidation pathway.

● It reacts with oxygen in the air as follows:

$$CH_3O• + O_2 \longrightarrow CH_2O + HO_2• \quad (R25)$$

Both of the products in reaction R25 deserve comment. First, HO$_2$• is, of course, a further potential source of ozone, via the reaction (R13) that also regenerates hydroxyl:

$$HO_2• + NO \longrightarrow HO• + NO_2 \quad (R13)$$

Other RO• radicals can react in a similar way to CH$_3$O•, again generating HO$_2$• together with an aldehyde — analogous to the second product, formaldehyde (CH$_2$O), in reaction R25. Just as methane is the simplest hydrocarbon, so formaldehyde is the simplest example of the family of aldehydes — usually represented by the generic formula RCHO (see Figure 5.9).

Like formaldehyde (Figure 3.17b in Activity 3.1), other aldehydes are subject to attack by HO• radicals. Such attack continues the hydrocarbon oxidation pathway, and can lead eventually to carbon monoxide — the additional *in situ* source of CO identified in Figure 4.9 (Section 4.3). However, aldehydes are also implicated in the formation of the powerful eye irritants that are, perhaps, the most obviously unpleasant feature of exposure to photochemical smog. In addition, formaldehyde and acetaldehyde are both toxic and potential carcinogens (causing cancer).

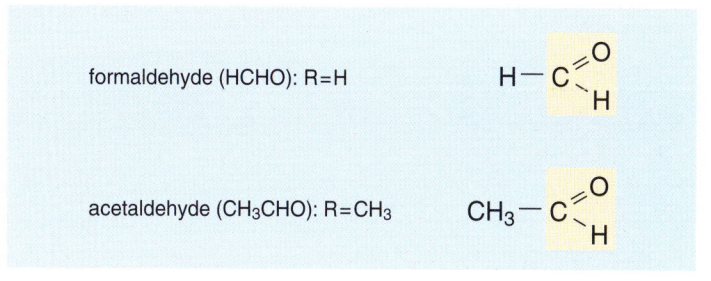

Figure 5.9 Structural formulae of formaldehyde and the next member of the aldehyde family, acetaldehyde. The characteristic feature of an aldehyde is highlighted in colour, and this is reflected in the generic formula RCHO.

Another undesirable effect of photochemical smog is the reduced visibility that arises because light is scattered by airborne particles or *aerosols*. This can be a hazard in itself. However, the chemical composition of these aerosols is also cause for concern: some of the aromatic compounds found are known to be carcinogens. The processes that lead to aerosol formation are complex, and still not fully understood. We shall not pursue the matter further, but it is interesting to note that some areas have long been prone to what appears to be a *natural* variant of urban photochemical smog. Blue hazes are formed in summer over certain forested areas, such as the Smoky Mountains (in the USA) and the Blue Mountains (in Australia; Figure 5.10). This phenomenon has been ascribed to the interaction between HCs released by the trees (the smell of a pine forest is due to such biogenic HCs) and ozone naturally present in the troposphere.

Figure 5.10 The Three Sisters rock formation in the Blue Mountains National Park at dawn. The sheer sandstone peaks rise nearly 700 m above the surrounding eucalyptus forests. Oil from these trees sometimes leads to the blue mists that give the park its name.

5.3 The role of thermal inversions

Having dealt with the chemistry of photochemical air pollution, we now look briefly at the influence of weather and geography. Clearly, the amount of sunshine is an important factor in ozone generation. However, the vicissitudes of the weather come into the equation in other ways as well. High wind speeds help to disperse air pollutants rapidly, and so reduce their impact on local air quality. The roughness of the ground also produces turbulent mixing of surface air, and this too promotes the dispersion of pollutants.

But the key factor in many serious pollution episodes is the extent of *vertical* mixing in the atmosphere. This is linked to the stability of the lower atmosphere in a given location — which, in turn, depends on how the local temperature profile (the **environmental lapse rate**, **ELR**) compares with the **adiabatic lapse rate** (**ALR**). Two possible situations are shown schematically in Figure 5.11.

○ Do you recall the significance of the ALR from earlier in the course?

● A rising parcel of air expands as the surrounding pressure falls, and as it expands it cools. The ALR is the expected decrease in temperature with height. For a parcel of dry air, it has a value of about 10 °C km^{-1}, or 1 °C for every 100 m. The value is somewhat lower for moist air, because the heat released by condensing water vapour slows the rate of cooling.

Figure 5.11 Schematic representation of the actual temperature profile above a given location (the ELR, red lines) compared with the ALR (blue lines) for (a) unstable and (b) stable conditions.

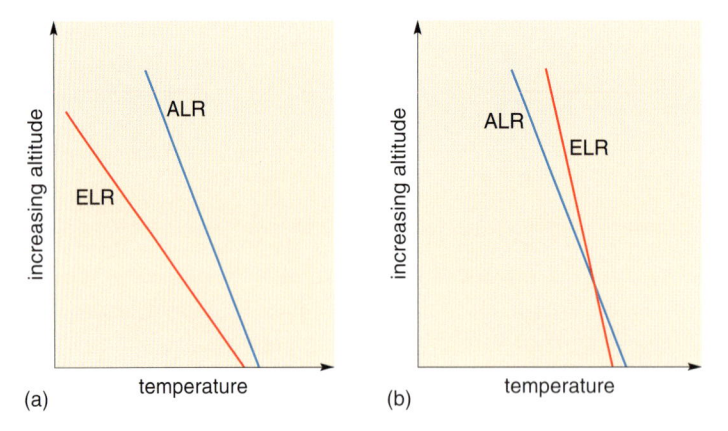

Try the following question to check that you understand what is meant by a 'stable' or 'unstable' atmosphere.

Question 5.2

Explain why the situation depicted in Figure 5.11a represents an unstable atmosphere, conducive to convection and rapid vertical mixing, whereas that depicted in Figure 5.11b represents a stable situation. (*A couple of short paragraphs*)

An unstable situation is common during daytime in summer. Heating of the surface increases air temperatures close to the ground, and temperature then decreases with height more rapidly than the ALR. Strong vertical mixing helps to disperse pollutants released at the surface.

By contrast, stable conditions necessarily arise if there is a thermal inversion at some height above the surface — a region of the *troposphere* through which temperature actually increases with increasing altitude, before starting to fall again. *The formation of thermal inversions in the lower atmosphere is one of the most important meteorological factors contributing to air pollution problems in urban areas.*

The most common cause of a **low-level inversion** (Figure 5.12) is overnight radiative cooling of the ground. Under clear night skies, the air close to the surface becomes colder than the air above, generating a stable inversion layer. Under calm conditions, such radiation inversions can be as low as 100 m or so above the ground. If winds are strong, however, the cold surface air is forced upwards by mechanical turbulence, and moderately stable conditions can then extend to some height in the atmosphere, effectively trapping ground-level emissions. After sunrise, heating of the surface gradually erodes the inversion until the unstable daytime profile (Figure 5.11a) is re-established, allowing any trapped pollutants to be dispersed.

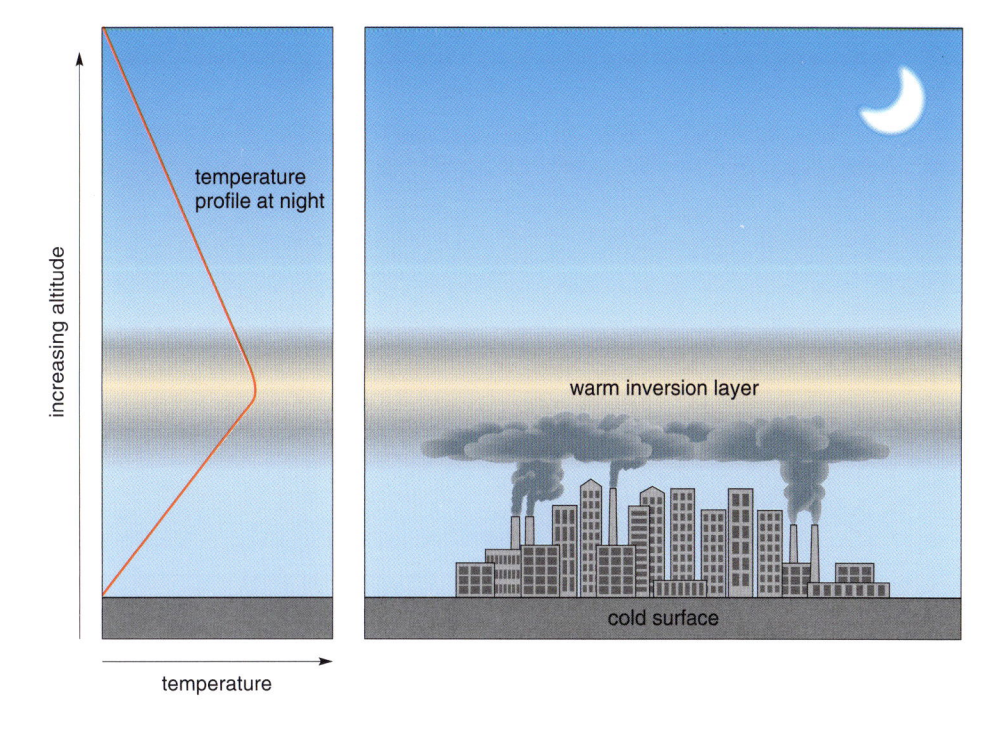

Figure 5.12 Schematic representation of a low-level radiation inversion.

○ In winter, radiation inversions can lead to fog formation, and are often slow to break-up the following day. Can you suggest why?

● Solar heating is weak in winter, and made more feeble still by the radiation scattered back from the upper surface of a bank of fog. The ground does not warm up.

Such conditions persisting for several days resulted in the disastrous sulfurous smog in London in the winter of 1952.

○ Mexico City suffers from serious pollution problems. It is situated in a valley between high mountains. Why would this geographical location be likely to exacerbate the problem of low-level inversions?

● If you have a frost-pocket in your garden, you will know that at night cold, dense air flows downslope under the influence of gravity. In a similar way, the air in a valley bottom can become cooler than the air immediately above, and this again inhibits vertical mixing.

Larger scale movements of air may also give rise to inversions, but higher up in the troposphere, typically at a height of 2 or 3 km — this situation is depicted in Figure 5.13 and is known as a **subsidence inversion**. It often occurs during anticyclonic (high pressure) conditions at the surface. Recall that air spirals downwards and outwards within an anticyclone. Just as a rising parcel of air expands and cools, so air subsiding at the centre of an anticyclone is compressed and progressively warms with decreasing height. Because turbulence is almost always present near the ground, this lowermost portion of the atmosphere is usually prevented from participating in the general subsidence. Thus, an inversion develops aloft between the lower turbulent zone and the subsiding warmer layers above. Vertical mixing of surface air is then limited to the region below the inversion — known as the boundary layer or mixed layer.

Figure 5.13 Inversions aloft often develop in association with slow-moving centres of high pressure, where the air aloft subsides and warms by compression.

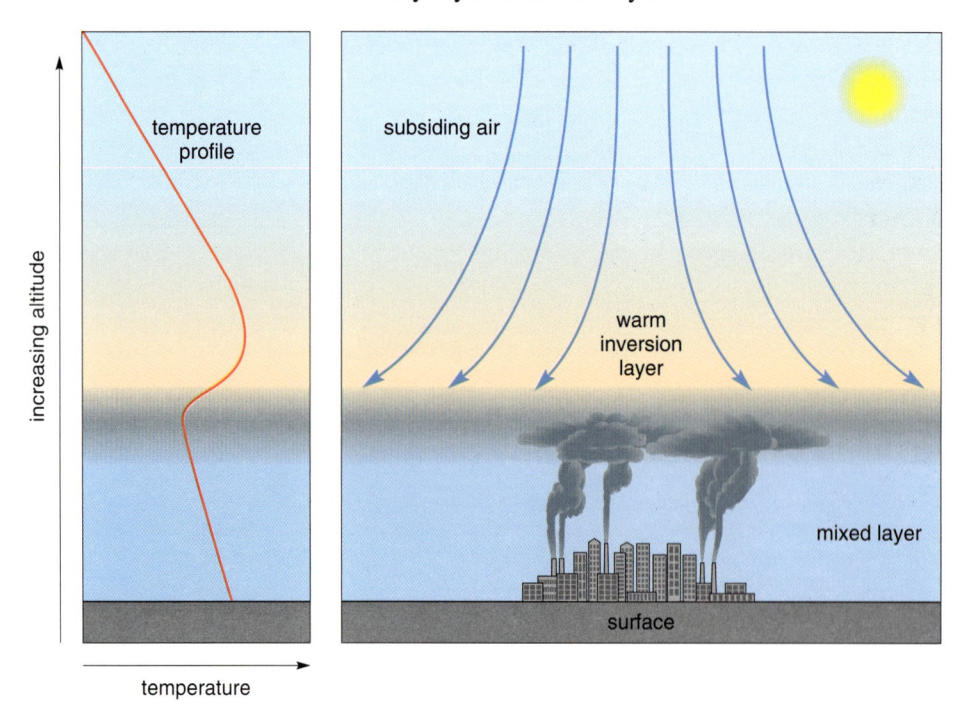

Subsidence inversions can occur over most regions during anticyclonic conditions. But they are a regular feature in the subtropical high-pressure belts (at about 30° latitude north and south), where air rising at the ITCZ eventually subsides (the Hadley circulation, revisited in Section 4.2). Some of the most polluted cities in the world are found at these latitudes — including Los Angeles, Mexico City and Athens in the North, and São Paulo (Brazil) in the South. In the Los Angeles Basin, the situation is exacerbated by the wall of mountains to the north and east (Figure 5.6), and by cool air brought in at low levels by onshore winds that have passed over cold ocean waters before reaching land (Figure 5.14)

Figure 5.14 Fog rolling into San Francisco Bay — a further manifestation of the way air moving inland from the Pacific Ocean is chilled by the cold California Current that flows south off the west coast of the USA.

5.4 A regional perspective

The picture of a city holding its pollution trapped in a mixed layer below a temperature inversion would seem to suggest that enhanced ozone levels are an exclusively urban problem. Figure 5.15 gives the lie to this idea, and reveals the broad spatial extent of the problem in North America. Recall that ozone concentrations in 'clean' surface air are in the range 5–30 ppbv (Table 2.1). But summer afternoon ozone concentrations in excess of 80 ppbv are found over southern California, eastern Texas, the industrial mid-west and the mid-Atlantic eastern states, roughly reflecting the distribution of population. Moreover, within a given region there is little difference in ozone concentrations between cities, suburbs and nearby rural areas. Indeed, concentrations are often higher at rural sites *downwind* of metropolitan areas. The processes that generate ozone take time, and continue while polluted urban air is being transported downwind.

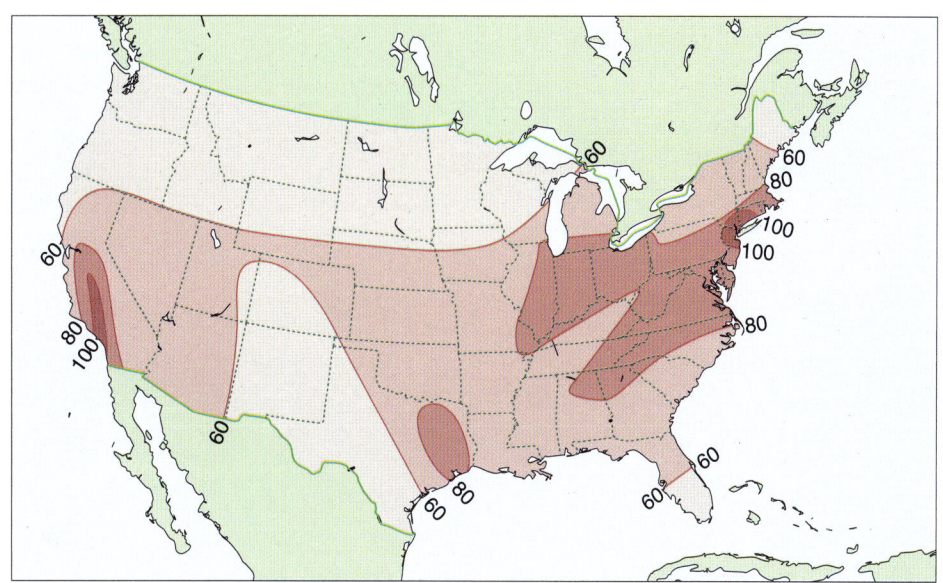

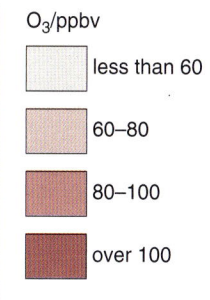

Figure 5.15 Ninetieth percentiles of summer afternoon ozone concentrations measured in surface air over the United States. 'Ninetieth percentile' means that concentrations are higher than this 10% of the time.

The monitoring of ozone levels in Europe has revealed that here too ozone pollution is a *regional*, rather than an urban, issue. Under warm, dry anticyclonic conditions in the summer months, elevated-ozone episodes can cover large parts of Europe. Indeed, the realization that ozone (and its precursors) can be involved in *long-range* transport was first made in Europe.

Just as the tropopause acts as a 'leaky lid' to the lower atmosphere, so too polluted air can escape the mixed layer below an inversion lower down. This can happen if the thermal inversion is a relatively weak affair. Fair weather cumulus clouds bubbling up on a summer afternoon can pump pollutants by convection above the mixed layer. The updraught of a thunderstorm is even more effective. Once in the free troposphere, higher wind speeds and lower temperatures can result in ozone and its precursors being carried for hundreds or even thousands, of kilometers downwind, before being mixed down to the surface.

○ Why should lower temperatures be a factor in the long-range transport of ozone precursors, such as HCs?

● Reactions often go faster at higher temperatures (Box 4.1). So lower temperatures can increase the lifetimes of reactive HCs (as they do for CH_4), allowing them to exert their influence further from the point of emission.

In the UK, summer ozone concentrations tend to be higher in southern England than elsewhere, reflecting the effect of polluted air masses brought in from continental Europe. This shows up in the pattern of ozone concentrations across the UK for the hot, sunny summer of 1995 in Figure 5.16. Noticeable too are the relatively high values associated with upland areas in Wales, northern England and Scotland. This reflects the influence of a second factor: dry deposition to the ground or vegetation acts as a sink for ozone in surface air (recall Question 4.4). This process is more important in sheltered, lowland areas than for exposed sites at higher altitudes. As a result, ozone levels can often be higher at elevated locations.

○ Perhaps the most striking feature of Figure 5.16 is the relatively *low* level of ozone associated with major conurbations, such as Greater London, Birmingham, Manchester, etc. Why is this surprising?

● Throughout, we have stressed that ozone formation in the troposphere depends on the availability of NO_x — the major source in the UK being the high traffic density in towns and cities!

Part of the explanation for this puzzling observation again goes back to the influence of reaction R12 — the direct reaction between NO and O_3. In discussing the smog episode in Figure 5.7 (Section 5.2.1), we said that this reaction delays the build-up of O_3 until most of the free NO has been converted to NO_2. Something similar happens when winds bring air *that already contains high levels of ozone* into an urban area. Then the input of fresh supplies of NO can act as an 'ozone scavenger', *reducing* O_3 concentrations in the immediate vicinity.

Meteorological conditions in the UK rarely favour episodes of seriously enhanced ozone concentrations. To date, the highest recorded value was 258 ppbv at Harwell (near Oxford) during the exceptionally hot summer of 1976;

Ozone air pollution

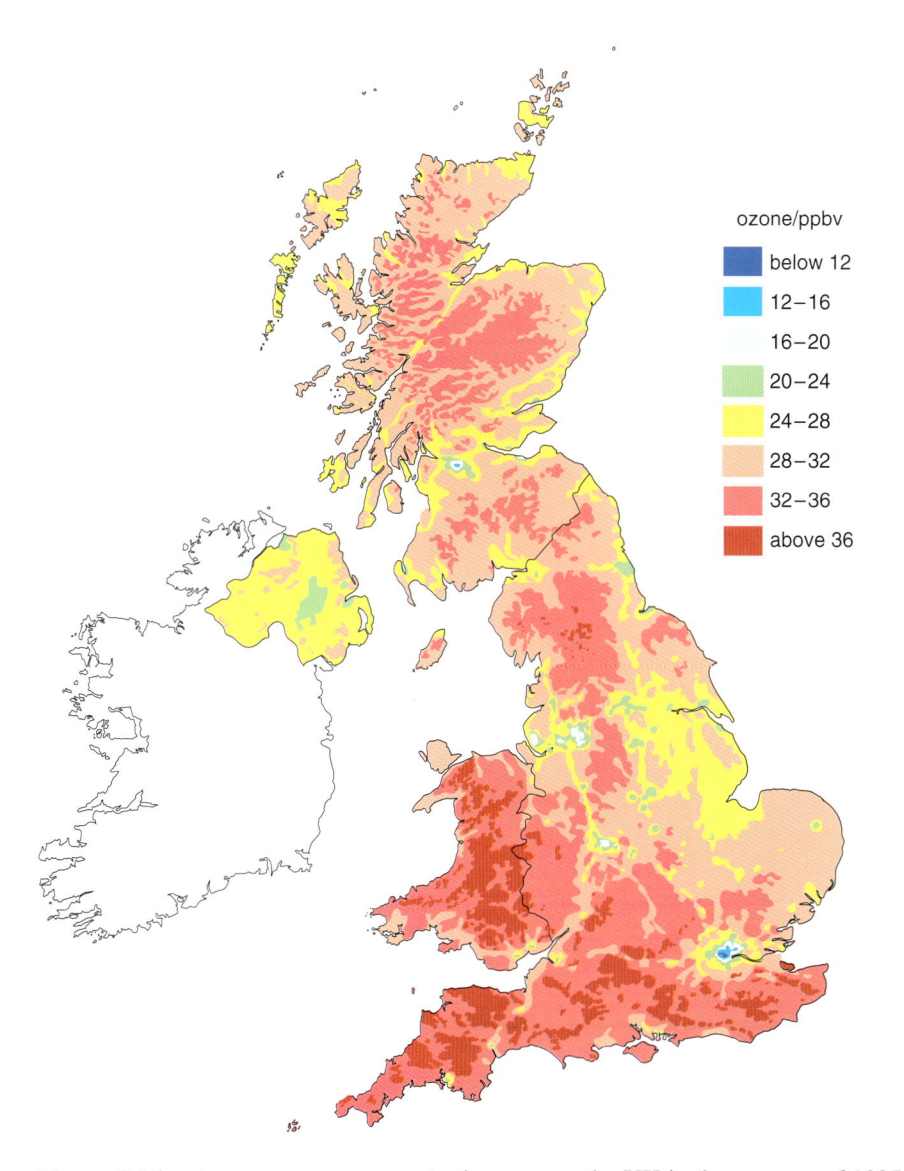

Figure 5.16 Average ozone concentrations across the UK in the summer of 1995. During this summer, the South Downs had the highest recorded O_3 levels, making it one of the most polluted places in Britain.

compare this with the peak of over 450 ppbv in the Los Angeles Basin evident in Figure 5.7. Within a European context, the UK and Scandinavia are the least affected by such episodes; unsurprisingly, this is more of a problem for countries further south. There are also intriguing differences between Europe and the USA, suggestive of an atmospheric regime over America that is more conducive to ozone formation: more on this in Section 5.6.

5.5 Ozone exposure: air quality standards

None of the constituents of photochemical smog is a desirable addition to the atmosphere, but here we concentrate (albeit briefly) on the effects of exposure to ozone. We consider these effects under two broad headings.

5.5.1 Effects on human health

Contrary to popular belief in Victorian times, a 'whiff of ozone' is not a health-giving tonic (Figure 5.17); ozone is a serious irritant of the respiratory system. However, the relationship between ozone exposure and a real impairment of human health is not clear-cut.

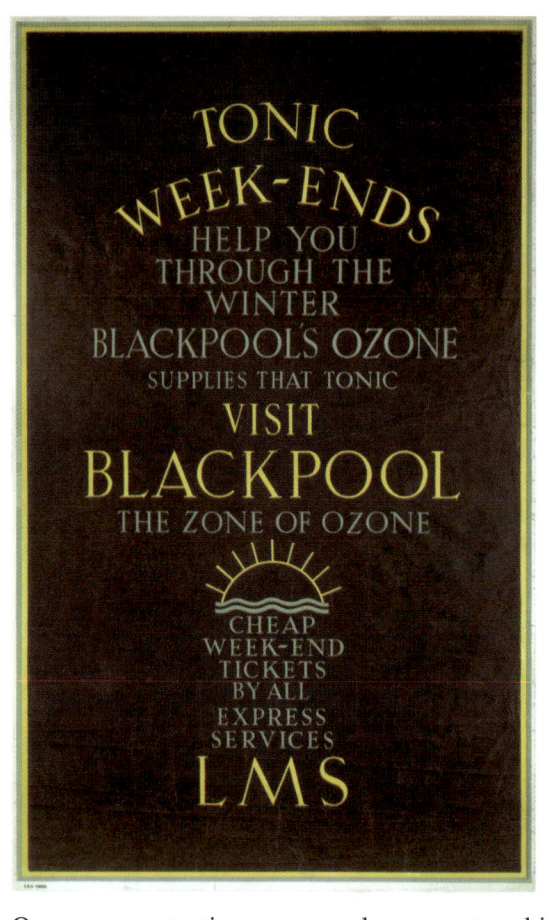

Figure 5.17 A London–Midlands–Scottish railway poster for Blackpool from 1935. Well into the twentieth century, many British coastal resorts were still being marketed on the supposed curative properties of their ozone-laden sea breezes. The 'whiff of ozone' at the seaside is more likely to be the smell of sulfur compounds from decaying seaweed (or sewage!).

Ozone concentrations commonly encountered in the UK are known to produce inflammation of the airways: typically, coughs, wheezing, throat dryness and increased mucous production are reported. Studies on healthy adults, exposed on a short-term basis to higher but realistic ozone concentrations, have consistently shown effects such as reduced lung function, pulmonary inflammation and increased sensitivity to other respiratory irritants and allergens (from pollen, dust mites, etc.). In healthy subjects, these effects are reversible within a few hours. However, studies in the USA have demonstrated that *long-term* exposure to the higher ozone levels commonly found there, is associated with a chronic decline in lung function. In general, people of all ages become more sensitive to ozone when they are active outdoors. Vigorous physical activity (sport, jogging, heavy work) causes people to breath faster and more deeply, allowing ozone to penetrate deeper into the parts of the lungs that are more vulnerable to injury.

There is no evidence that ozone exposure *causes* asthma or other chronic respiratory diseases, such as emphysema and bronchitis. However, these diseases do make the lungs more vulnerable to the effects of ozone. Individuals with these conditions generally experience discomfort more quickly, and at lower ozone levels, than less-sensitive groups. Certainly, asthmatic attacks are exacerbated at high ozone levels — above about 250 ppbv. For conditions more typical of the UK, a recent report concluded that ozone pollution has a detectable, but relatively small, effect on the provocation of existing asthma. So far, there is little evidence to suggest that the elderly or people with heart disease have heightened sensitivity.

Given the brief discussion above, it is clear that there are difficulties in setting appropriate **air quality guidelines** for ozone exposure — difficulties that are compounded by the high background levels. The World Health Organization (WHO) guidelines for human health, which incorporate safety margins, are 76–100 ppbv for 1-hour exposure, and 50–60 ppbv for 8-hour exposure. At present, the US federal standards are somewhat higher than this: a maximum 1-hour and 8-hour average of 120 ppbv and 84 ppbv, respectively. The current UK ozone standard, part of the National Air Quality Strategy adopted in January 2000, is 50 ppbv, measured as the mean over *any* continuous 8-hour period.

In the UK, surface ozone concentrations are monitored continuously at 24 urban and 17 rural sites. Data from the network are made available to the public via an air quality telephone line, the internet and weather bulletins. In the latter context, a set of criteria, shown in Table 5.3, are used to classify ozone levels into bands. Continuous monitoring also allows determination of the number of exceedences throughout the year at each site in the network; that is, the number of times ozone levels exceed the UK standard given above.

Table 5.3 UK air quality guidelines for ozone, expressed as hourly averages in ppbv.

Band	O_3 concentration
low	less than 50
moderate	50–89
high	90–179
very high	180 or more

Activity 5.2

Figure 5.18 shows the number of exceedences each year over the period 1986–1997 for six sites in the UK — two in cities and the rest in rural locations. The period shown includes two years with unusually hot summers: 1990 and 1995.

Figure 5.18 Annual exceedences of UK ozone air quality standards for six sites; for use with Activity 5.2.

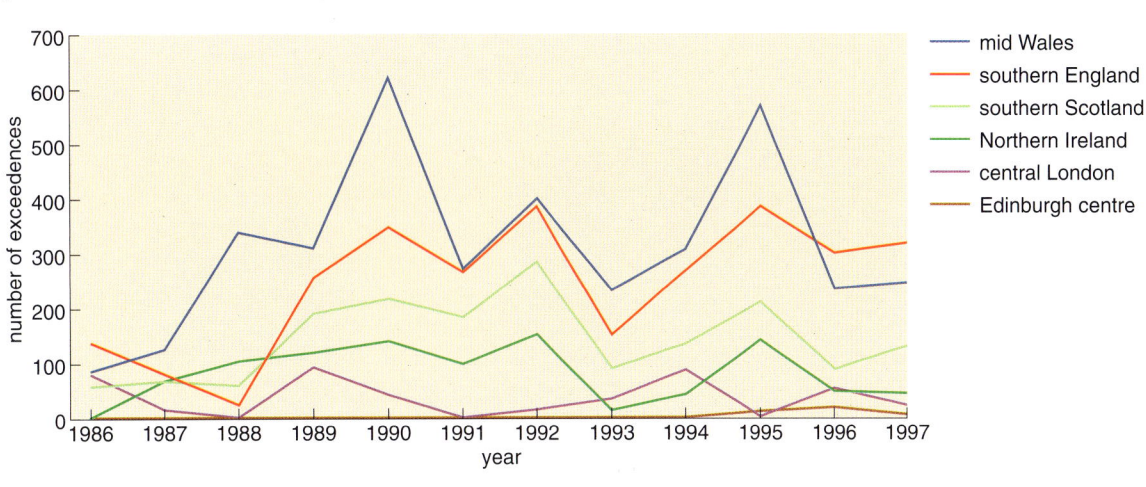

(a) Discuss briefly the factors that probably contribute to the different levels of exceedence found at the various sites in Figure 5.18, and to the way they vary from year to year. (*1–2 short paragraphs*)

(b) Is there any evidence in Figure 5.18 for a long-term trend in annual exceedences in the UK? Explain your answer. (*One sentence*)

5.5.2 Effects on plants

Visible damage to plant leaves (Figure 5.19) was one of the earliest signs of photochemical pollution in the Los Angeles Basin; it was soon established that ozone was the main phytotoxic agent. Since then, ozone has caused widespread injury to agricultural and horticultural crops, and to natural and managed forests, not only in California but also in many other US states prone to high-ozone episodes. For example, the eastern USA has suffered extensive damage to tobacco crops. Hundreds of thousands of trees have died in the mixed conifer forests of southern California. In terms of plant injury, ozone has proved to be the most costly air pollutant in the USA, possibly accounting for 90% of the financial losses.

Figure 5.19 Characteristic symptoms of ozone injury to plants: (a) brown necrotic lesions (due to tissue death) on the under surface of potato leaves; (b) tip necrosis and chlorosis (yellowing) on white pine. Other typical symptoms on plant leaves include tiny tan flecks or a 'bronzed' appearance.

Plants are vulnerable to air pollution via the **stomatal pores** on the leaf surface. As well as regulating the uptake of CO_2, and loss of water, these pores provide the main entry route for gaseous pollutants. Uptake of O_3 is thought to induce changes in cell membrane permeability and the production of free radical species, leading to cellular and biochemical changes in the leaf tissue. The result is visible leaf injury (or even leaf drop) and impaired photosynthesis, producing reduced vigour and growth, and loss of yield in crop species.

Plants vary markedly in their sensitivity to ozone exposure. Unfortunately, the most sensitive plants include many important crops (barley, oats, corn and wheat, as well as tobacco and cotton), fruits and vegetables (e.g. citrus fruit and grapes; potato, tomato, peas and beans), together with certain types of tree (e.g. beech, birch and many pine species). Field observations and growth experiments in open-top chambers (Figure 5.20) have established that ambient ozone levels across Europe can result in visible injury to many commercially grown crops, together with marked reductions in yield.

Figure 5.20 Typical array of open-top chambers during a field study to measure effects of ozone on crop yield.

Studies have shown that prolonged exposure to as little as 50–100 ppbv ozone can have a significant effect on yield for sensitive plants. Such exposure is common in many parts of continental Europe, and across vast tracts of the USA (Figure 5.15). But short-term exposure to higher ozone levels has an even greater impact. These findings have led to the adoption of the critical level concept in assessing the risk of damage to different types of vegetation — usually classified as 'crops', 'forests and woodlands', and 'semi-natural vegetation'. Critical levels are defined using the idea of an **accumulated ozone time** (**AOT**) — a measure of ozone exposure through the *accumulated time during a growing season for which the hourly ozone concentration exceeds a given threshold*. Currently, the threshold is taken to be 40 ppbv, leading to the **AOT40 index**. Thus, an exposure of 1 hour at 90 ppbv would give an AOT40 of 50 ppbv-hours.

○ Calculate the AOT40 for exposure to ozone at 50 ppbv for 4 hours per day throughout a growing season of 3 months.

● The exposure is 10 ppbv above the threshold for $3 \times 30 \times 4 = 360$ hours. So the AOT40 is 3600 ppbv-hours.

Recommended critical levels for Europe are an AOT40 of 3000 ppbv-hours over 3 months for crops and semi-natural vegetation, and 10 000 ppbv-hours over 6 months for forests. In the UK, it has been estimated that the second standard is exceeded in about 23% of the wooded area, mainly in southern Britain. By contrast, the critical level for other types of vegetation is exceeded throughout most of England and Wales, and in many parts of Scotland — representing some 76% of the area covered in semi-natural vegetation, and over 90% of that devoted to arable crops.

Estimates based on the AOT40 index are, then, indicative of a widespread *risk* to vegetation from ozone exposure in the UK. Whether, and to what extent, that risk translates into actual damage is still uncertain. There is a pressing need for more work in this area, where difficulties are compounded by the need to take account of factors such as the timing of high ozone episodes, and the vagaries of the weather from year to year. Other important issues include the possibility of secondary effects in natural plant communities, resulting from the differing sensitivities to ozone of the species present, and increased susceptibility to other forms of air pollution (e.g. acid rain) and insect damage.

Topic 2 Atmospheric Chemistry and Pollution

Bearing in mind that the UK is one of the European nations that is least affected by ozone air pollution, it is clear that this is an important issue throughout the developed world.

Question 5.3

Figure 5.21 shows data from several European experiments to determine the effect of ozone exposure on the grain yield from spring wheat. Use the figure to estimate the percentage loss in grain yield at the critical level for crops quoted above.

Figure 5.21 Relative grain yield of spring wheat as a function of ozone exposure, expressed as an AOT40 for 3 months.

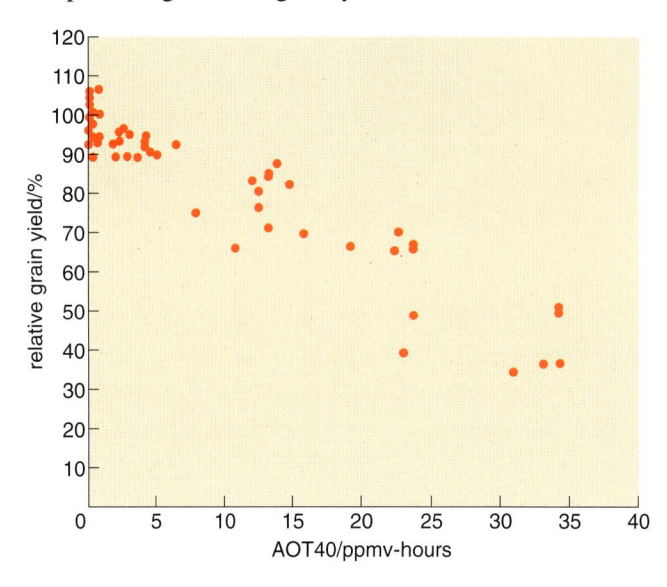

5.6 Controlling ozone pollution

In principle, control of ground-level ozone concentrations can be accomplished by reducing anthropogenic emissions of the main precursor pollutants — reactive HCs and NO_x. In practice, the most effective control strategy can be elusive, and it has proved difficult to bring ozone levels into compliance with air quality standards. In this closing section, we focus on the science behind these difficulties — and touch only briefly on the other factors involved, whether technological, or socio-economic and political.

5.6.1 Control strategies

To simulate the complexities of the ozone budget in polluted environments — and hence calculate ozone concentrations, and predict their response to different control measures — the appropriate tool is a computer model. Successive generations of models of varying degrees of sophistication have been developed. Some of the essential elements can be appreciated from the depiction of a simple **box model** in Figure 5.22. This invites you to think of the air mass over a given locality as a well-mixed 'box' of air, into which precursor pollutants are emitted and undergo the reactions (outlined in Section 5.2.2) that act as sources and sinks of ozone within the box.

Central to any model is, then, a specification of this set of reactions — and how the rate of each process depends on the concentrations of the species involved, and on temperature or sunlight (for the photochemical steps). To simulate the situation

Figure 5.22 Schematic representation of the essential features of a simple box model capable of simulating ozone formation in a polluted environment.

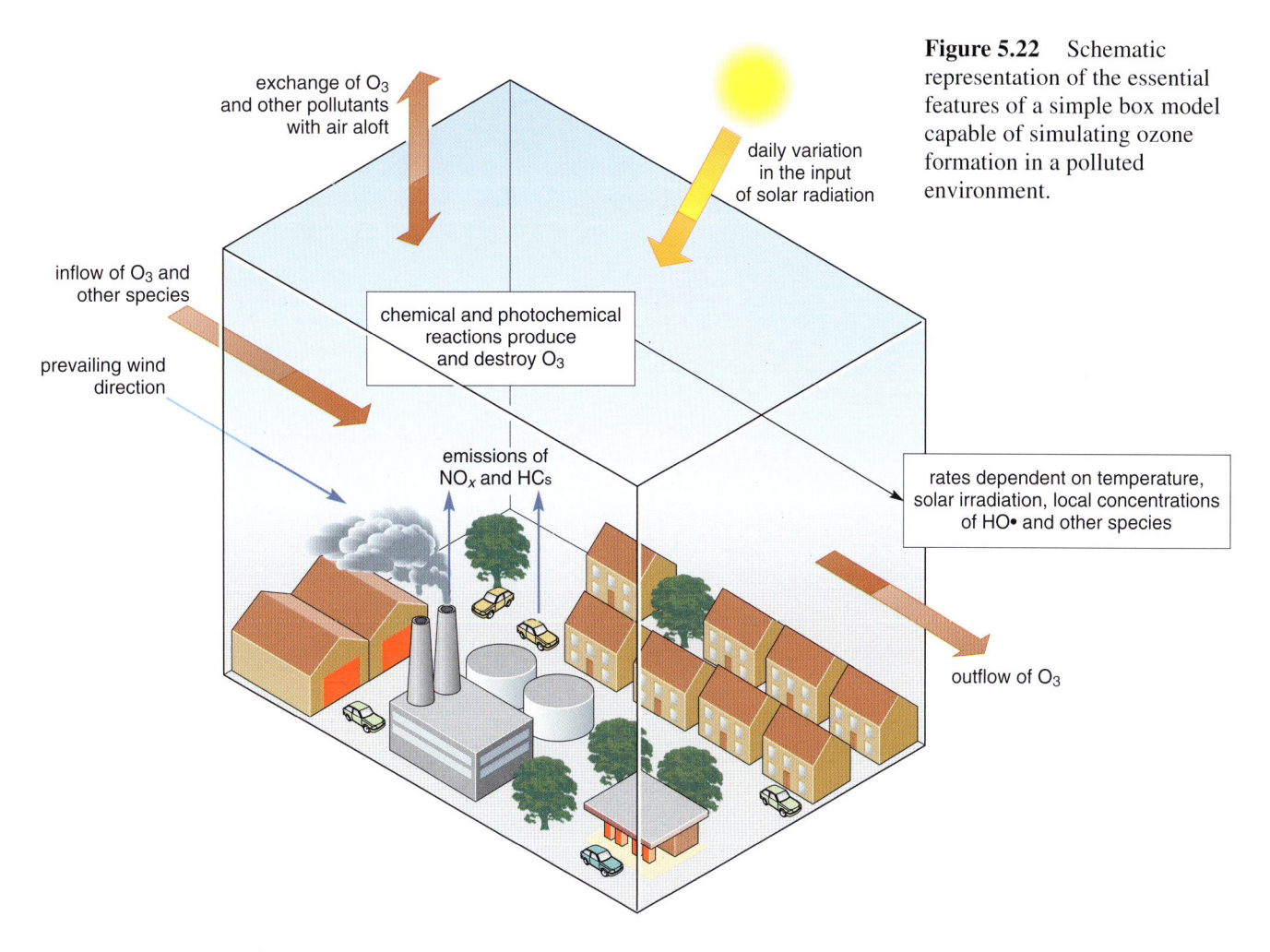

in a particular place, local conditions and meteorological variables must also be specified: diurnal variations in the input of precursor pollutants, for example, and in temperature and solar irradiation; transport of ozone and other key species into and out of the box by prevailing winds; and so on.

The output from a model can only ever represent an approximation of the situation in the real atmosphere. Nevertheless, modelling studies have helped to identify one important issue in the formulation of a successful control strategy — the importance of the *relative* concentrations of NO_x and HCs in the air mass over the area to be protected. To summarize the key findings:

- At high [HC]/[NO_x] ratios, O_3 concentrations increase with increasing NO_x, but are relatively insensitive to the concentration of HCs. This situation is known as the **NO_x-limited regime**.
- At high [NO_x]/[HC] ratios, O_3 concentrations increase with increasing HCs, but *decrease* with increasing NO_x. This situation is known as the **hydrocarbon-limited regime**.

Insight into the counterintuitive dependence of O_3 on NO_x under the second of these regimes can be gained from the information collected in Figure 5.23.

The left-hand side of the figure is a reminder that most NO_x is emitted as NO, that net O_3 production depends on peroxy + NO reactions, and that the supply of

Figure 5.23 Schematic representation of some of the key reactions involved in O₃ production and loss in polluted environments.

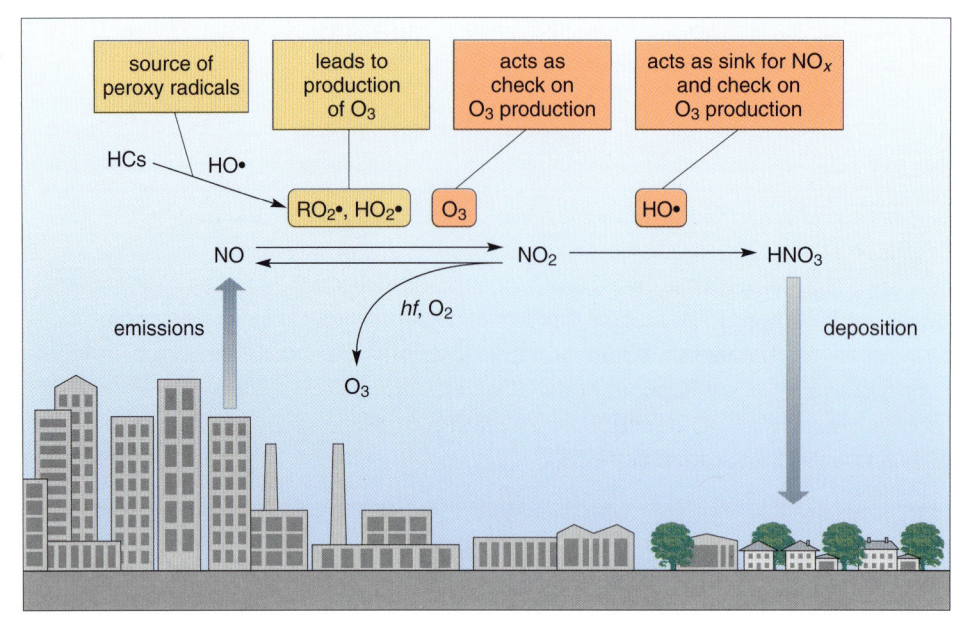

peroxy radicals is initiated by HO• attack on HCs. In the hydrocarbon-limited regime, the high input of NO acts as a check on the build up of O₃ in two ways — by 'ozone scavenging' via the direct NO + O₃ reaction (noted earlier, in Section 5.4), and through the influence of the resulting high concentration of NO₂. The latter can divert HO• from the attack on HCs by forming HNO₃, as shown on the right of Figure 5.23. This effectively short-circuits O₃ production by reducing the supply of peroxy radicals. With this regime in place, increasing the input of NO simply increases the effectiveness of these checks on O₃ production.

○ Why should ozone concentrations increase with increasing HC concentrations under these circumstances?

● Increasing the input of HCs allows them to compete more effectively for HO•, increasing the supply of peroxy radicals and shifting the balance in favour of increased ozone production.

Take it far enough, and an increase in HCs can shift the chemical regime from hydrocarbon to NO_x-limited. With a plentiful supply of HCs, one can, in effect, only form as much ozone as there is NO to be oxidized to NO₂ and subsequently photolysed.

The difficulty faced by pollution control agencies is evident from the two bullet points on p. 117. An effective strategy against ozone pollution depends on a knowledge of whether the chemical regime in the ambient air is NO_x-limited or hydrocarbon-limited: if the former, controls on HC emissions are of little benefit for reducing ozone; if the latter, cuts in NO_x emissions can be actively counterproductive, resulting in an *increase* in ozone.

The difficulties are compounded by the fact that ozone pollution is a regional problem. The hydrocarbon-limited regime (high $[NO_x]/[HC]$ ratios) is typical of the air in heavily polluted urban centres, where vehicle emissions of NO are high. But as the air mass moves downwind over suburban and rural areas, where the input of NO is lower, the chemical regime tends to shift over to NO_x-limited.

○ Look back at Figure 5.23. What chemical process will also act to reduce the concentration of NO$_x$ as air is transported away from an urban centre?

● Conversion of NO$_2$ (and hence NO$_x$) to HNO$_3$, which is rapidly removed by wet (or indeed, dry) deposition.

Recent research has identified yet another factor in the shift to a high [HC]/[NO$_x$] regime in downwind rural areas: natural emissions of reactive *biogenic* HCs from trees and other vegetation. In this context, the most important such compound appears to be isoprene (structure **6** in Box 5.2). As a quick glance back at Table 5.2 (Section 5.4) will confirm, isoprene reacts very rapidly with HO•; its atmospheric oxidation can produce large amounts of ozone. Indeed, estimates suggest that 5 ppbv of isoprene is as effective in generating ozone as 50 ppbv of a typical mix of HCs released by human activities.

Large emissions of isoprene, stimulated by high temperatures and extensive forest cover, help to explain why the atmosphere over the USA appears to be especially conducive to ozone formation. Inventories drawn up in the 1990s indicate that isoprene emissions in the USA are larger than the sum of *all* anthropogenic HCs — sufficient by itself to make ozone production NO$_x$-limited everywhere except large metropolitan areas. Of course, these natural emissions of HCs would not realise their ozone-producing potential were it not for the widespread anthropogenic sources of NO$_x$ — the message emphasized in Activity 5.1.

Question 5.4

One early move to combat ozone pollution in California (implemented in 1966) was effective in reducing HC emissions from motor vehicles, but led simultaneously to a large increase in NO$_x$ emissions. Suggest why this had the desired effect of reducing ozone levels in downtown Los Angeles, but resulted in increased ozone pollution in downwind rural areas. *(A few sentences)*

No wonder ozone pollution has proved to be such an intractable problem — when efforts to protect one area can exacerbate the situation elsewhere!

5.6.2 Controlling emissions

Much of the regulatory effort in recent decades has targeted the major source of ozone precursors — vehicle emissions. Unsurprisingly, the State of California led the way. Moves there during the 1960s were followed by legislation at the level of the US Federal government, through amendments to the US Clean Air Act in 1970. Since then, progressively more stringent limits have been set for emissions of HCs, NO$_x$ *and* CO (a hazardous pollutant in its own right, in addition to its role in the chemistry of the background atmosphere).

In the context of the discussion above, it is interesting to note that the initial focus in the USA was on reducing HC emissions — a strategy underpinned by modelling studies which indicated ozone production to be hydrocarbon-limited across the country. It is now recognized that the inventories of HC emissions fed into the models were in error: they both underestimated anthropogenic sources and took no account of natural sources. With the changed understanding that ozone production is, in fact, primarily NO$_x$-limited in the USA (except in large urban areas) has come a comparable effort to control NO$_x$ emissions.

Topic 2 Atmospheric Chemistry and Pollution

Box 5.3 The three-way catalytic converter

If the hot exhaust gases from a petrol engine contain the right amount of O_2, then the ingredients of chemical reactions that can destroy all three targeted pollutants are present. These reactions include:

- The complete combustion of HCs (to give $CO_2 + H_2O$), which was meant to occur in the engine.
- The destruction of CO, by reaction with O_2:

  $2CO + O_2 = 2CO_2$

- The mutual annihilation of CO and NO:

  $2NO + 2CO = 2CO_2 + N_2$

The problem is that these reactions take place too slowly in the exhaust system, and the pollutants are swept out before significant conversion (to CO_2, N_2 and steam) can occur.

This is just the sort of situation that can be put right by finding an appropriate **catalyst** — a substance that can speed up a reaction without undergoing changes itself. The catalysts currently used in the three-way catalytic converter (Figure 5.24) are the precious metals platinum and rhodium (and sometimes palladium as well).

Inside a metal canister, the core of the converter consists of a ceramic material containing a honeycomb of fine channels. The walls of these channels are coated with a highly porous material (mainly alumina), impregnated with particles of the metal catalysts where the conversion reactions occur as the exhaust gases pass through. Efficient conversion occurs only if the air/fuel ratio is kept within strict limits. This is achieved via automatic adjustments in response to signals from an O_2 sensor placed in the exhaust stream.

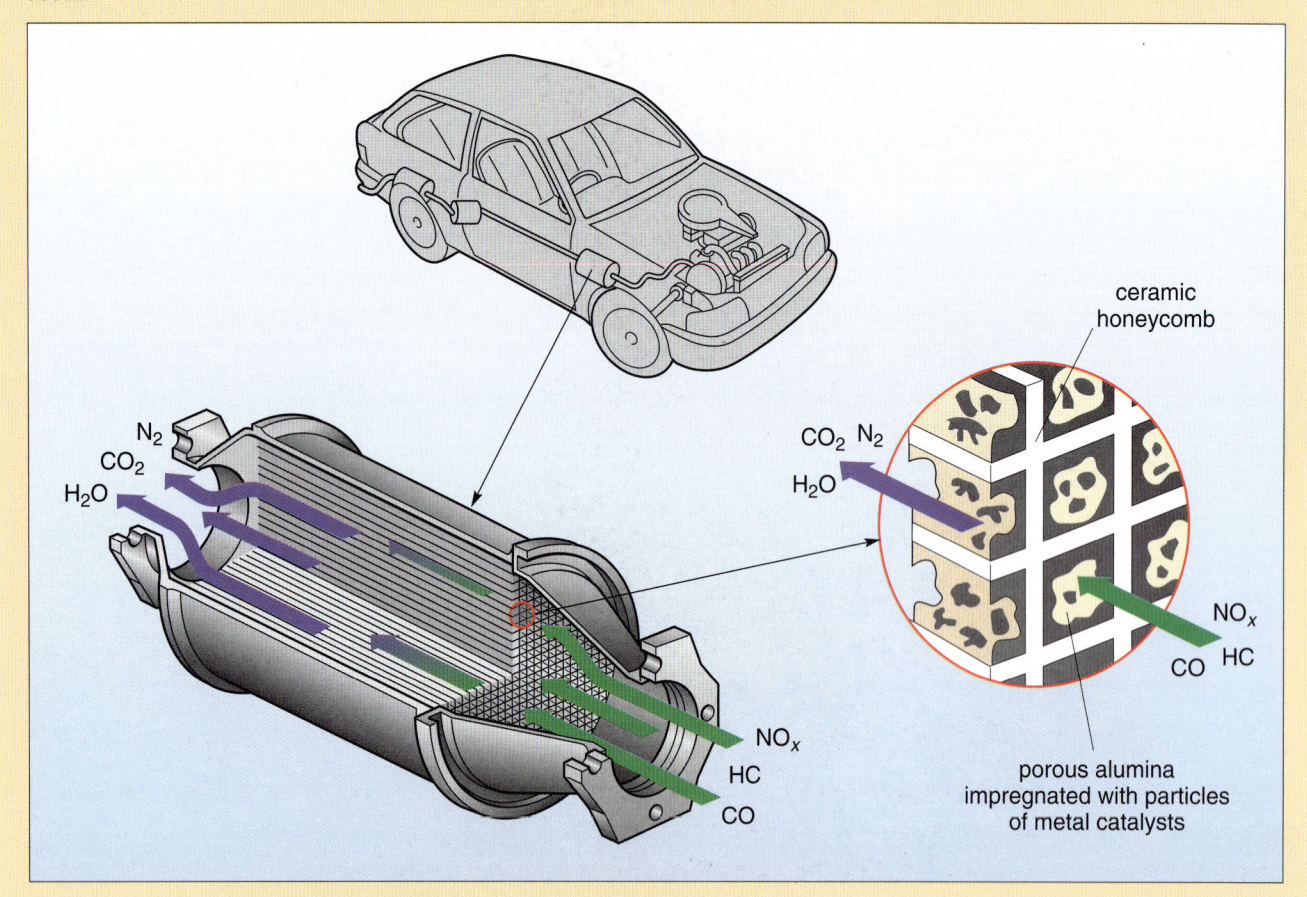

Figure 5.24 Schematic representation of the three-way catalytic converter.

By the mid-1990s, the US federal standards for petrol-fuelled vehicles represented a reduction from uncontrolled emission levels of over 90% for all three targeted pollutants. Even more stringent limits have now been adopted, not only in the USA but also in the European Union (EU) and elsewhere in the industrialized world. Standards have also been set for commercial and diesel-fuelled vehicles.

To meet the mandatory standards for petrol vehicles, the present strategy is to destroy the pollutants before they exit from the tail-pipe. The trick is to fit a device, known as the **three-way catalytic converter** (Box 5.3), into the exhaust system. Control is achieved by promoting reactions that have the net effect of *simultaneously* oxidizing CO and HCs (to CO_2 and $CO_2 + H_2O$, respectively) and reducing NO (to harmless N_2). For each pollutant, a conversion efficiency of 90% or more is achievable.

It is evident from Figure 5.25 that the catalytic converter (mandatory in the EU since 1993) is very effective at reducing emissions of ozone precursors — and over a wide range of speeds. Unfortunately gains made, as new catalyst-equipped vehicles replace old cars on the road, are offset by ongoing traffic growth. In the UK alone, car numbers are predicted to increase from 20 million in the early 1990s to over 50 million by 2025. Moreover, vehicle exhausts are not the only anthropogenic sources of O_3 precursors (Table 5.1, Section 5.2.2). Evaporative loss of HCs can occur at every stage in the production, storage and distribution of petrol — and during refuelling at the pump. Solvent usage, industrial processes and chemical manufacture are further significant sources of HCs. And so are coal- and gas-fired power stations for NO_x.

Figure 5.25 Emission levels of (a) CO, (b) HCs and (c) NO_x from petrol-engined vehicles as a function of speed, with (red) and without (blue) a three-way catalytic converter.

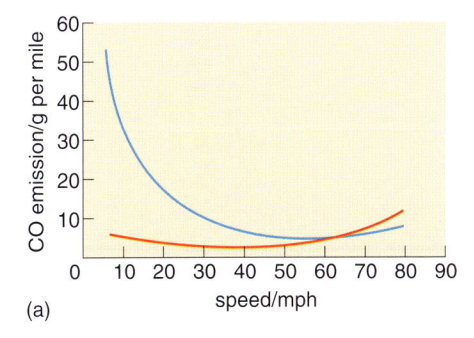

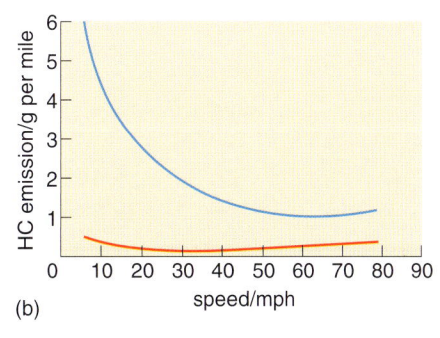

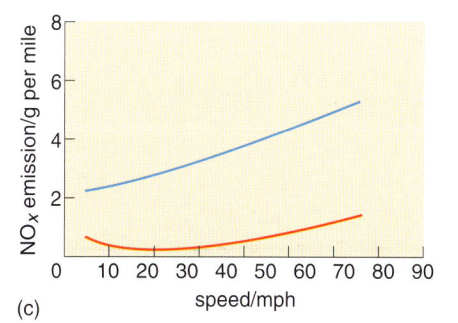

Efforts are being made to control emissions from all these sources. But it is an uphill battle. In the USA, for example, it is estimated that anthropogenic emissions of HCs fell by 38% between 1970 and 1997, while NO_x emissions went up by 11% — still no small achievement considering the population grew by 31% over this period and vehicle miles travelled increased by 127%.

5.6.3 Trends in ozone air quality

Have the regulatory efforts to date been translated into improved air quality?

○ What sort of information is needed in order to detect a downward trend in O_3 in a given locality that can be attributed to reduced emissions of the precursor pollutants? (Recall your thoughts on Activity 5.2)

● The basic requirement is a reasonably long record of O_3 measurements, together with some means of 'filtering out' the influence of other factors that affect O_3 levels — notably changing weather conditions from year to year. So records of these conditions are also needed.

One place that amply fulfils these criteria is the Los Angeles Basin (Figure 5.6). Routine monitoring of ozone and other air pollutants has been in place at stations across the region since 1976. Figure 5.26 shows the record of annual violations of the current US federal standards for ozone for the period up to 1999. This analysis is based on 'raw' ozone measurements, but weather-adjusted studies confirm the significant downward trend in ozone pollution that is evident in Figure 5.26. And the 'smog season' is now shorter then it used to be, being largely confined to May–September. Up to the late 1980s, it was common to have days exceeding the federal standards as early as February, and well into November.

Figure 5.26 Number of days each year (as a percentage) when the current US federal standards for O_3 concentrations were exceeded somewhere in the LA Basin, 1976–1999.

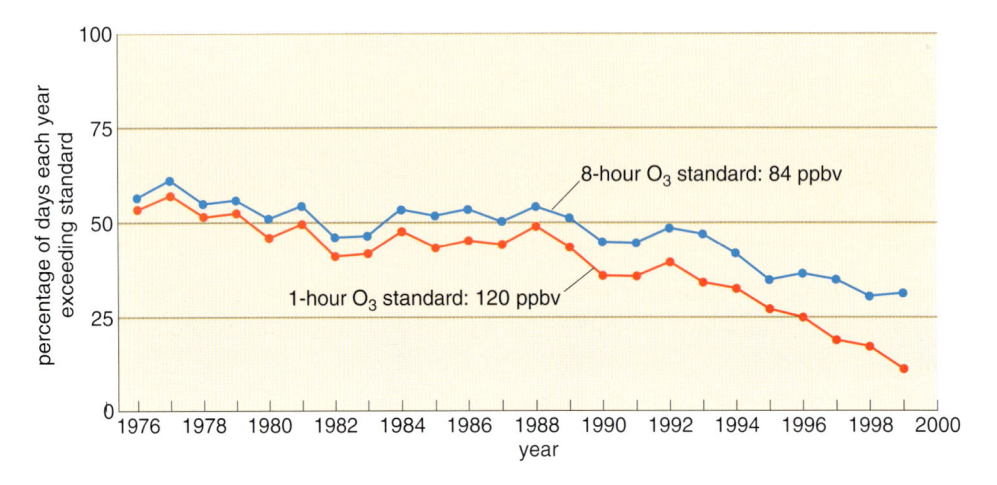

At the time of writing, analyses of long-term trends in ozone levels elsewhere in the USA indicate a real improvement for some of the other serious offenders (e.g. the New York City metropolitan area). But no significant amelioration is yet apparent in large areas of the country where the ozone standards are violated. A similar mixed success has been reported in Europe and Japan, and elsewhere in the developed world.

It is apparent that further reductions in ozone pollution will require yet tighter emission controls. In the USA for example, California has already put in motion legislation intended to *eliminate* vehicles as a serious source of ozone precursors. Much research effort is currently being devoted to the technology required to meet this laudable objective (Figure 5.27).

In conclusion: with all the technical and economic resources at its disposal, the developed world is beginning to make inroads into the problem of ground-level ozone pollution. Meanwhile in many developing countries, elevated ozone levels are just part of the growing air pollution that follows in the wake of population growth and rapid industrialization. Recall too that South America, Africa and parts of Asia are home to a source of ozone precursors that has nothing to do with traffic or industrial activity.

○ What source is this? (Refer back to Figure 4.9 and your thoughts on Activities 4.1 and 5.1.)

● The deliberate burning of biomass — linked to tropical deforestation and the widespread seasonal burning of grasslands in the subtropics.

Figure 5.27 Part of a 'solar hydrogen' demonstration plant in Germany. To the left is an array of solar photovoltaic cells: electricity generated by these cells is used to split water into H_2 and O_2. The hydrogen is liquefied and stored in large tanks (to the right), providing a liquid H_2 filling station for a test car converted to run on this fuel. Since burning H_2 produces H_2O, this option would at once eliminate the HC problem. It requires further development to become economically competitive with other cleaner-burning, if still carbon-based, fuels — such as liquefied petroleum gas (LPG, mainly propane) and compressed natural gas (CNG).

Such activities contribute 22% of the annual input worldwide of CO, and some 15% of all NO_x emissions — setting the scene for O_3 formation in the huge plumes generated by these fires (Figure 5.28). On a regional scale, the burning in Africa is believed to produce as much ozone as urban industrialized areas. Any attempt to bring this source of ozone pollution under control will need to address a very different set of issues.

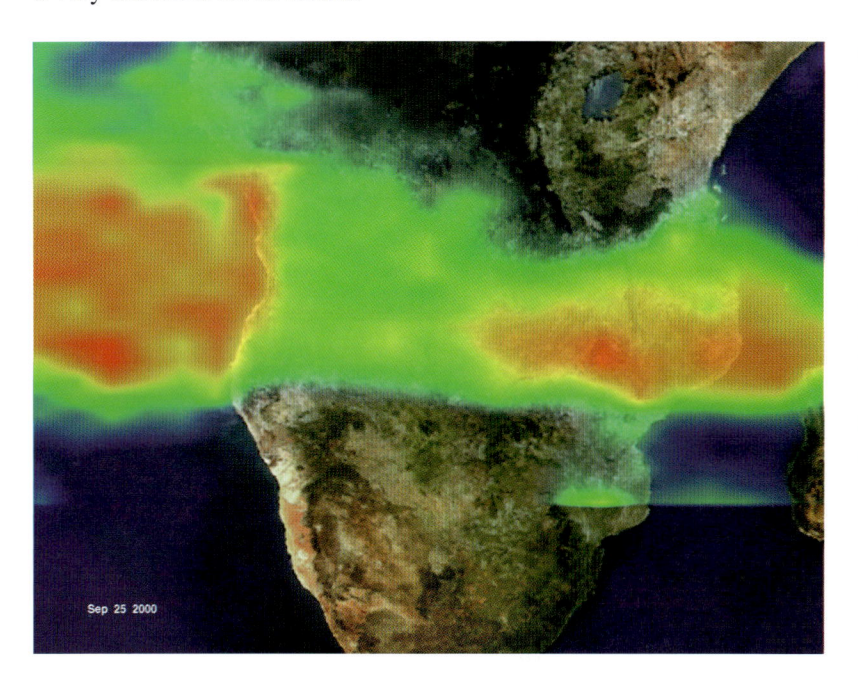

Figure 5.28 In September 2000, at the same time as fires were raging across southern Africa (Figure 2.3), satellite-borne instruments detected high levels of tropospheric ozone (shown in orange and red) over the region. The ozone levels were frequently similar to those found in air pollution alerts in major US cities. At this time of year, prevailing circulation patterns tend to trap and concentrate gaseous emissions from the fires. Clear skies result in the photochemical transformations that generate ozone, which can then be carried out over the Indian Ocean (to the east) and southern Atlantic (to the west).

Question 5.5

Consider the introduction of electric vehicles powered by conventional rechargeable batteries as part of the strategy to further reduce emissions of ozone precursors. What factors would need to be considered in assessing the likely effectiveness of this option? (*Answer in a few sentences*)

5.7 Summary of Section 5

1. Background levels of tropospheric ozone have increased over the past century, alongside the growth in anthropogenic sources of the precursor gases, CH_4, CO and NO_x (Activity 5.1).

2. Enhanced ozone concentrations are associated with photochemical (Los Angeles type) smog. Conditions conducive to smog formation are intense solar irradiation and the trapping of pollutants by thermal inversions, either near the ground (low-level inversions) or aloft (subsidence inversions, common under high-pressure conditions).

3. Smog ozone chemistry mirrors that in the background troposphere, but the dominant source of peroxy radicals is oxidation of reactive *non-methane* HCs (especially unsaturated compounds), rather than CH_4 and CO. Vehicle exhausts are the main source of NO_x and HCs, but there are stationary sources (power stations, solvent use, etc.) as well.

4. Ozone pollution is a regional problem — due in part to the transport of ozone and its precursors away from urban sources, and in part to the input of biogenic HCs (especially isoprene) from vegetation in rural and forested areas.

5. The pattern of summer ozone concentrations across the UK is influenced by several factors: weather conditions, including air circulation patterns over Europe; dry deposition of ozone; and ozone scavenging in urban centres (Activity 5.2).

6. Air quality standards for ozone are established using human health criteria. They aim to protect against both acute effects (e.g. provocation of existing asthma) and possible long-term chronic effects (e.g. reduced lung function).

7. High ozone levels in the USA have caused significant damage to crops and forests resulting in major financial losses. Estimates based on the AOT (accumulated ozone time) criterion indicate a widespread risk of damage to crops and natural plant communities in the UK — one of the European nations that is least affected by ozone pollution.

8. The most effective strategy against ozone pollution depends on whether the chemical regime for ozone production is:

    • Hydrocabon-limited (high $[NO_x]/[HC]$ ratios) — conditions typical of large urban centres; or

    • NO_x-limited (high $[HC]/[NO_x]$ ratios) — conditions typical of downwind rural areas, and more remote regions.

9. Regulatory efforts to reduce vehicle emissions of HCs and NO_x have been partly offset by population growth and increased car usage. Ozone remains a pervasive air pollution problem in many parts of the developed world, and is a growing problem elsewhere — through rapid industrialization and widespread biomass burning.

Learning outcomes for Topic 2

Now that you have completed this topic, you should be able to:

1. Show understanding of the terms 'oxidation' and 'reduction', and their significance in the context of atmospheric chemistry. (*Questions 2.2 and 2.3*)

2. Define, calculate, and/or comment on the significance of, the atmospheric lifetime of a specified trace gas. (*Activity 4.1; Questions 4.1–4.5 and 5.1*)

3. Show familiarity with the kinds of process (emissions, chemistry, deposition) and other factors (transport, atmospheric lifetime) that can influence the abundance and geographical distribution of a trace gas (especially, CH_4, CO and NO_x), and/or interpret observed variations in its atmospheric concentration. (*Activities 4.1 and 5.1; Questions 2.1, 2.5 and 4.5*)

4. Demonstrate understanding of the factors that determine, and the important effects of, the absorption of incoming solar ultraviolet radiation by different atmospheric constituents. (*Activities 3.1 and 5.1; Question 3.1–3.4*)

5. Demonstrate knowledge and understanding of:

    (a) the role of ozone in maintaining the oxidizing power of the troposphere;

    (b) the main natural sources and sinks of tropospheric ozone, including the interpretation of relevant reaction cycles and the roles of CH_4, CO and NO_x in this context;

    (c) the kinds of human activities that can perturb background levels of tropospheric ozone.

    (*Activities 3.1 and 5.1; Question 3.4*)

6. Apply the knowledge and understanding referred to in 3 and 5 to the problem of ozone air pollution in (and downwind from) polluted environments, and hence show understanding of:

    (a) the mechanism involved in its formation, including the influence of anthropogenic emissions and other factors;

    (b) the factors to be considered in formulating an effective control strategy.

    (*Activity 5.2; Questions 5.2, 5.4 and 5.5*)

7. Show familiarity with the harmful effects of ozone exposure, and the significance of data on ozone air quality. (*Activity 5.2; Question 5.3*)

Topic 2 Atmospheric Chemistry and Pollution

Comments on activities

Activity 2.1

Your entries may well differ from those in Table 2.3, but a careful reading of Box 2.4, in the context of material elsewhere in Section 2, should have allowed you to extract the key points.

Table 2.3 A comparison between the combustion of methane and its atmospheric oxidation.

Combustion process	Atmospheric process
burns with blue flame	no flames
occurs at high temperature (over 1200 °C)	occurs at (much lower) atmospheric temperatures
overall, results in rapid oxidation of CH_4 to CO_2	overall effect the same, but happens very slowly
needs energy (provided by a spark or flame) to get it going	no information
occurs in a series of discrete steps, initiated by the rupture of a C–H bond in CH_4	must involve something similar, but no information about how it starts
intermediates in overall process include CH_2O and CO; both usually have a fleeting existence	as combustion process, but acts as an important atmospheric source of CH_2O and CO (Section 2.3)
other intermediates include highly reactive molecular fragments	no information

Activity 3.1

(a) The first step is to identify the net effect of the reaction cycle in Figure 3.17a. Following the line of reasoning suggested in Box 3.4 (*cf* Question 3.4), this gives:

$$CH_4 + O_2 + NO + O_2 + NO = H_2O + NO_2 + CH_2O + NO_2$$

or $\quad CH_4 + 2O_2 + 2NO = CH_2O + H_2O + 2NO_2$ \hfill (RA1)

Thus, each trip around the cycle — each molecule of CH_4 that is oxidized to formaldehyde (CH_2O) — produces two molecules of NO_2, and hence potentially two molecules of O_3, via the photolysis of NO_2 (reaction R11), followed by the ($O + O_2$) reaction (R2). Adding these two steps, their net effect can be summarized as:

$$NO_2 + O_2 = NO + O_3 \quad \text{(R11 + R2)}$$

Repeating the sequence twice over thus has the following net effect:

$$2NO_2 + 2O_2 = 2NO + 2O_3 \quad \text{(RA2)}$$

Adding equations RA1 + RA2 then gives the required overall reaction:

$$CH_4 + 4O_2 = CH_2O + H_2O + 2O_3 \quad \text{(R18)}$$

(b) (i) Starting at the top and moving clockwise, the second step is reaction of the CHO• radical with atmospheric O_2, as:

$$CHO• + O_2 \longrightarrow HO_2• + CO \quad \text{(RA3)}$$

126

This is the step that produces carbon monoxide, CO, *and* generates the all-important peroxy radical, $HO_2\bullet$

(ii) Following the line of reasoning in part (a), the net effect of the cycle in Figure 3.17b can be written down directly as:

$$CH_2O + O_2 + NO = CO + H_2O + NO_2 \quad \text{(RA4)}$$

Including the net effect of NO_2 photolysis and ozone formation (R11 + R2), this becomes (as required):

$$CH_2O + 2O_2 = CO + H_2O + O_3 \quad \text{(R19)}$$

(c) As established in Question 3.4, under NO_x-rich conditions, the net effect of the CO-oxidation cycle in Figure 3.16 is reaction R15:

$$CO + 2O_2 = CO_2 + O_3 \quad \text{(R15)}$$

So one molecule of O_3 is produced per molecule of CO oxidized.

(d) According to parts (a) to (c), in a NO_x-rich environment, the oxidation of one molecule of $CH_4 \longrightarrow CH_2O \longrightarrow CO \longrightarrow CO_2$ yields:

2(R18) + 1(R19) + 1(R15) molecules of O_3; that is, 4 molecules of O_3 (as shown in reaction R17 in the text).

(e) Accounting in terms of molecules is equivalent to accounting in terms of *moles*, so from parts (c) and (d):

oxidation of 4×10^{13} moles of CO yields 4×10^{13} moles of O_3

oxidation of 3×10^{13} moles of CH_4 yields $4 \times 3 \times 10^{13}$ moles of $O_3 = 12 \times 10^3$ moles.

On this basis, the global production rate of tropospheric $O_3 = (4 + 12) \times 10^{13} = 16 \times 10^{13}$ moles per year.

This estimate is based on analysis of the CO and CH_4 oxidation pathways under 'high-NO_x' conditions. But attack by $HO\bullet$ converts NO_2, and hence NO_x, to nitric acid, which is rapidly rained out in the troposphere. As a result, a proportion of the CO and CH_4 released at the surface is oxidized in NO_x-poor environments, and proceeds *without ozone formation* — or even leads to ozone destruction, via other possible fates (e.g. reaction R9) for peroxy radicals (as discussed in Section 3.3.1).

The analysis above takes no account of the influence of this 'low-NO_x' regime, and so would be expected to overestimate the global production rate of O_3.

Activity 4.1

The data in Figure 4.11 provide a snapshot of the global distribution of CO on two days in 2000 — 30 April (top) and 30 October (bottom). As such, they reflect the influence at these times of regional surface sources of this relatively short-lived gas — predominantly, from the burning of fossil fuels and biomass (forests and grasslands); Figure 4.9. With this in mind, the most striking feature of the 30 April, 2000 image is that levels of CO are much higher in the Northern Hemisphere than in the south — broadly in line with the geographical distribution of densely populated industrialized nations. In addition, forest fires probably made a significant contribution to the peak CO levels over Southeast Asia at this time.

The seasonal impact of CO released from forest and grassland fires burning in South America and southern Africa is evident in the October 30, 2000 image. Now the peak CO levels (more than 220 ppbv) are over these regions, and concentrations through much of the Southern Hemisphere are comparable to those in the north — reflecting the transport of CO away from the source regions.

Activity 5.1

Introduction

Higher background levels of O_3 in the troposphere must reflect a shift in the balance between *in situ* production and loss. That balance depends on the availability of NO_x (NO + NO_2) for the following reasons.

Discussion

- Photolysis of NO_2 is the only *in situ* source of tropospheric ozone, but most NO_x is emitted as NO.
- So conversion of NO to NO_2 is a key factor, and this is directly influenced by the concentrations of peroxy radicals, generated during the oxidation of CO and CH_4.
- Human activities have led to a doubling in CH_4 concentrations over the past century or so, and now contribute some 80% of all CO *emissions* (from Figure 4.9a).
- An enhanced source of peroxy radicals will enhance O_3 production in a NO_x-rich environment, but lead to ozone loss in a NO_x-poor one (via reactions such as $HO_2\bullet + O_3$).

Conclusion

The key to higher ozone concentrations in the background troposphere must, then, lie with the enhanced and more widespread emissions of NO associated with fossil fuel combustion and extensive biomass burning.

Activity 5.2

(a) In broad terms, the different levels of exceedence found at the various sites in Figure 5.18 echo the pattern of summer ozone concentrations across the UK shown in Figure 5.16. Overall, the sites in southern England and mid-Wales have exceeded the standard more often than those in Scotland and Northern Ireland, reflecting the influence of polluted air brought in by winds from continental Europe. Also, the urban sites have by far the lowest range of exceedence numbers, because of the ozone scavenger effect, where abundant local sources of NO act as a check on the build up of ozone.

The other factor that shows up in Figure 5.18 is the influence of the weather conditions in a given year. 1990 and 1995 experienced hot sunny summers, and show high exceedences, especially in Wales and southern England.

(b) The short answer is 'no'. The variability from year to year is so large, that it is impossible to discern any overall trends over the 11-year period shown in Figure 5.18.

Answers to questions

Question 2.1

The main natural sources are:

- respiration of land plants — CO_2 released from the controlled 'burning' in oxygen of simple organic compounds made by plant photosynthesis;
- respiration of decomposers and detritivores (worms, fungi, bacteria, etc.) as they break down dead organic matter (biomass) in soils, leaf moulds and sediments; and
- outgassing from the oceans.

Further minor sources include volcanic activity and the oxidation of organic carbon in rocks and sediments exposed to the atmosphere.

The major sinks for atmospheric CO_2 are:

- plant photosynthesis on land; and
- dissolution in the oceans (which drives the oceanic carbon cycle).

Question 2.2

The order is: NH_3, N_2, N_2O, NO, NO_2, NO_3^-.

Moving from left to right, the proportion of hydrogen in the molecule decreases, and the proportion of oxygen increases.

Question 2.3

It is both! In the process of anaerobic decomposition, some of the carbon in dead organic matter (biomass) is oxidized to CO_2, and some is reduced to methane (CH_4). (Any oxidation process is always accompanied by a reduction process (Box 2.3), and *vice versa*. Thus, for example, the activities of denitrifying bacteria result in the carbon in biomass being oxidized to CO_2, while the nitrogen in NO_3^- is reduced to N_2 (or N_2O or NO).)

Question 2.4

The water molecule has just two oxygen–hydrogen bonds, so the only possible *molecular* fragment is OH, formed by breaking one of these bonds, as depicted in Figure 2.10. The other fragment is a hydrogen atom — itself a reactive species.

Figure 2.10 The rupture of an O–H bond in a water molecule.

Question 2.5

(a) It is a simplification because the atmosphere also contains other more reduced carbon compounds — notably CH_4 and CO (together with trace amounts of formaldehyde (Figure 2.5), and indeed many other organic compounds of biogenic origin, as we shall see later on). The most obvious justification is that CO_2 is far more abundant than any of the other carbon species: its mixing ratio is over 200 times that of CH_4, for example (see Table 2.1). (The traditional emphasis on CO_2 is also justified through its importance to the biosphere, and the very large amounts cycled through the atmosphere each year.)

(b) The assertion would be justified provided the atmospheric oxidation of CO (whatever *its* source) acts as only a minor *in situ* source of CO_2 — compared with the amounts emitted at the Earth's surface. (This is indeed the case: CO oxidation accounts for only some 0.4% of the total annual input of CO_2 to the atmosphere.)

(c) As regards methane, the discussion in the text hopefully turned your thoughts to human food production — specifically rice cultivation (mentioned in the caption to Figure 2.4), and rearing grazing animals (cows, sheep, etc.) — and leakage of natural gas. The deliberate burning of forests and grasslands adds to this natural source of both CH_4 and CO. It may also have occurred to you that increased emissions of CH_4 are likely to augment the *in situ* atmospheric source of CO.

We shall see later (Section 4.3) that human activities now dominate the inputs to the atmospheric budgets of these two gases.

Topic 2 Atmospheric Chemistry and Pollution

Question 3.1

Using Equation 3.5:

For violet light:

$$E = \frac{1.20 \times 10^5}{400} \text{ kJ mol}^{-1} = 300 \text{ kJ mol}^{-1}$$

For red light:

$$E = \frac{1.20 \times 10^5}{700} \text{ kJ mol}^{-1} = 171 \text{ kJ mol}^{-1}$$

Question 3.2

Reading across to the vertical axis from the point labelled O_2 in Figure 3.6 gives a wavelength of about 250 nm. The shorter-wavelength (higher-energy) radiation absorbed by O_2 ($\lambda < 240$ nm) does, therefore, impart enough energy to break the oxygen–oxygen bond in O_2: photolysis is possible.

Question 3.3

According to Figure 3.6, absorption of wavelengths longer than 300 nm could lead to the photolysis of just two of the atmospheric constituents included there: ozone (O_3) and nitrogen dioxide (NO_2).

Question 3.4

To answer this question, you could write down an equation for each step in the mechanism (starting at the top and moving clockwise, that would give reaction R7, following by R8 and then R13), and then add together all of the reactants and products (as we did in Section 3.3.1).

However, it is quicker and simpler to just note the species that enter the cycle at various points (CO, O_2 and NO) and those that leave it (CO_2 and NO_2). Since these must be the reactants and products, respectively, in the overall reaction (Box 3.4), that checks with reaction R14.

For the next bit, you probably *did* need to write down the reactions, and then add things up:

$$CO + O_2 + \cancel{NO} = CO_2 + \cancel{NO_2}$$
$$\cancel{NO_2} + hf \longrightarrow \cancel{NO} + \cancel{O}$$
$$\cancel{O} + O_2 \longrightarrow O_3$$

Net: $CO + 2O_2 + hf = CO_2 + O_3$

Note that the 'hf' (the symbol for a photon) is not included in the overall reaction (R15) given in the question. It is usual to omit this symbol when writing the equation for an *overall* reaction like this: including it (which would be OK) is a reminder that one of the steps in the mechanism (R11 in this case) is a *photochemical* process.

Question 4.1

(a) The balance of water entering and leaving the atmosphere each year implies that, on this timescale, the *total* amount of water in the atmosphere remains relatively constant: the system is in a steady state. If this were not the case, we would expect to observe a long-term trend in the total amount of water in the atmosphere — a gradual increase or decrease over a period of years.

(b) Earlier in the course, we defined the residence time for water in the atmosphere, or in any other 'box' or 'reservoir' (the oceans, say, or an individual lake), as the *average* length of time that a water molecule stays in the reservoir. It can be calculated from the equation:

$$\text{residence time} = \frac{\text{total amount in reservoir}}{\text{rate of loss}}$$

Note that in a steady state (as here) rate of loss = rate of input, so information about either would suffice. Using the values in Figure 4.1 gives for H_2O in the atmosphere:

$$\text{residence time} = \frac{15 \times 10^{15} \text{ kg}}{505 \times 10^{15} \text{ kg yr}^{-1}} = 3.0 \times 10^{-2} \text{ yr}$$

In days, this is $365 \times 3.0 \times 10^{-2} = 11$ days (to two significant figures).

(Recall that this short mean residence time is why the *actual* moisture content of the air is highly variable — sensitive to relatively local variations in the flow of water to and from the underlying surface. Keep this in mind as you study Section 4.2)

Question 4.2

To estimate the atmospheric lifetime of CH_4 from Equation 4.4 (lifetime = $1/k'$), the first step is to calculate the value of k', as:

$$k' = k \times [\text{HO} \bullet] = \{6.2 \times 10^{-15} \text{ (molecules cm}^{-3})^{-1} \text{ s}^{-1}\} \times \{1 \times 10^6 \text{ molecules cm}^{-3}\}$$

$$= 6.2 \times 10^{-9} \frac{\text{molecules cm}^{-3}}{\text{molecules cm}^{-3}} \text{ s}^{-1} = 6.2 \times 10^{-9} \text{ s}^{-1}$$

Then lifetime = $\frac{1}{6.2 \times 10^{-9}\ s^{-1}}$ = 1.61 x 10^8 s

In years this becomes

$\frac{1.61 \times 10^8\ s}{60 \times 60 \times 24 \times 365\ s\ y^{-1}}$ = 5.1 years

Question 4.3

According to Equation 4.2: rate of loss = $k' \times$ [CH$_4$]

But k' = 1/lifetime (from the hint given in the question), so this becomes:

rate of loss = [CH$_4$]/lifetime

So, to estimate the rate of loss of CH$_4$ (per year), you would need to know its mean lifetime (with respect to removal by reaction with HO•) in years (given as about 12 years in the text) and its globally averaged tropospheric concentration. If the latter is expressed in the unit (molecules cm^{-3}), then to estimate the mass of CH$_4$ removed each year, you would also need to know the total volume of the troposphere (in cm^3) and the mass of a molecule of CH$_4$.

Question 4.4

For ozone loss via dry deposition, we can use Equation 4.3 to write:

lifetime = $\frac{[O_3]}{\text{rate of loss}} = \frac{[O_3]}{k \times [O_3]} = \frac{1}{k}$

Substituting the value of the 'rate constant' k given in the question gives:

lifetime = $1/(1 \times 10^{-5}\ s^{-1})$ = 10^5 s = $\frac{10^5}{60 \times 60}$ hours = 28 hours

So the lifetime of ozone with respect to dry deposition in rural areas is about a day.

Question 4.5

Hopefully the hint in the question directed your attention to the modified rate constant $k' = k \times$ [HO•] that appears in Equation 4.4. Thus, the lifetime of CO depends not only on the value of k, but also on the concentration of hydroxyl radicals. From point 5 in the Summary of Section 3, the primary source of HO• is ozone photolysis, followed by reaction of the O* atoms so formed with water vapour (reactions R5 + R6 in Section 3.3.1). So we might expect concentrations of HO• to be highest in the tropics (where the moisture content of air and input of uv radiation are both high), increasing the value of k' and so shortening the lifetime of CO. Observations confirm this line of reasoning.

By contrast, the air at high latitudes in winter is often relatively dry, and the input of uv radiation is low. Both factors would be expected to lower [HO•], and so lengthen the lifetime of CO.

Question 5.1

All of the HCs in Table 5.2 will have shorter atmospheric lifetimes than both CO and methane. Compared to CH$_4$, for example, the lifetime is lower by a factor of around 10^3 for ethene, and well over 10^4 for isoprene — reflecting the greater reactivity of these unsaturated hydrocarbons.

Question 5.2

In Figure 5.11a, the actual temperature of the atmosphere (the red line) decreases with height above the surface more rapidly than the ALR (the blue line). A parcel of air heated by contact with the ground may become slightly warmer and less dense than its surroundings (as shown) and start to rise. As it moves upwards, it will cool at the ALR. Because the rising parcel of air cools more slowly than the surrounding air, it will continue to be warmer and less dense than its surroundings. Thus upward movement continues and is magnified. The situation is unstable and conducive to convection, with updraughts in one location being balanced by downdraughts elsewhere. Rapid vertical mixing ensues.

By contrast, in Figure 5.11b temperature decreases with height less rapidly than the ALR. Surface air that is slightly warmer than its surroundings again starts to rise and to cool at the ALR. But now the temperature difference between the rising parcel of air and its surroundings soon decreases to zero, and upward movement ceases. The atmosphere is stable because any vertical movement of air tends to be damped out.

Question 5.3

The critical level for crops in Europe is currently set at an AOT40 of 3000 ppbv-hours. But the values of AOT40 plotted in Figure 5.21 are given in the unit ppmv-hours (note the label on the horizontal axis). Since 1 ppbv = 10^{-3} ppmv (Box 2.1), the critical level becomes an AOT40 of 3000×10^{-3} ppmv-hours = 3 ppmv-hours. The data in Figure 5.21 are rather scattered, but a reasonable estimate of the corresponding relative grain yield is 90–95%. Hence the percentage loss in grain yield is about 5–10%.

This example highlights the fact that setting a critical level for ozone exposure incorporates a judgement about the 'acceptable' level of potential damage.

Question 5.4

With a high density of traffic in downtown LA, the chemical mix of pollutants in the ambient air (i.e. high [NO_x]/[HC] ratios) should make ozone production hydrocarbon-limited. Under these circumstances, a simultaneous *decrease* in HC emissions and *increase* in NO_x emissions would be expected to result in lower ozone concentrations, as observed. By contrast, an increase in the concentration of NO_x in the air transported downwind would be expected to lead to increased ozone production under the NO_x-limited regime typical of rural areas in the USA — again, as observed.

Question 5.5

In principle, an electric-powered vehicle produces zero emissions *on the road*. However, the full 'pollution audit' would need to include the emissions produced in generating the electricity used to recharge the batteries — especially NO_x from coal and gas-fired power stations. On top of this are the factors likely to influence the public acceptability, and hence uptake of such vehicles: performance characteristics (e.g. range, speed, etc); ease and cost of recharging (compared with conventional fuels); economic incentives (or penalties, linked to tightening legislation); etc.

Acknowledgements for Topic 2 Atmospheric Chemistry and Pollution

Grateful acknowledgement is made to the following sources for permission to reproduce material in this book:

Figures 1.1, 2.3, 4.11, 5.28: NASA; *Figure 1.2*: Kunsthaus Zurich; *Figures 1.3, 5.26*: South Coast Air Quality Management District; *Figure 2.1*: Picture David Whillas, CSIRO Atmospheric Research; *Figure 2.2*: Peter Francis; *Figure 2.4*: Ecoscene; *Figure 2.7*: Centrica, plc; *Figure 3.10*: Dean/Greenpeace/SPL; *Figure 3.11*: NOAA; *Figure 4.3*: University of Berne, Institute of Geography; *Figure 4.4*: National Academy Press; *Figure 4.5*: Academic Press (London) Ltd.; *Figure 4.7*: Simon Fraser/Northumbrian Environmental Management Ltd/SPL; *Figures 4.8, 5.1*: SPL; *Figures 4.10, 5.16, 5.18*: AEA Technology; *Figure 5.2*: Commission of the European Communities, Brussels, 1993; *Figure 5.4*: Camera Press Ltd; *Figure 5.5*: Connor Caffrey/SPL; *Figure 5.6*: © 2002 GlobeXplorer International; *Figure 5.7*: John Wiley & Sons, Inc.; *Figure 5.10*: Photo courtesy of Tom Howder; *Figure 5.11*: Swiss Federal Research Institute for Agricultural Chemistry and Environmental Hygiene, Liebefeld-Bern; *Figure 5.14*: Photo by Al Greening. © Golden Gate National Parks Association; *Figure 5.15*: Fiore, A. M. *et al.*, *Geophysics Research*, **103**, pp. 1471–1480. Copyright © 1999 Princeton University Press. Reprinted by permission of Princeton University Press; *Figure 5.17*: National Railway Museum/Science and Society Picture Library; *Figure 5.19a*: Photograph supplied by G. J. Holmes; *Figures 5.19b, 5.20*: Photograph supplied by A. Heagle; *Figure 5.21*: Swiss Federal Dairy Research Station, Liebenfeld; *Figure 5.27*: Martin Bond/SPL.

Every effort has been made to trace all the copyright owners, but if any has been inadvertently overlooked, the publishers will be pleased to make the necessary arrangements at the first opportunity.

TOPIC 3
WETLANDS and the CARBON CYCLE

Nancy Dise

1	**Introduction**	**137**
2	**What is a wetland?**	**139**
2.1	Hydrology and water quality	139
2.2	Soils	144
2.3	Biota	147
2.4	Wetlands of the world	152
2.5	Summary of Section 2	155
3	**Wetland types**	**157**
3.1	Peatlands	157
3.2	Marshes	163
3.3	Swamps	170
3.4	Linking peatlands, marshes and swamps through hydrology	177
3.5	Wetland transitional communities	178
3.6	Wetland classification in Britain	180
3.7	Summary of Section 3	182
4	**Carbon cycling processes in a wetland**	**185**
4.1	Carbon gains and losses in a 'model' wetland	185
4.2	Processes of carbon storage and accumulation in wetlands	199
4.3	Summary of Section 4	202
5	**Carbon cycling in wetlands and the global picture**	**204**
5.1	Global carbon storage by wetlands	204
5.2	Global carbon accumulation in wetlands	208
5.3	Global carbon release by wetlands	210
5.4	Summary of Section 5	211

6 Impacts of wetland management and climate change on carbon cycling 213
6.1 Drainage and carbon storage 213
6.2 Rice and greenhouse gases 214
6.3 Wetland restoration and carbon balance 215
6.4 Climate change 215
6.5 Summary of Section 6 217

Learning outcomes 219

Answers to questions 220

Acknowledgements 224

Introduction

> … [The] path zigzagged … among those green scummed pits and foul quagmires which barred the way … lush slimy water-plants sent an odour of decay and a heavy miasmatic vapour into our faces, while a false step plunged us more than once thigh-deep into the dark quivering mire, which shook for yards in soft undulations around our feet. Its tenacious grip plucked at our heels as we walked, and when we sank into it it was as if some malignant hand was tugging us down into those obscene depths, so grim and purposeful was the clutch in which it held us …
>
> Arthur Conan Doyle, *The Hound of the Baskervilles*

Wetlands have always had an image problem. For millennia, people have considered them wasteland, places to fear, avoid and if possible 'improve' by ditching, burning or draining. Literature abounds with colourful descriptions of swamps, bogs and mires and the dire fates of their victims.

Yet humans have also recognized that wetlands and wetland products can be of immense value. The great civilizations of Assyria, Babylon and Egypt were built upon the fertile, regularly flooding alluvial floodplains of rivers of the Near East. Papyrus reed grown in marshes, immortalized for all time the Egyptian civilization of thousands of years ago (Figure 1.1), and rice, a wetland plant, nourishes more people than any other food. Peat, cut for centuries in the great northern peat bogs of Siberia, Canada, Finland and Ireland, has provided fuel for generations. Forestry, agriculture and horticulture have exploited the natural fertility of wetland soils for thousands of years.

Today, the attitude that wetlands are wastelands is gradually being replaced by the recognition that wetlands collectively comprise some of the most important biomes on the planet. In addition to their previously recognized local values — providers of fertile soil, foodstuff and fuel — our growing understanding of the Earth's biosphere and its complex web of interactions has revealed some key global functions of wetlands. It's now clear, for instance, as we deplete fishery stocks the world over, that wetlands are the nurseries for enormous numbers of commercially important fish and the organisms they feed upon. It is also becoming clear, as we take inventory of our planet's stock of living things, that wetlands support a disproportionate amount of the Earth's species, either as 'permanent residents' or migrating visitors such as many waterfowl. We are discovering, as we produce and release ever more technological products into the environment, that wetland soils can be effective filters of many pollutants, including metals such as cadmium, zinc and arsenic. Wetlands are also being increasingly protected, or even recreated along river banks, where they can remove excess fertilizer and pesticides in runoff from agricultural land before these pollutants ever reach streams and rivers.

Figure 1.1 Papyrus (*Cyperus papyrus*), used by the ancient Egyptians for making paper.

It may be hard to imagine, but it is likely that none of these is the most important role of wetlands in the biosphere. This role has only recently been appreciated: that wetlands are key players in the global balance of carbon. Here, however, wetlands show a contradictory nature. At the same time as they store massive amounts of carbon, wetlands also release greenhouse gases. Indeed, wetlands are a major source of the two most important greenhouse gases after carbon dioxide: methane (CH_4) and nitrous oxide (N_2O). As we begin to recognize that global warming is not just a theory but is actually occurring, which of these two faces of wetlands will dominate: carbon accumulator or greenhouse gas source?

In this topic we first describe what makes a wetland, defining in Section 2 the characteristics most wetlands have in common. In Section 3 we describe the major different types of wetland — their unique characteristics, distribution, some aspects of their element cycling, and some of the threats to their existence. Section 4 marks the switch to biogeochemistry, looking at the role of wetlands in the global carbon budget. Understanding the processes governing carbon biogeochemistry in wetlands plays a key role in our efforts to limit the potential effects of human-induced climate change. In Section 4 we delve in detail into the gains, losses and transformations of carbon in a single wetland. Section 5 looks at wetland carbon cycling on a global scale; here you will calculate whether or not wetlands make a difference to global warming by storing and accumulating carbon. Finally, Section 6 describes the effects of human activity, including climate change, on wetlands and on the way wetlands interact with the rest of the biosphere.

What is a wetland?

Like other biomes, wetlands cannot be defined with exclusive precision, but in general all wetlands share a few general characteristics:

- They are areas in which the water-table is at or near the surface for a significant part of the year.
- Their soils tend to be low in oxygen and are often rich in organic matter.
- Their vegetation is adapted to the conditions of at least occasional waterlogging and oxygen deficiency.

There are many definitions of wetlands, including many very long ones. Let's consider the one offered by Professor Paul Keddy at Southeastern Louisiana University, USA:

> A wetland is an ecosystem that arises when inundation by water produces soils dominated by anaerobic processes and forces the biota, particularly rooted plants, to tolerate flooding.

This one sentence encompasses the cause (inundation by water), the primary, chemical effect (scarcity of oxygen), and the secondary, biological effect (adaptations to tolerate flooding and anoxia).

In the next three sections we examine the key characteristics of each of the factors — hydrology, soils and biota — that are shared by the majority of the world's wetlands. In Section 2.4 we step back for an overview of the world's major wetland regions before turning in Section 3 to explore the major wetland types in detail.

2.1 Hydrology and water quality

2.1.1 Sources of water

Water is the overriding feature of wetlands, determining almost all their other properties. Many wetlands form the physical and hydrological gradient between **upland** (i.e non-wetland) environments and open water. Clearly, for water to accumulate the amount of water entering the ecosystem must exceed the amount of water leaving the system — that is, precipitation and water inflow from the surrounding area must exceed water loss through evaporation, transpiration, seepage and runoff. Climate and terrain are the major factors determining wetland hydrology. In addition to evaporation and transpiration (considered together as evapotranspiration), the slope of the land in areas with wetlands is usually gradual, decreasing the loss of water through runoff.

About half the world's wetlands are found in cool, northern and boreal regions of the world, where low temperatures reduce the rate of water loss through evaporation. Cool temperatures and daylight limitations also shorten the growing season, so that there are fewer days in the year when plants are at peak activity.

○ How could the activity of plants affect the amount of water accumulating in an ecosystem?

● Active plants use energy from sunlight to 'fix' carbon into carbohydrate, using water and carbon dioxide. Water is taken up through the roots but most of it is usually lost to the atmosphere (transpired) through stomata in leaves, which are open while taking in carbon dioxide. Thus high plant activity often reduces the amount of water accumulating in soil because it promotes water transfer from soil to atmosphere.

Many wetlands also occur in tropical and subtropical regions of the world. These have high rates of water loss through evapotranspiration. For water to accumulate, warm-climate wetlands must have high rates of rainfall, high inflow of water from surrounding water bodies, or both. They often occur along broad, flat, river floodplains underlain by clay or compacted organic matter which impedes drainage.

The amount, the flow rate and the chemical composition of the water in a wetland determine many of its major physical and chemical characteristics.

The amount of water

By definition, the water-table in wetlands lies close to the surface, and is often above the surface. It may be stable, or fluctuate dramatically over a year. The most dramatic fluctuations are in wetlands that are influenced by large, seasonal rivers. The water-table in some Amazon floodplain swamps, for instance, can vary from several metres above the surface to several metres below the surface over the cycle from rainy to dry seasons.

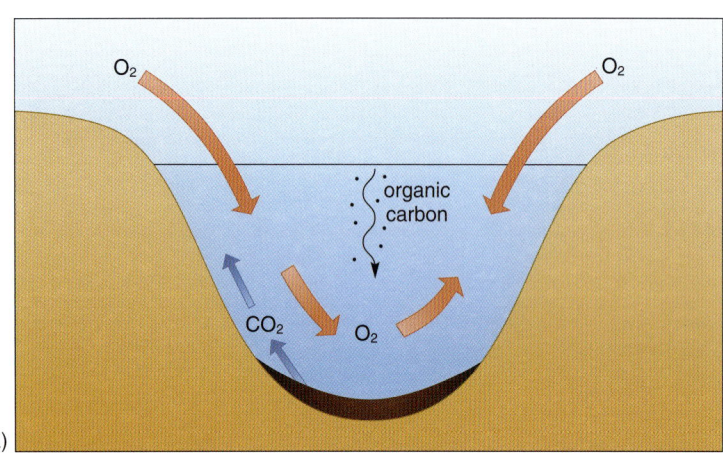

In wetlands, the level of the water-table roughly marks the transition between an oxygen-rich environment and an oxygen-depleted environment. Oxygen diffuses through water about 10 000 times slower than it diffuses through air. But not all water-saturated environments are depleted in oxygen — the water in many lakes and in the oceans is oxygen rich. Here, the supply of organic matter from the surface is low in comparison to the amount of water; in addition, wind, waves and currents increase the rate of diffusion of oxygen, and phytoplankton release oxygen into the water. Thus although deeper sediments in these bodies of water may become anaerobic (oxygen-free), the main water column usually stays oxygenated (Figure 2.1a).

Figure 2.1 (a) Many surface waters are aerobic because they have low rates of input of organic matter. (b) Anaerobic conditions occur in wetlands due to a high rate of organic matter supply in relation to the rate of diffusion of oxygen.

Compare this with wetlands, where the large amounts of organic carbon produced each year are available for decomposition. Oxygen becomes depleted in these waterlogged environments when the rate of supply of organic carbon exceeds the rate of supply of oxygen to decompose that carbon aerobically (Figure 2.1b).

The rate of flow of water

Water flows through wetland soils as percolating rainfall or meltwater, as water flowing laterally below the surface or as water flowing over the surface. Since these external sources of water are often fully oxygenated, rapidly flowing water can allow oxygen to reach plant roots and often permeate deep into wetland soils. Flowing water can also flush out the products of anaerobic decomposition, which can build up to toxic levels.

Rainfall percolates through wetlands gradually during slow drizzles or rapidly during torrential downpours. Snowmelt or floodwater can provide a large amount of oxygen-rich water over a short time period. Throughflow, as the word implies, describes water that flows laterally through the upper layers of soil, above the water-table. This water generally comes from the immediate surrounding area. Groundwater, which in some wetlands flows through deeper, saturated soils, can be a fairly constant source of water to some wetlands. Throughflow and groundwater can be grouped together as subsurface lateral flow.

Water flow can be very slow in wetlands where there is little connection to the regional water-table (groundwater surface), for example in wetlands that are 'perched' above the groundwater surface (Figure 2.2a). Accumulated organic matter in these wetlands can serve as a kind of plug, impeding downward percolation of water and creating a 'local water-table' above the regional water-table. Wetlands in standing water also have very slow rates of water flow (Figure 2.2b). Conversely, flow can be rapid in wetlands that are connected to groundwater flow or that receive runoff from catchments upstream (Figure 2.2c). In extreme cases, water throughflow in wetlands can be very rapid, for example when a fast-flowing river overflows its banks.

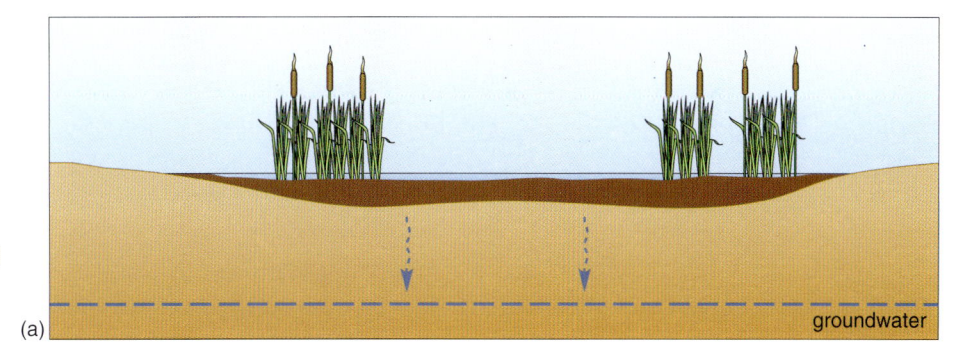

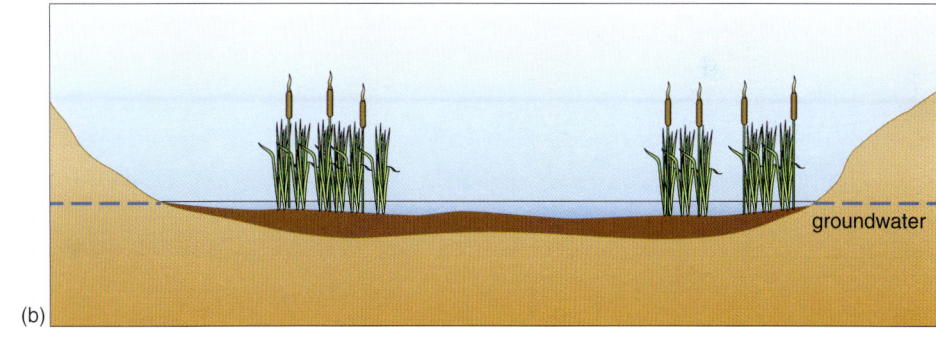

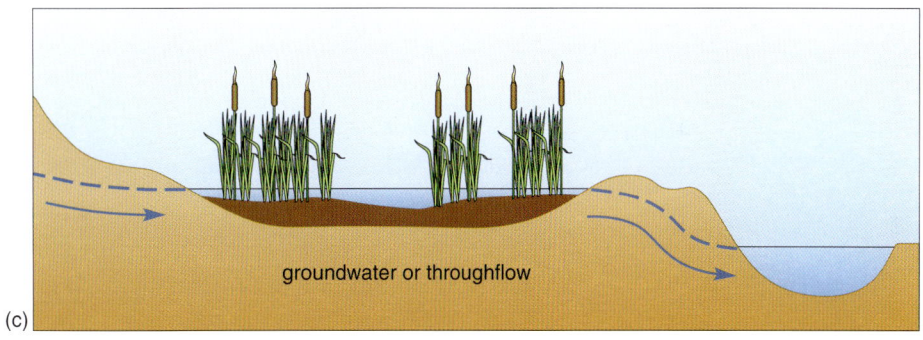

Figure 2.2 Water flow through three different wetlands: (a) a perched wetland; (b) a basin wetland; (c) a wetland downstream from a catchment. Arrows in (a) and (c) show the main direction of water flow.

Water flow is often seasonal. High-latitude wetlands (e.g. those in boreal and Arctic regions) may have one period of rapid water flow in spring, when the snowpack melts, with minimal flow rates the rest of the year. In high latitudes in winter, much of the water in a wetland may be tied up in snow, and throughflow may be minimal. Many wetlands in floodplains, particularly in tropical regions, show strong hydrologic seasonality of flooding and soil exposure linked to alternating wet and dry seasons. Flow can also be pulsed over shorter time intervals, for example in wetlands receiving tidal inputs of water.

Water quality

In addition to oxygen, water flowing through wetland soils can carry dissolved organic carbon and organic nitrogen, major ions such as calcium (Ca^{2+}), magnesium (Mg^{2+}), potassium (K^+), sodium (Na^+), nitrate (NO_3^-) and sulfate (SO_4^{2-}), and more complex organic compounds. Many of these 'dissolved solids' are important plant nutrients, and different sources of water typically contain different concentrations of dissolved material.

Precipitation generally has low levels of dissolved solids, with a relatively low pH (usually around 5 or 6). In contrast, throughflow water and groundwater entering a wetland have usually been in contact with mineral soils in upstream catchments, giving them higher concentrations of major ions and higher pH. Throughflow water draining upstream grasslands or agricultural land can bring with it nutrients such as NO_3^- and phosphate (PO_4^{3-}) from these ecosystems.

Because groundwater and throughflow water generally are richer in basic cations and nutrients than is precipitation, plant productivity is generally higher in wetlands receiving a high proportion of subsurface flow. Vegetation is often also richer and more diverse in these wetlands, and is dominated by vascular plants. Conversely, wetlands that primarily receive water from precipitation are usually nutrient poor and dominated by mosses.

Throughflow or groundwater that passes through soils affected by human activities can also, however, bring an excess of nutrients, pesticides and herbicides, or trace metals that may potentially be toxic to wetland biota. Water flowing through estuaries also brings salts that wetland biota must be adapted to in order to survive.

Table 2.1 summarizes the major sources of water to wetlands.

Table 2.1 Major components of hydrologic budgets of wetlands.

Component	Characteristics
precipitation	low in dissolved solids, relatively low pH many tropical/subtropical regions have distinct wet and dry seasons many high-latitude regions have distinct snowmelt water pulses in spring
surface flow (flooding)	can be seasonal, e.g. flooding of river banks, inundation by tidal water or snowmelt water from surrounding catchments
throughflow	can bring in oxygen, nutrients, dissolved carbon and basic cations such as Ca^{2+} and Mg^{2+}, but also pollution and salts
groundwater	can be a major source of oxygen and basic cations, water pH can be high, can provide a steady baseline water supply, but not always present
evapotranspiration	highly seasonal — peaks in summer, low rates in winter

2.1.2 The wetland landscape

The geomorphology of the landscape plays a crucial role in the hydrology and productivity of a wetland. As described above, a perched wetland is essentially isolated from throughflow and the regional groundwater, resulting in a nutrient-poor, often hydrologically variable environment, and vegetation adapted to these extreme conditions (Figure 2.2a).

A wetland in a basin (Figure 2.2b) loses water mainly through slow seepage, and evaporation. It may be nutrient-poor if its source of water is very low in dissolved solids, or it may become nutrient-rich if it is receiving water that has accumulated soluble nutrients and organic matter. Low levels of water flushing through a wetland can allow the build-up of products of anaerobic decomposition, reducing productivity.

A wetland at the bottom of a catchment receives runoff water from the immediate surrounding area (Figure 2.2c). This water is nearly always more nutrient-rich and usually less temporally variable than rainwater, and thus supports a richer and more productive vegetation. However, because they are closely linked with the upstream catchments, these wetlands are vulnerable to any alterations upstream. Logging, building, or intensified agriculture, for instance, can flush nutrients, toxins or particulate matter through a downstream wetland, potentially causing dramatic species changes and even the death of the biota. In less extreme cases, a downstream wetland can filter out such material before it reaches the stream or lake.

Many wetlands are formed on floodplains (Figure 2.3). The **floodplain** of a river or stream is the land bordering that body of water that may periodically be flooded. Floodplain wetlands can range from wet grasslands along small streams with extreme and variable hydrology, to wide marshes or forests along shallow-graded, winding rivers and streams.

Although some classification systems consider floodplain wetlands a type of wetland in themselves, they encompass a broad variety of different communities depending upon factors such as the regional climate, the local soils and the hydrology of the river. All floodplain wetlands, however, share the characteristics of being strongly influenced, physically and chemically, by the adjacent stream

Figure 2.3 Floodplain wetlands: (a) temperate wetland dominated by the reed *Phragmites* sp.; (b) a tropical swamp dominated by trees of the genus *Melaleuca*, Australia.

(a) (b)

or river. Their communities must be adapted to highly variable conditions of flooding and drying, fluctuations in oxygen levels, periodicity in nutrient or water supplies, and sometimes destructive floods.

The floodplain forms a physical and hydrological link between the river and the adjacent upland. When not modified by ditching, the link between the upland, floodplain and river is a continuum. A narrow strip of floodplain wetland is frequently the only non-cultivated land in otherwise developed areas. Floodplain wetlands are thus an important refuge for animals and rare plants, providing habitat diversity, abundant water and migration corridors for wildlife.

2.2 Soils

Wetland soils are defined in the International Soil Classification system as **hydric soils**: 'soils that formed under conditions of saturation, flooding or ponding long enough during the growing season to develop anaerobic conditions in the upper part'.

The common feature of wetland soils is that they are saturated during at least part of the year, and that this saturation leads to anaerobic conditions. Beyond this they can vary enormously, from relatively thin, mineral-rich, high pH soils in some marshes to deep (tens of metres) carbon-rich, acid peats.

Wetland soils may be classified as either organic or mineral. **Organic soils** contain at least 20% organic carbon. Organic wetland soils (Figure 2.4) contain a high proportion of plant material in various stages of decomposition, which is often less decomposed than in organic non-wetland soils. In some wetland soils, the decomposition rate is so limited that much of the original plant material can still be identified, even deep in the soil.

Some wetland soils (especially those where water loss in floodwater, throughflow or groundwater is low) are extremely rich in organic matter. A handful of peat (described further in Section 3), for example, dried and chemically analysed, contains about 50% carbon by mass. The high carbon content of some of these soils has, for centuries, made them valuable as fuel, and fuel peat is still harvested on a commercial scale in some parts of the world.

Organic soils lend certain characteristics to the wetland and the wetland plants found in them. They tend to be richer in nutrients such as nitrogen and phosphorus than mineral soils, but many of these nutrients may be tied up in organic material that is difficult for microbes to decompose, and thus is unavailable to plants. Organic soils generally have a high cation exchange capacity (CEC; see Block 2, Part 2 *Earth*) although they also have a low pH, so that the exchange sites are usually saturated with hydrogen ions.

Figure 2.4 Organic wetland soil, in this case from a fen. The fibrous nature of the lower soil is due to partially decomposed vegetation, and roots.

○ Why do organic soils have a high cation exchange capacity?

● Organic matter contains a high proportion of acidic functional groups (e.g. –COOH and phenolic –OH) that can lose or gain hydrogen ions, giving a high pH-dependent charge to the surface.

Soils containing less than 20% organic carbon are generally classified as **mineral soils**. Mineral soils often have a higher pH than organic soils due to the weathering of minerals containing basic cations. Since organic matter is the main source of cation exchange capacity, mineral soils usually show a lower CEC than do organic soils. However, with a generally higher pH and a higher concentration of basic cations from weathering, mineral soils have a higher proportion of the CEC occupied by basic cations (i.e. higher base saturation). They are often a good source of Ca^{2+} and Mg^{2+} to plants, although a relatively poor source of nitrogen and phosphorus, which are mainly derived from organic soil components.

Wetland mineral soils are often distinguished from non-wetland mineral soils by the presence of characteristic features associated with inundation by water. These features include **gleying**, bluish-grey clay-rich patches which may occur in a distinct horizon (Figure 2.5). Gleying indicates the loss of more oxidized forms of iron, Fe(III), and manganese, Mn(IV) or Mn(III), which together give soil its typical red–brown colour. The more reduced forms of these metals (Fe^{2+} and Mn^{2+}, respectively) generally form soluble compounds, and either leach out of the soil, leaving behind the natural colour of the parent material, or remain in the soil, leaving a characteristic blue–grey colour. Thus the colour of a gleyed horizon can derive from either the colour of reduced ions remaining in the soil, or the colour of the uncoated sand and silt particles from which iron and manganese have been removed.

Mineral wetland soils may also show **mottling**: discrete areas or spots of gleying occurring together in the soil profile with areas of red–brown oxidized soil. Mottling indicates that oxidizing and reducing conditions have occurred intermittently, corresponding with periods of unsaturated and saturated conditions.

The presence of both oxidized and reduced states suggests that the energy released by decomposing organic matter is transferred through organisms in wetland soils through more diverse pathways than in non-wetland soils, where nearly all of the decomposition occurs through aerobic respiration. However it is important to realize that in wetland soils aerobic respiration occurs as well, since most wetlands are not entirely saturated all of the time. Oxygen can rapidly enter wetland soils whenever the water-table drops below the surface.

Figure 2.5 Mineral wetland soil showing gleyed lower horizon, with characteristic blue–grey colour.

Topic 3 Wetlands and the Carbon Cycle

○ What is another way in which oxygen can enter a wetland soil?

● Water moving through a wetland as percolation, throughflow or groundwater may have high concentrations of dissolved oxygen.

Because of the large energy yield from aerobic respiration, aerobic organisms dominate where oxygen is present. So efficient is this process, compared to anaerobic respiration, that in many wetlands most of the decomposition occurs aerobically, even if only a few centimetres of the soil is oxygenated, or if the soil is exposed to oxygen during only a short time during the year.

When all the available oxygen is respired, however, different organisms take over the role of organic matter decomposition. To oxidize the remaining organic carbon for energy, these organisms use oxidizing agents other than O_2, including NO_3^-, SO_4^{2-}, long-chain organic molecules, and carbon dioxide (CO_2). The organic carbon is cycled through a chain of microbial and chemical transformations, and reduced products such as nitrous oxide (N_2O), hydrogen sulfide (H_2S), ethanoic acid (CH_3COOH, known informally as acetic acid) and methane (CH_4), respectively, are formed.

The production, accumulation and release of these reduced compounds is one of the major distinguishing characteristics of wetlands. Scrape down a few centimetres in the sediment of a stagnant marsh and your nose will immediately detect this 'distinguishing characteristic'! These compounds can build up to concentrations that would be toxic to organisms not adapted to them.

But this is far from the end of the story, for some of the waste products are *themselves* a source of energy. In fact, it is likely you use one of them to heat your home in winter.

○ Which one of the above products of anaerobic decomposition is commonly used for home heating?

● Methane, CH_4, is the main constituent of *natural gas*.

Many of the waste products of anaerobic respiration can, in turn, be used by whole different populations of microbes for energy. But there is a catch — oxygen is usually required. In a wetland with a water-table below the surface, and therefore oxygen in contact with the soil, a large proportion of the reduced gases produced in deeper anaerobic layers may be oxidized as they diffuse upward and reach the aerobic soil. However, in fully saturated wetlands a large proportion of the gases may escape oxidation in the soil, and so be released to the atmosphere.

This ability of wetlands to transform nutrients and metals into different oxidation states, with different toxicities and different physical forms (particulate/dissolved/gaseous), mediated by specially adapted populations of aerobic and anaerobic microbes, causes wetlands to be critical in biogeochemical cycles, especially those of carbon, nitrogen and sulfur.

Wetland soils are also critical in the cycling of phosphorus, but in a different way. Unlike carbon, nitrogen and sulfur, there is no major gaseous phase for phosphorus, and the supply in precipitation is very low. Nearly all the phosphorus coming into a wetland arrives in sediment, plant litter, or dissolved form in throughflow, groundwater or floodwater. Therefore chemical cycling of phosphorus is dominated by local or regional-scale sources and sinks, rather than global ones. However, wetlands can strongly influence the cycling of phosphorus by the dissolution of compounds such as $FePO_4$ and $Ca_3(PO_4)_2$, which releases phosphorus as phosphate ion (PO_4^{3-}) into the soil. So the oxidation–reduction status of the soil (the **redox** status), the pH, and the concentration of ions such as Ca^{2+} and Fe^{2+}/Fe^{3+} are vital to the cycling of this nutrient.

2.3 Biota

Wetland organisms are faced with three major challenges:

- Oxygen limitation: as described above, waterlogged conditions limit oxygen availability.
- Toxicity: reduced forms of metals such as Fe^{2+} and Mn^{2+}, and species such as H_2S, CH_4, NH_4^+, or ethene, can reach levels that would be toxic to non-adapted species. Acidity may also be high in organic wetlands, and salinity may cause problems for biota in wetlands exposed to the sea.
- Nutrient limitations: whereas some ions or compounds may build up to toxic levels, some vital nutrients such as NO_3^-, PO_4^{3-} and Ca^{2+} may be in critically low supply in wetlands.

2.3.1 Microbial adaptations

Oxygen limitation is a *requirement* for many wetland bacteria. These organisms have developed the capacity to use compounds other than oxygen for respiration. Some of the bacteria that respire anaerobically belong to the earliest life forms that appeared on Earth. They evolved and dominated the Earth during the first two billion years of its history, when the atmosphere contained little oxygen. It took over a billion years of photosynthesis to enrich the atmosphere with enough oxygen to allow the development of aerobic life and drive the anaerobic bacteria to oxygen free refuges, such as the sediments at the bottom of lakes or wetland soils. Many wetland bacteria are poisoned by oxygen; others can respire aerobically, but switch to anaerobic respiration when oxygen becomes limiting. Most are out-competed by the energetically more efficient aerobic bacteria when oxygen is freely available.

The microbial community in wetlands, as in other ecosystems, is the planet's ultimate recycling factory. Many have developed the capacity to resist the toxicity of the reduced end products of metabolism, or to 'pass on' these end products to yet other microbes. For instance, H_2S and CH_4, produced by some anaerobic bacteria, can be oxidized by other bacteria in aerobic zones in the wetland soils. The oxidation of these compounds prevents their build-up in the soil. Since different microbes can use a wide variety of substrates for obtaining energy, including non-carbon substrates such as H_2S and NH_4^+, a shifting availability of various chemical species favours different populations of bacteria. The microbial community can respond rapidly to changing concentrations of different compounds and ions as long as other conditions for growth are adequate.

Plants, animals and microbes living under saline conditions, such as in saltmarshes or mangrove swamps, must contend with the high osmotic pressure of the soil solution, which would cause fatal diffusion of water out of cells if not counteracted. The major microbial adaptation is the maintenance of a slightly higher osmotic pressure inside the cell than outside, through accumulation of ions such as K^+, or organic molecules like glycerol.

2.3.2 Multicellular organism adaptations

Multicellular organisms cannot survive for long periods of anaerobic respiration. Wetland organisms, especially plants, cope with anaerobic conditions, metal toxicity and nutrient limitations by various adaptations.

○ Why is adaptation especially important for plants?

● Unlike animals, plants can't move to more benign areas when conditions are unfavourable.

This rather obvious fact has broad implications. Being literally rooted to the spot, plants must be able to tolerate the environmental conditions that occur at the site over the reproductive span of the plant, which may be a single season, a year, several years, or hundreds of years.

Oxygen limitation

For many wetland plants, a common adaptation to waterlogging is the development of tissue containing air spaces (called aerenchyma) in the cortex of the stems and roots (Figure 2.6a). The air spaces allow the stem to act as a kind of honeycombed straw, allowing oxygen to diffuse from the stem into the roots. Thus the roots no longer depend upon oxygen dissolved in the soil solution to meet their respiratory demands, as is the case with non-wetland plants.

Over 50% of the volume of roots and stems in wetland plants may be air, as opposed to a few per cent in non-wetland plants. Some plants can even build up a slight pressure gradient in young leaves, allowing air to be actively 'pumped' into the roots (Figure 2.6b). The movement of oxygen to the roots through passive or pressurized flow allows aerobic root respiration to occur in a largely anaerobic environment. Other wetland plants simply stick their roots out from the sediment into the air.

Unlike vascular plants, mosses do not have specialized structures for bringing oxygen to their roots. Because they have no vascular system for bringing water from their roots to their leaves, they also need to be in more or less constant direct contact with water. They therefore tend to grow in areas where the water-table is close to the surface.

In addition to getting oxygen to their roots, plants need to get CO_2 to their leaves for photosynthesis. Many submerged wetland plants have adapted to the low diffusion of CO_2 through water by producing large, often single leaves that float on the surface. These collect sunlight and atmospheric CO_2 for photosynthesis,

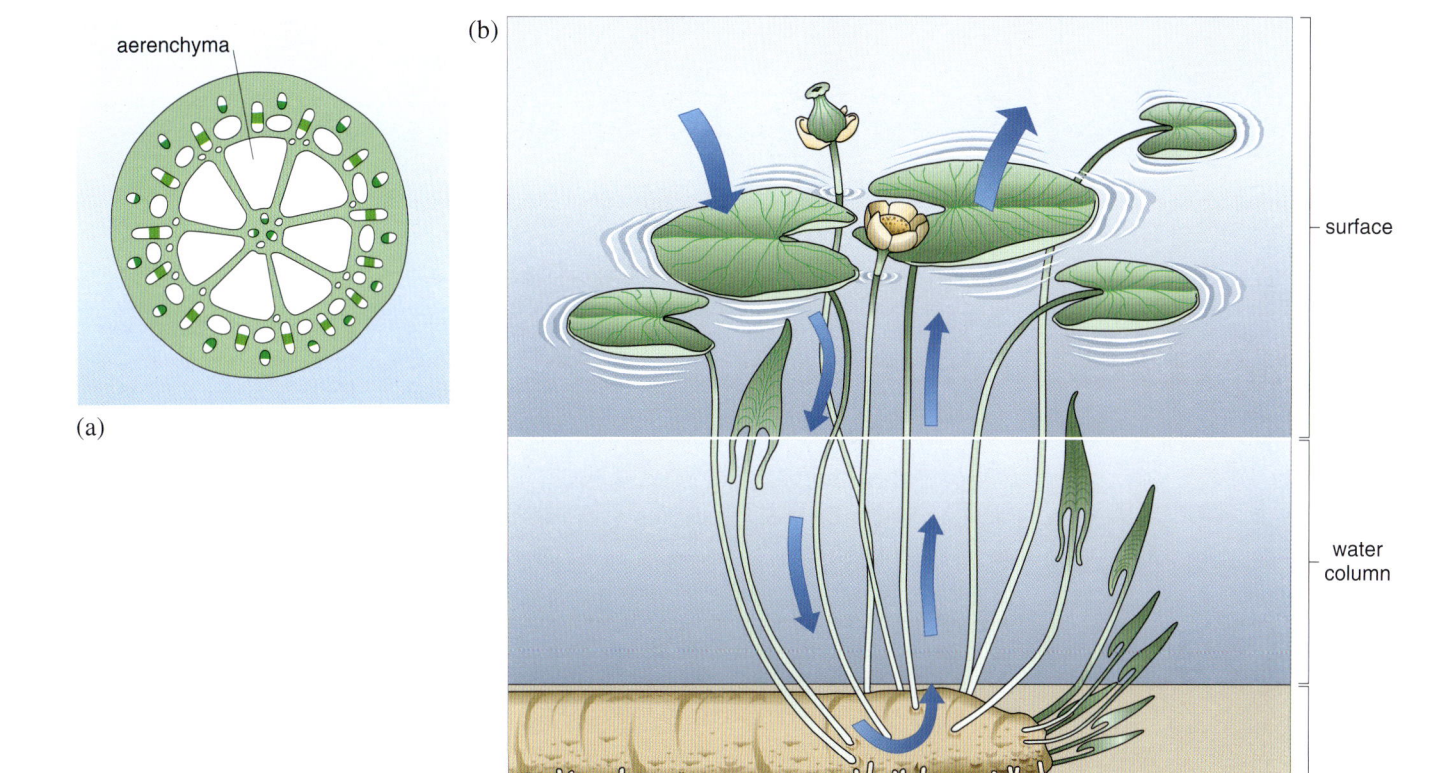

Figure 2.6 Plant adaptations to life in a waterlogged environment: (a) enlarged cross-section of a wetland plant stem, showing air spaces (aerenchyma); (b) pressurized gas flow in water lilies.

in addition to O_2 for respiration. Submerged wetland plants also use the buoyancy of water to their advantage: they do not need to produce strong structural tissue to hold the plant upright.

Aquatic animals that spend some or all of their lives in wetlands often show unique behavioural or physiological adaptations, especially to conditions of oxygen limitation. These include mechanisms to increase the efficiency of oxygen collection or of circulation, or reduced oxygen demand of tissues and organs. Some species of fish living in shallow floodplain pools in the Amazon, for instance, are able use their lower lip, swim bladders or stomachs to remove oxygen from swallowed air. Other adaptations to low oxygen levels include high concentrations of respiratory pigments, allowing more efficient collection of oxygen. This adaptation is seen in some invertebrates, such as clams and worms.

Many animals exhibit behavioural adaptations, such as migration (Figure 2.7) or going into a low-activity dormant phase when conditions are harsh. Perhaps surprisingly, plants may also show certain 'behavioural adaptations' to adverse

Figure 2.7 African mudskipper (*Periopthalamus sobrirus*), a fish that can survive periods of drought, by migration.

wetland conditions. These mainly concern the timing of seed production or germination, since seeds and seedlings are often the life stage most vulnerable to low O_2 levels. Seed production may coincide with the seasonal pattern of low water-tables, and the seeds themselves may be able to withstand long periods of inundation, sprouting quickly when water levels decline. Wetland plants may also produce floating seeds, allowing wide distribution. Some wetland plants, such as many mangroves, produce small 'plantlets' complete with leaves and tiny roots instead of seeds, giving the progeny a head start in establishing themselves (Figure 2.8).

Nutrient deficiency

Organisms that survive in nutrient-poor wetlands must be adapted to conditions of nutrient deficiency. Plant adaptations include thick leaves and woody stems, which reduce any leaching of nutrients out of the vegetation, and evergreenness, which reduces the loss of nutrients in shed leaves. Other adaptations to nutrient deficiency include:

- high root biomass and deep rooting to access a greater area of potential nutrients;
- nutrient translocation, to move valuable nutrients into parts of the plant that are not shed in autumn or are not at high risk of being eaten, such as stems and roots;
- 'plasticity' in nitrogen use — the ability to use different nitrogen species, including NO_3^-, NH_4^+ and perhaps organic nitrogen; and
- nitrogen-fixing bacteria on roots, which increase levels of nitrogen available to the plant.

These organisms are almost invariably stress-tolerant, often growing slowly; when conditions are particularly harsh they may shut down their metabolism almost entirely.

The requirements for light and nutrients in mosses are relatively low in comparison with vascular plants. They are well adapted to wetlands with low nutrient availability.

Toxic compounds

Some of the oxygen released to the roots of plants may 'leak' out of the roots into the surrounding soil, producing local aerobic zones (Figure 2.9). Reduced compounds that build up in the wetland soils may be oxidized in these areas, effectively detoxifying them. These local aerobic zones can also support a rich population of aerobic microbes which can oxidize methane, reduced nitrogen compounds and hydrogen sulfide for energy. Often a sizeable proportion of the gases produced anaerobically is not released to the atmosphere because of oxidation in aerobic microsites in an otherwise anaerobic environment.

Oxygen diffusion from roots can often be detected even long after the plant itself has disappeared, as veins of red, orange or brown in the soil where oxidized iron and manganese leave a trace of the former outline of the root system.

Figure 2.8 Mangrove shoots colonizing an exposed mudflat.

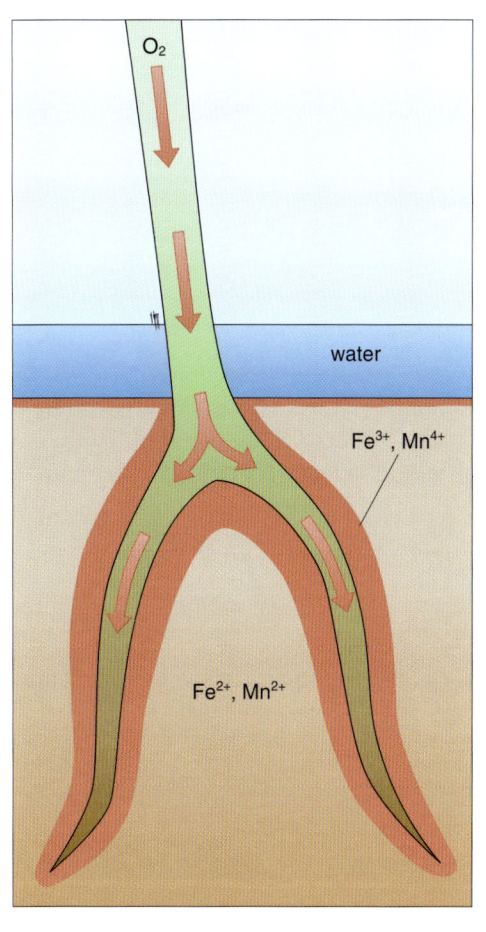

Figure 2.9 Local aerobic zone around root system of a vascular wetland plant, showing red patches of oxidized soil. The figure also shows a narrow aerobic zone at the water–sediment interface.

Some reduced and potentially toxic compounds are inevitably taken up by the plant. Within the plant, these can be dealt with in a number of ways. One strategy is to accumulate and isolate these compounds in areas of the plant such as vacuoles (fluid-filled vesicles in cells), vascular support tissue, or even senescing cells or tissues. Thus locked out of harm's way, the compounds do not influence the metabolism of the healthy cells and tissues. Some wetland plants can also biochemically convert dissolved forms of toxic compounds into gaseous forms that then diffuse out of the plant. Many wetland plants and animals are simply more tolerant to higher levels of potentially toxic compounds than their non-wetland counterparts.

To prevent cellular water loss, plants living in saline environments employ the same strategy as do microbes: in some cells, particularly those close to the root–water interface, they maintain high osmotic concentrations. Other strategies often involve specialized cells. These include:

- active pumping of salt out of cells;
- developing resistant 'barrier cells' in areas such as the cortex where water passes through, thus preventing diffusion of salt into other cells;
- active excreting of salt through leaves or roots; or simply
- enhanced physiological tolerance to high salt levels.

Many plants living in saline wetlands have adopted the C4 biochemical pathway of photosynthesis, which allows plants to fix CO_2 more efficiently than those using the more common C3 mechanism. In comparison with their C3 counterparts, C4 plants use enzymes with a higher affinity for CO_2, have lower respiration rates and can fix CO_2 over a wider range of temperatures and light intensities. Since the stomata must be open to bring in CO_2 for photosynthesis, and a large amount of water can be lost through the open stomata, all these measures that allow more efficient use of CO_2 also allow the plant to conserve water. Water conservation may sound a bit counter-intuitive in a wetland! However, water is lost from cells exposed to a saline solution, and thus a saline wetland can be a physiological desert. Water conservation is therefore an important strategy of wetland plants in saline environments.

2.4 Wetlands of the world

2.4.1 Global distribution

There have been several attempts to estimate the total global extent of wetlands. These rely upon various sources, including regional surveys, national surveys, digital maps and international databases. Although there is large variability in the intensity of surveying and the definition of wetlands, a reasonable estimate is that the total global extent of 'natural' wetlands is about 6×10^6 km^2, and that of rice paddies is about 1.5×10^6 km^2. Wetlands thus cover about 5% of the land surface of the Earth (Figure 2.10). The global distribution of natural (non-rice agriculture) wetlands broadly divides into northern latitude (>45° N in North America and >60° N in Eurasia) peatlands, and tropical/subtropical (30° N–30° S) marshes, swamps and floodplain wetlands. (We shall describe these terms in greater detail in Section 3.)

2.4.2 Losses and protection

The current area of the world's wetlands is probably only about half that which existed before people began modifying them. The major human activities on northern peatlands have included drainage for agriculture and forestry, mining for fuel and horticulture, and perhaps climate change. Tropical and subtropical wetlands are impacted by drainage for agriculture, forestry and mosquito control. Many temperate marshes have been lost due to drainage for property development

Figure 2.10 Wetlands of the world.

□ major wetland area   □ area with abundant wetlands

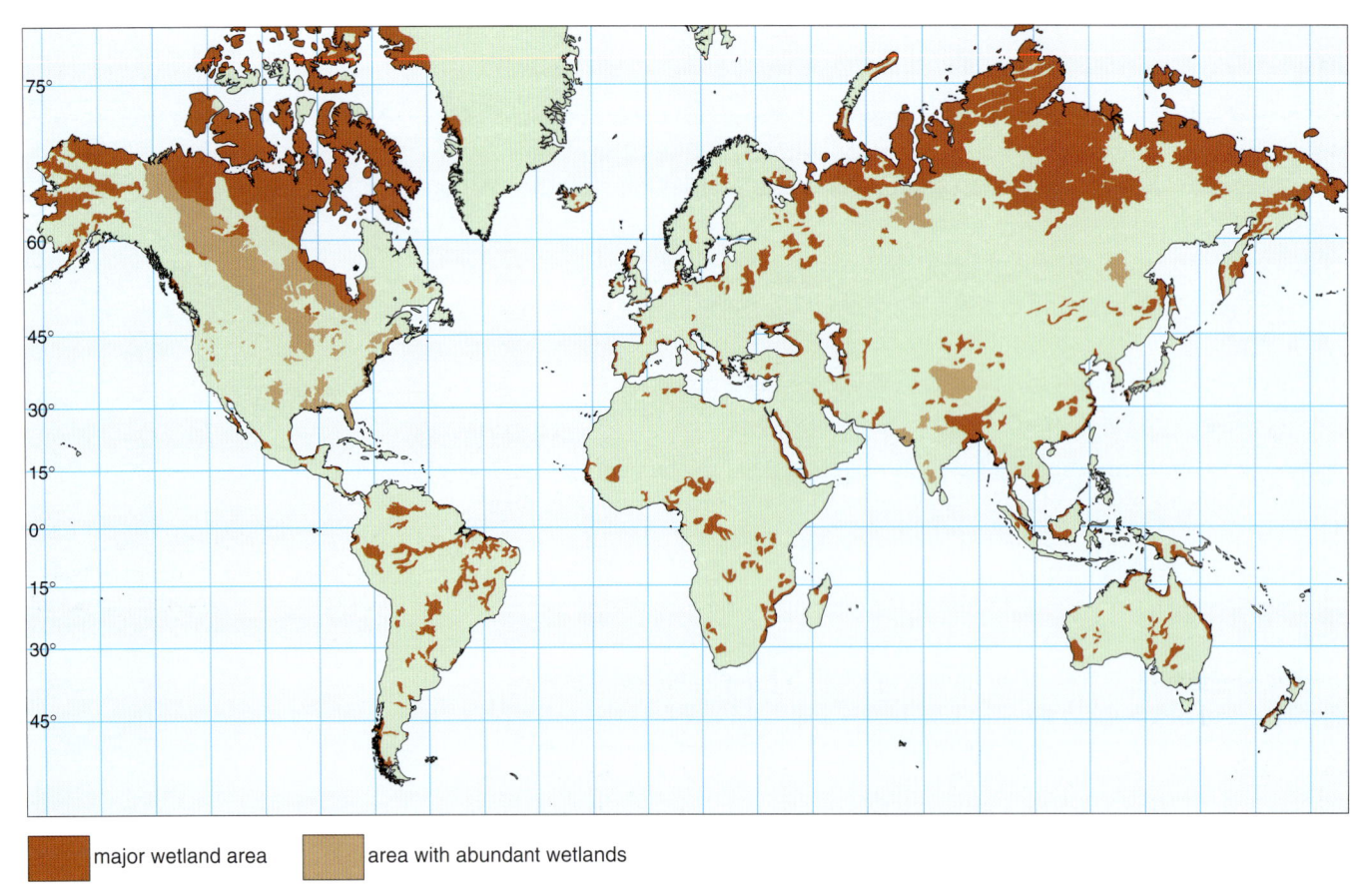

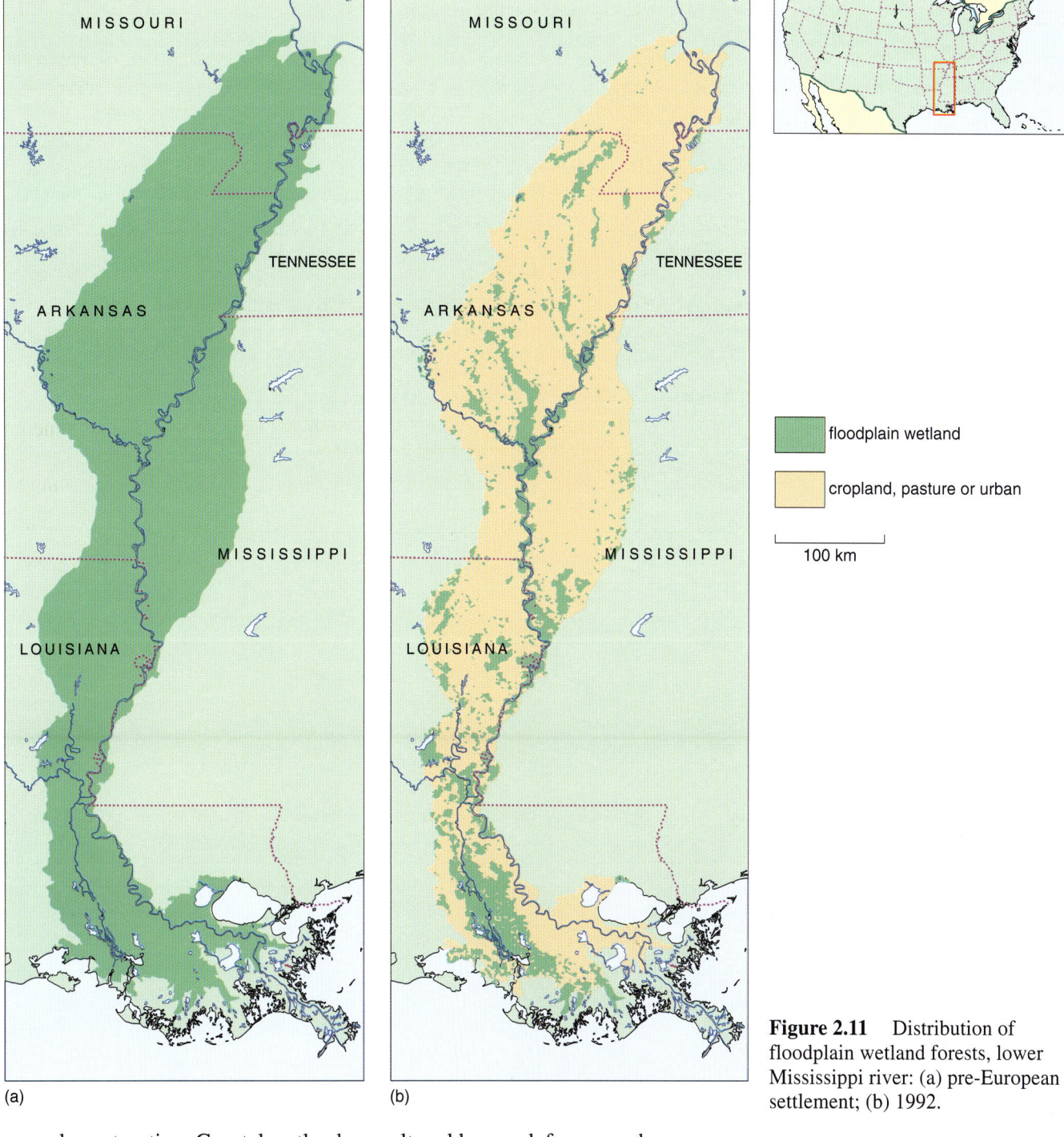

Figure 2.11 Distribution of floodplain wetland forests, lower Mississippi river: (a) pre-European settlement; (b) 1992.

or road construction. Coastal wetlands are altered by sea defences such as seawalls, and by dredging.

Most of the wetlands in the developed nations of the world were lost long ago, as people drained and cultivated nearly all the 'useless swampland' that could be used for agriculture. Nearly 80% of the floodplain wetlands that once lined the Mississippi river, for instance, are gone (Figure 2.11). Another example is the

Topic 3 Wetlands and the Carbon Cycle

'Great Black Swamp', a mosaic of 4000 km² of marsh and swampland in the midwestern United States southwest of Lake Erie, well documented by the travails of many pioneers, settlers and occasional armies who had to slog thigh-deep through it in the 1700s and 1800s. Today it is pasture, cropland and housing developments. Much of the wetland loss in Europe occurred hundreds of years ago and is undocumented.

A repeat of this process is facing the wetlands of the tropical and subtropical regions of the world today. The Panatal, for instance, is an area of about 140 000 km² of floodplain marsh and swampland, roughly the size of England, located almost in the exact geographic centre of South America (Figure 2.12). During the wet season the rivers flood their banks and support a rich population of wetland vegetation. In the dry season the vegetation reverts to a dry savannah. The Panatal is legendary for its wildlife, including jaguar, ocelot, giant otters, wolves, snakes, deer, more than 540 species of fish and about 650 species of birds, including the Jabiru, the largest flying bird in the Western Hemisphere.

The Panatal has long supported low levels of ranching and use by native populations, but increasing population pressure has led to intensification of grazing and regulation of water levels, massive poaching (up to a million animals a year), and large-scale agriculture, deforestation and mining. Agriculture and industry have in turn caused the contamination of rivers with herbicides, pesticides and metals. But the biggest threat to the area is the Hidrovia project, a scheme that will dredge a 3440 km-long waterway through the region, including 1670 km straight through the Panatal. The project will greatly reduce the annual flooding of the wetlands, allowing expansion of agriculture as well as further development of forestry and mining. Although agreed to by five South American countries in 1988, the project is still being actively debated.

The Panatal is symptomatic of the pressures on wetlands in developing areas of the world today — the encroachment of ever more people, and the application of new technology in draining, agriculture, forestry and mining. Often these are backed by national or multinational development programmes.

Despite past loss and current development of wetlands, their value has become increasingly recognized, resulting in laws and agreements in most countries to protect wetland habitats. Coordinated international efforts in wetland conservation began at the Convention of Wetlands of International Importance, held in Ramsar, Iran in 1971, and universally referred to as the Ramsar Convention.

Figure 2.12 Section of the Panatal, Mato Grosso State, Brazil.

The importance of wetlands for waterfowl habitat was primarily recognized in the convention, although more recently biodiversity, preservation of rare or endangered species, refuges, nurseries and fisheries, and non-biological properties like limnology and hydrology, have also been recognized. The Ramsar convention resulted in a global treaty, endorsed by 130 countries in 2002, under which members were obliged to develop national wetland policies and identify sites for a list of wetlands of international importance for protection, based on rareness or uniqueness of type, or on the above characteristics. In 2002 there were 1112 of these Ramsar wetland sites identified throughout the world, totalling an area of nearly 90×10^6 hectares (1 ha = 10^4 m^2). The UK has 149 Ramsar sites, the most of any country.

2.5 Summary of Section 2

1. A wetland is an ecosystem that arises when inundation by water produces soils dominated by anaerobic processes, thus supporting species that are adapted to flooding.

2. The hydrologic budget of a wetland is made up of precipitation, surface flow, throughflow, groundwater and evapotranspiration. The amount and the source of the water to a wetland together determine its major physical and chemical characteristics.

3. Water flowing through wetland soils can carry with it oxygen, dissolved organic carbon and nutrients, and carry away toxic compounds. For this reason, productivity is generally higher in wetlands that are dominated by throughflow or groundwater. Conversely, wetlands that are dominated hydrologically by precipitation are usually nutrient poor.

4. Wetlands may be dominated by either organic or mineral soils, but share the common feature of being saturated during at least part of the year, leading to anaerobic conditions. Waterlogging in wetland soils leads to characteristic features such as gleying or mottling and the presence of reduced compounds of carbon, nitrogen and/or sulfur.

5. Oxygen limitation, nutrient limitations and toxic compounds are the major challenges facing wetland organisms. Vascular wetland plants show characteristic adaptations to these conditions, including aerenchyma to facilitate oxygen transport to the roots, strategies to conserve nutrients, and root aeration, which oxidizes potentially toxic reduced compounds near the roots.

6. Aquatic animals that spend some or all of their lives in wetlands often show unique adaptations to low oxygen levels. These include mechanisms to increase the efficiency of oxygen collection or of circulation, reduced oxygen demand by tissues and organs, or behavioural changes such as dormancy or migration when conditions are harsh.

7. Wetlands, including rice paddies, cover about 5% of the land surface of the Earth. Global wetland distribution broadly divides into northern latitude peatlands, and tropical/subtropical marshes, swamps and floodplain wetlands.

8. Wetlands have been extensively impacted by humans in the past and in the present, primarily due to drainage for agriculture.

Question 2.1

Describe how oxygen limitation due to inundation by water determines some major differences between:

(a) Wetland soils and non-wetland soils.

(b) Wetland vascular plants and non-wetland vascular plants.

Question 2.2

Rain gauges, soil wells, weirs and other hydrologic instruments are installed in two wetlands to establish their water budgets. After five years of data collection, the results are analysed. Summaries of the proportion of water received from precipitation, throughflow and groundwater are shown in Table 2.2.

Table 2.2 Proportional contributions of water sources to two wetlands.

	Wetland A (%)	Wetland B (%)
precipitation	90	10
throughflow	9	70
groundwater	1	20

Based purely on the data in Table 2.2, infer the major characteristics of wetlands A and B, and compare them to each other, in terms of nutrients, soil fertility, soil pH and vegetation.

Wetland types

> billibong, muskeg, mire, moor, pakihi, pocosin, carr, slough, flark, hochmoor, lagg, quagmire, swale, yazoo, paalsa, pingo…

Despite the common characteristics of wetness, anaerobic soils and adapted organisms, wetlands show a remarkable variety. They may be dark forests tens of metres high, open grasslands or low, mossy carpets. They may be submerged by several metres of water, cut by braided rivers or streams, or show no standing water at all. They may be alive with insect life, birds and mammals or poor in fauna. They may be frozen for ten months of the year, or be steaming jungles.

This enormous variety, and the long history of interest in wetlands from people living in or near them, has spawned a bewildering array of names for wetlands around the world. These regional, or in some cases, 'continental'-scale, names often evoke distinct communities with recognizable vegetation, hydrology, climate, soil and associated fauna. They continue to be widely used, even by scientists working within the field, and are valuable for describing specific communities. Problems only occur as scientific communication becomes global, for in an international context common terms do not always convey the same meaning.

There is no universally recognized classification system yet for wetlands, although most current texts for an international audience have adopted the North American system, and there is movement to standardize the terms, at least across international boundaries. In the words of Paul Keddy on this topic, '*In an era of international flights, e-mail and global telephone linkages, scientific dialects are no longer acceptable. Let us hope that teachers will try to bequeath to their students one standard terminology*'.

In the above spirit we have also adopted the North American terminology for our global tour of wetlands and their properties. Recognizing the value for a largely British audience of distinguishing well-identified wetland communities in Britain, however, we describe these in Section 3.6.

In the North American system, wetlands can be described as **peatlands**, **marshes** and **swamps**. In the following sections we describe each in turn.

3.1 Peatlands

Peatlands are wetlands that accumulate significant amounts of partially decayed plant material, or **peat**. Peat formation requires significant organic matter production, in excess of decomposition, and restricted or impeded drainage, which limits the removal of dissolved or particulate carbon.

Peatlands are the largest major type of wetland, covering a total of about $3.4 \times 10^6 \text{ km}^2$. Although they can be found wherever drainage is impeded and anaerobic conditions prevail, peatlands are distributed primarily in the sub-boreal, boreal and subarctic zones of the world, north of about 45° N latitude in North America, and about 60° N latitude in Eurasia. This includes large areas of Canada, Alaska, Siberia and northern Scandinavia (Figure 2.10). Here, cool temperatures limit plant growth rates and therefore transpiration, and also limit the rate of evaporation. Drainage is also often poor in peatlands. Thus precipitation in high-latitude regions often exceeds water removal, leading to waterlogged soil and anaerobic conditions. The combination of cool temperatures and anaerobic soil leads to reduced rates of decomposition. Because peatlands tend to be hydrologically dominated by precipitation rather than throughflow or groundwater, organic matter accumulates, forming peat.

3.1.1 Properties and development of peatlands

There are two major types of peatland. **Bogs** are isolated from the groundwater and receive little throughflow water; their main source of water is precipitation (Figure 3.1). Since groundwater generally contains much higher concentrations of calcium, magnesium and other nutrients than rainwater does, bogs have low pH, low nutrients and relatively low species diversity.

One genus of moss, *Sphagnum*, is highly successful in bogs, and often dominates the understorey. Bogs also contain other mosses and sedges, and low, woody shrubs with thick leaves to reduce nutrient loss, called **ericaceous** shrubs (a family of acid-tolerant plants — including rhododendrons, heathers, etc.). Adaptations in ericaceous shrubs, such as thickened leaves and woody stems, also reduce herbivory by making plant parts more difficult to digest. The conditions of a high water-table, acid soils and nutrient-poor, acid vegetation cause the build-up of large amounts of peat. Bogs generally have very deep peat, up to many tens of metres.

Fens (Figure 3.2) receive a significant proportion of groundwater or throughflow water, but not so much that the accumulation of peat is prevented. The input of nutrients and basic cations from water flowing through fens leads to a higher pH and greater overall nutrient levels than in bogs. Consequently species diversity in fens is usually higher. The amount of water supplied from groundwater and runoff, and the chemical composition of that water, largely determines the species found in a fen. Fens with only a moderate or irregular supply of throughflow water may be dominated by *Sphagnum*, like bogs (sometimes called 'poor fens' or 'transitional fens'). With increasing nutrients they may become dominated by other mosses, sedges and grasses.

Peatlands form in two major ways: terrestrialization and paludification (Figure 3.3). **Terrestrialization** (Figure 3.3a), occurs when shallow lakes are gradually filled by the accumulation of organic matter deposited on the lake floor and from the encroachment of peatland plant vegetation from the sides. The organic matter comes from internal lake production, such as algae, from material that has fallen into the lake from the shore (leaves, etc), and from organic matter or sediment carried in streams and rivers that feed the lake. Herbaceous plants such as reeds and sedges, and mosses such as *Sphagnum*, can form a mat floating on the lake surface, which gradually thickens as these plants die and others grow on top of them. The

Figure 3.1 Peat bogs. (a) Bog formed over impervious clay soil, Sussex, UK. (b) Forested bog in the Queen Charlotte Islands, British Columbia, Canada.

thick mat can eventually become colonized by other plants, including shrubs and trees, forming a **quaking bog**. These are so called because the surface undulates, or quakes, when walked upon. Eventually the lake fills from both above and below. Lakes are temporary landscape features; most have a lifespan of only a few thousand years before they are completely filled.

Many of the vast peat bogs and fens of the boreal zone were formed from terrestrialization of the millions of small ponds and lakes left behind as the glaciers retreated at the end of the last glaciation. These include so-called **kettle hole bogs**, in which a block of ice left behind by a retreating glacier formed a deep hole which filled with water and then, eventually, peat (Figure 3.4), as well as much larger bogs and fens.

Figure 3.2 Calcareous fen in Oxfordshire, UK.

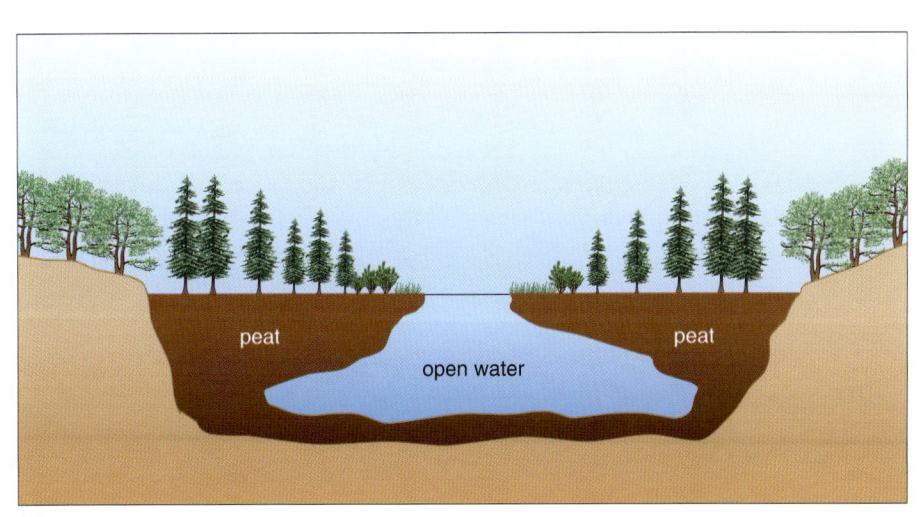

(a)

(b)

− − − water-table

Figure 3.3 Formation of peatlands by (a) terrestrialization and (b) paludification.

Figure 3.4 'Kettle hole' pond in tundra. Ponds such as that shown on the right-hand side of the picture could have formed as glacial ice retreated, eventually filling with peat to form bogs.

The other way that peatlands form is through **paludification** (Figure 3.3b), in which terrestrial vegetation is blanketed by peatland vegetation due to a change in the site hydrology, leading to persistent soil saturation. This can occur if the climate changes, for instance becoming cooler and/or wetter. As a consequence, evapotranspiration is slowed.

○ Why would cool temperatures slow the rate of evapotranspiration?

● Cool temperatures reduce the atmosphere's capacity to hold water vapour, and therefore its demand for water is decreased.

○ What consequence would the above have on water accumulation?

● Since evapotranspiration is a major mechanism of water removal, slowing its rate would increase water accumulation.

A consequence of a climate change toward cooler and wetter conditions may therefore be a rise in the water-table.

Paludification can also occur if the local hydrology is altered such that water permanently floods previously dry land, for example if a river course is changed through flood scouring, or if segments of a river are dammed due to activities of humans or beaver. Logging can lead to paludification since trees remove large amounts of water through evapotranspiration.

Higher water-tables, and the resulting reduction in the diffusion of oxygen into soils, lead to local conditions of anoxia. If the zone of prolonged anoxia is also the rooting zone, this can kill fine roots and eventually whole trees. The organic matter from dead vegetation and roots creates a high demand for oxygen by bacteria engaged in aerobic decomposition, so oxygen is even more rapidly depleted. With aerobic respiration rates too low to decompose all of the organic matter, it begins to accumulate.

As more and more partially decomposed organic matter accumulates as peat, the vegetation becomes more and more isolated from the mineral soil. In extreme cases, with the accumulating peat impeding downward percolation of water, a local water-table can develop, isolating the soil from groundwater and throughflow water (Figure 2.2a). As the source of nutrients becomes dominated by precipitation, the nutrient levels in the soil and soil solution decline and the ecosystem becomes more acid. Plants that are not adapted to these conditions do not survive, but mosses, particularly *Sphagnum*, thrive. The highly organic, acid peat becomes colonized by peat vegetation.

The transformation from an upland forest to a peat bog as described above can take many decades, or even hundreds of years. But once formed, *Sphagnum* bog communities may persist for many thousands of years.

3.1.2 Biota

Peatland surface vegetation is dominated by bryophytes — plants that do not have vascular tissue — and in particular, mosses. Mosses must have periodic access to free water, but do not need to support large quantities of structural tissue. Therefore their requirements for light and nutrients are relatively low.

Sphagnum is exceptionally well adapted to life in bogs due to a number of physiological features, including:

- a unique ability of its leaves to exchange cations, leading to a high capacity to retain nutrients; and
- a highly water-absorbent anatomy, similar to a sponge, allowing it to retain moisture.

In fact, some of the above characteristics create conditions that exclude other plants:

- The high cation exchange capacity means that *Sphagnum* exchanges hydrogen ions for basic cations in the environment. The basic cations are adsorbed onto the leaves of the *Sphagnum* and can be used by it. The hydrogen ions are released to the soil, acidifying it. This reduces colonization by other plants, and also inhibits soil fauna.
- The water-absorbent anatomy creates even more waterlogged conditions in the environment, which excludes some other plant and animal species.
- *Sphagnum* can also produce and release organic acids and inhibitory compounds to the outside environment. Although the significance of these compounds is not clear, they may further suppress the growth of competitive vascular plants and restrict some soil fauna. In addition, these compounds may retard bacterial action, slowing decomposition.

The combination of anoxic acid conditions and the inhibitory compounds released by *Sphagnum* is responsible for the extremely slow rates of decomposition in bogs, which are sometimes dramatically illustrated by the discovery of a 'bog body' (Figure 3.5).

Figure 3.5 The mummified head of Tollund Man, dated 220–240 BC, found in a bog in Jutland, Denmark.

Topic 3 Wetlands and the Carbon Cycle

(a)

(b)

(c)

Figure 3.6 Peatland plants: (a) cotton grass (*Eriophorum* spp.), (b) leatherleaf (*Chamaedaphne calyculata*), (c) Labrador tea (*Ledum groenlandicum*).

Other common peatland plants include members of the **sedge family** such as *Eriophorum* spp. (Figure 3.6a) and *Carex* spp. Ericaceous shrubs are also common peatland plants, especially in bogs. These include plants such as leatherleaf (*Chamaedaphne calyculata*) and Labrador tea (*Ledum groenlandicum*) (Figure 3.6b and c) in North America, and bilberry (*Vaccinium myrtillus*) and bog myrtle (*Myrica gale*) in both North America and Europe. Black spruce (*Picea mariana*) and tamarack (*Larix laricina*) are common in forested bogs. All of these plants are adapted to low levels of oxygen and nutrients through the various strategies described in Section 2.3.2.

Some peatland plant species have adapted to critically low levels of nutrients such as nitrogen by becoming 'carnivorous' (Figure 3.7). The presence of **carnivorous plants** indicates very low levels of available nitrogen and phosphorus in the environment. These remarkable plants have developed adaptations to attract, trap and digest animals, primarily insects and spiders, although small reptiles and amphibians have been known to fall victim to some of the large tropical carnivorous plants. The digested animals provide nutrients that are deficient in the soil.

A common carnivorous plant in peat bogs, the sundew (*Drosera* spp.; Figure 3.7a), has small stalks at the end of its leaves; these are often brightly coloured and sticky with nectar, which attracts insects. Unfortunately for the insects, the nectar also contains adhesive compounds and digestive enzymes. Insects that land on the stalks become stuck and are slowly digested. The movement of the insect often stimulates nearby stalks to curve over and around the victim, further entrapping it.

The pitcher plant (several genera, most commonly *Sarracenia* spp.; Figure 3.7b) features specially adapted leaves curled in the shape of a pitcher, in which rainwater collects and into which the plant secretes digestive enzymes. Hairs lining the inside of the leaves are oriented so that insects entering the pitcher cannot climb out, but are funnelled into the water, where they drown and are then digested. Pitcher plants, however, are themselves host to a number of insects which feed predominantly or entirely upon them. Some have turned the tables, taking advantage of the pitcher plant's uniquely dangerous environment. The mosquito *Wyeomyia smithii*, for example, lays its eggs and completes its larval life stages in the mix of water and digestive enzymes in the pitchers, feeding on the carcasses of insects and spiders without any noticeable ill-effects!

Peatlands are generally less productive than other vegetation communities in the same climatic zone, and less productive than other wetlands. Because of this, and other features such as high acidity that limits soil detritivores, thickened leaves and woody stems, peatlands support relatively low levels of animal populations. Some insects such as mosquitoes may be common, as well as organisms that feed on them, such as dragonflies. Vertebrate populations tend to be low, although deer, moose, beaver and mink are found in the large peatlands in sparsely populated regions. Many bird species are seasonal visitors, although few require strictly peatland habitats. Wading birds including snipe, curlew and dunlin use peatlands for food and nesting.

3.2 Marshes

A marsh is a wetland community that is dominated by non-woody, vascular plants, and does not show a deep accumulation of peat. Marshes are characterized by a significant flow of subsurface water and often have some contact with the regional groundwater-table. This leads to a rich nutrient status. A significant amount of organic matter may be leached from marshes and support aquatic communities downstream. Marshes are usually more dominated by mineral soils than are peatlands.

Marshes are the most diverse of the Earth's wetlands. They may occupy large swaths of the landscape, like the 4000 km^2 Everglades in Florida, or be only a few metres across. They occur throughout the world, across the spectrum of climate zones. In northern latitudes marshes coexist with peatlands, depending upon factors like the amount of throughflow water. In tropical and subtropical latitudes, marshes coexist with swamps. Marshes are often fairly ephemeral features of the landscape, converting to peatlands, grasslands, or forests as they gradually fill.

We can consider two different types of marsh: **freshwater marshes** and **saltmarshes**. Saltmarshes are found worldwide along coastlines, especially in middle and high latitudes. They are often replaced by mangrove swamps in tropical and subtropical regions.

There are about 950 000 km^2 of freshwater marshes in the world, accounting for approximately 15% of the world's natural wetlands, but this estimate is highly uncertain because they are often categorized with other wetlands, such as floodplains, and they may be difficult to spot and count as they are often small and may disappear at some times of the year. Large areas of freshwater marshes occur in South America, particularly Paraguay, Brazil, Venezuela, and along the Amazon, as well as sub-Saharan Africa, eastern and western coastal Australia, Southeast Asia, Indonesia, China and India. Small patches of marshy land may dot otherwise urban or agricultural landscape, making them valuable refuges for wildlife in developed regions.

(a)

(b)

Figure 3.7 Carnivorous wetland plants: (a) sundew (*Drosera intermedia*); (b) pitcher plant (*Sarracenia purpurea*), showing pitcher-shaped leaves and two flowers.

(a)

(b)

(c)

Figure 3.8 (a) Coastal saltmarsh, Keyhaven, Hampshire, UK; (b) sand/shingle spit with saltmarshes and creeks, Norfolk, UK; (c) mudflat colonized by glasswort (*Salicornia* spp.), Devon, UK.

Saltmarshes (Figure 3.8) form along protected coastlines or in the deltas of major, slow-moving rivers as they empty into the sea. The major requirement for saltmarsh formation is shallow topography and protection from wave action, leading to the accumulation of sediment. These sediments come from reworked marine deposits, the remnants of organisms within the marsh or marsh water itself, or, in the case of delta marshes, from river silt.

Coastal saltmarshes are commonly found along the lee side of offshore islands, called **barrier islands**, or where the shore is otherwise protected from the full force of the sea. Such marshes often dominate the **intertidal zone**, below the high tide line but above the low tide zone. There are about 380 000 km^2 of saltmarshes in the world. Figure 3.9 shows their overall global distribution.

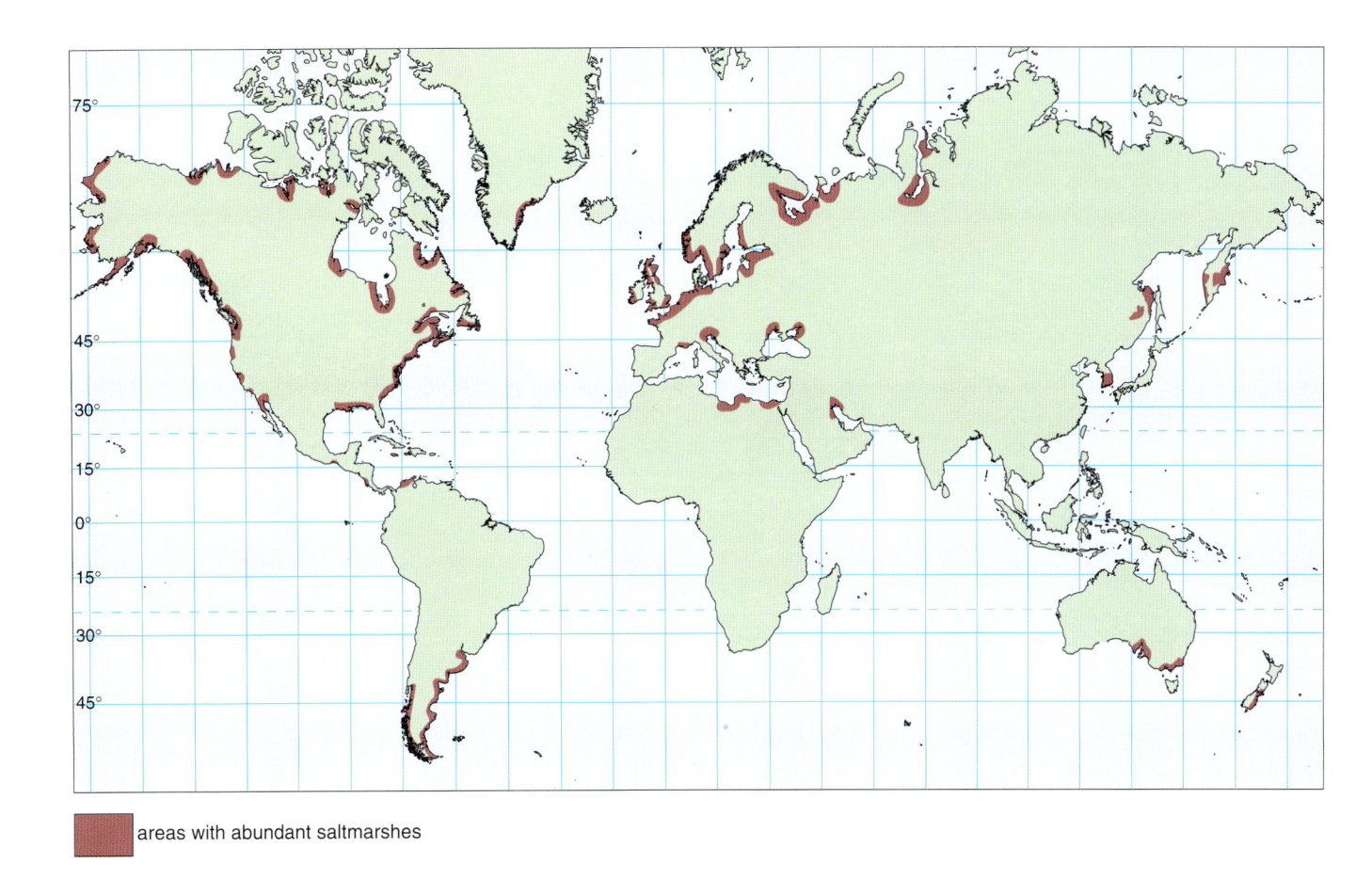

Figure 3.9 General distribution of saltmarshes in the world.

areas with abundant saltmarshes

3.2.1 Properties and development of marshes

Properties of marshes

In comparison with peatlands, marsh soils generally show a high pH, high available calcium, high base saturation and relatively high nutrient content.

- Where does this high level of nutrients originate?
- It must come from water passing through the marsh as throughflow or groundwater, or from seawater in the case of saltmarshes.

As a consequence, marshes often have extremely high productivity (Figure 3.10). However, high soil microbial activity often leads to rapid decomposition of the organic matter, and recycling of nutrients.

Figure 3.10 *Typha* marsh. Some marsh plants can grow to well over 2 m in height.

Water passing through a marsh also removes substances, such as toxic or reduced compounds, which might otherwise accumulate. The rate and frequency of water throughflow is therefore important for the maintenance and productivity of the ecosystem. Some marshes have an irregular hydrology, alternating between periods of flood and drought without any clear pattern; others have a more regular hydrology. Since connection to the regional water-table, throughflow water or seawater is important to the biogeochemistry of the marsh, changes in the regional hydrology brought about by factors such as construction of dams or sea walls, drainage of streams or rivers, or climate change, can impact strongly on marshes.

Perhaps the most regular hydrology is found in tidal saltmarshes (Figure 3.11). Tides flush accumulated salts and toxins out of the marsh and replenish nutrients. This regular supply of nutrients, together with often optimal conditions of water and climate, make saltmarshes among the most productive ecosystems in the world, with net primary production (NPP) almost as high as in commercial agriculture. Their productivity is limited by local effects such as erosion, soil anoxia, sulfide toxicity and salinity.

As with other wetlands, the limited amount of oxygen results in a significant amount of organic matter in marshes being broken down through anaerobic pathways. In saltmarshes, sulfate (SO_4^{2-}) plays an important role in carbon metabolism. The high level of sulfate available in seawater means that much of the carbon decomposition in saltmarshes passes through sulfate reduction (revisited in more detail in Section 4). In freshwater marshes, much of the carbon decomposition occurs through the production of methane, or through aerobic respiration if water levels decline temporarily or highly oxygenated freshwater passes through.

Development of marshes

Because of the great diversity of marshes, there is no generalized way in which they develop. A common pathway of freshwater marsh formation is terrestrialization: the infilling of a shallow lake with organic material (Figure 3.3a). Continued filling of the lake may result in the entire lake being transformed to a marsh, which may eventually (over a few thousand years) convert to a peatland or forest, depending upon factors such as the climate and the connection to regional hydrology.

An important type of freshwater marsh is the **prairie pothole marsh** (Figure 3.12). These are formed in the depressions left by glaciers in areas where there is a surplus

Figure 3.11 UK saltmarsh at low tide, dominated by the cordgrass *Spartina* spp.

of water, fertile soil and warm summers. Prairie pothole marshes grade northward into kettle-hole bogs, which are formed in the same way, but in environments where peat accumulates. They may be found wherever there are prairies or former prairies, a topography influenced by glaciers and a temperate climate, but are most commonly found in the northern USA and southern Canada.

Like freshwater marshes, saltmarshes are often ephemeral features. They are constantly sculpted by wind and waves, removed from some areas and built up in others as tides and currents shift. The oldest saltmarshes are only about 3000–4000 years old.

Figure 3.12 Prairie pothole marsh, western USA.

As a saltmarsh develops, it often forms an extensive network of slow-moving, meandering open water. These areas of protected water are home to the juvenile or larval stages of numerous open-water marine species of fish, molluscs, crustaceans such as crab or shrimp and other commercially valuable species. Thick vegetation and low oxygen levels often prevent large carnivorous fish and invertebrates from venturing into these backwaters, affording the young some protection. However, saltmarshes are home to numerous predatory wading birds such as herons and egrets.

Saltmarshes are dynamic systems in a continuing state of sedimentation and erosion, both vertically and along their edges. A state of 'dynamic equilibrium' persists in many saltmarshes in which, over a number of years, the inflow of sediment roughly balances the outflow of sediment. An excess of sediment deposition can eventually lessen the influence of the sea and tides, so that the influence of freshwater as surface flow, throughflow, groundwater and precipitation dominates, eventually leading to the development of a freshwater marsh. Many extensive marsh systems, such as the Everglades, are saltmarshes that grade into freshwater marshes in their upper reaches.

3.2.2 Biota

Freshwater marshes are often dominated by large rushes, reeds or sedges such as those in the genera *Typha, Phragmites, Scirpus, Cladium, Cyperus,* or *Carex*. These may be highly diverse in species composition, or may consist of nearly uniform stands. Shallow, still surface waters in marshes may also contain aquatic plants with floating leaves, such as the water-lily *Nymphaea*. Marsh plants are exposed to perhaps the widest variety of conditions of all wetland plants. Since marshes can be alternately flooded and dry, plants must be able to tolerate not only deficits in oxygen, but also oxidation and drying of substrate, herbivory, competition and the occasional fire during times of drought. Exposed to the movement of water, and sometimes, ice, marsh plants need to have strong anchorage, but also to be able to sacrifice above-ground vegetation when conditions are highly unfavourable and be able to regenerate rapidly from their roots.

Marsh plants have adapted to these conditions by producing dense growth that often arises from a large underground storage root, tuber or rhizome (Figure 3.13). The rhizome acts as a food store when conditions are poor, and the dense underground rooting system limits establishment by other plants.

Figure 3.13 Rhizomes of a freshwater marsh plant, *Typha latifolia*.

Figure 3.14 Saltmarsh plants. (a) Sea lavender, (b) glasswort (*Salicornia* spp.).

In addition to these stresses, saltmarsh plants (Figure 3.14) must also tolerate extremes in salinity, frequent flooding by tides in low-lying wetlands, or exposure and storms in higher wetlands. Saltmarsh plants have evolved strategies to cope with such stresses, but these strategies require energy. In order to cope with salinity, for instance, plants expend energy to increase the internal osmotic concentration of their cells. The different stresses in areas within a saltmarsh may lead to a clear zonation of plants adapted to differing conditions.

In coastal saltwater marshes in North America, the zone closest to the sea is often dominated by the smooth cordgrass *Spartina alterniflora*, giving way to the smaller and less robust saltmeadow cordgrass *S. patens* toward the upper marsh where conditions are less harsh (more fresh water, more sheltered conditions), and then to the saltmarsh rush *Juncus gerardii*. Other salt-tolerant plants commonly found in coastal saltmarshes include species in the genera *Puccinellia* (saltmarsh grasses) and *Salicornia* (glasswort). Saltmarshes in Europe are characterized by other species of *Spartina* (e.g. *S. anglica*, *S. townsendii*), and often show extensive areas of mudflats.

Marshes often support a rich diversity of animal life. The copious amounts of organic matter produced in freshwater marshes (Figure 3.15) support a large community of detritivores such as nematode worms, insect larvae and molluscs such as snails and freshwater clams. Flying insects like midges and mosquitoes, and their predators the dragonflies, often abound, as do birds that feed on the flying insects or on the submerged vegetation. Mammals that frequent freshwater marshes show a huge variety, such as the racoon, muskrat, deer, moose and bear in remote North American marshland, large hoofed mammals such as wildebeest, elephants, buffalo and hippos in Asia and Africa, and water voles and otters in Britain and central Europe. Amphibians, particularly frogs, provide a food chain link between detritivores and wading birds or mammals in freshwater marshes.

Fish may be locally abundant in deeper freshwater marshes and saltmarshes; shallow water marshes may become stagnant and show major fluctuations in temperature, which can kill fish. Many fish species spend their juvenile life stages in marsh waters, where the thick tangle of vegetation affords them protection from predators such as other fish.

Salt is a major constraint for higher animals in saltmarshes. However, saltmarshes support many detritus feeders and filter feeders such as crabs, amphipods and molluscs (Figure 3.16a, b). They are also very valuable wildlife habitat for large populations of wading birds such as herons and egrets, shore birds and numerous geese, ducks and migratory species (Figure 3.16c, d). Saltmarsh mammals are rare, although in North America the muskrat (*Ondatra zibethicus*) is locally common.

Wetland types

Figure 3.15
Freshwater marsh animals. (a) Grey heron, Norfolk, UK; (b) brown hawker dragonfly, Milton Keynes, UK; (c) sitatunga antelope, Kenya; (d) muskrat (*Ondata zibethicus*), California, USA. These species may also be found in other wetlands.

Figure 3.16 Saltwater marsh animals: (a) horseshoe crab (*Limulus polyphemus*), Florida, USA; (b) unidentified crab in mud; (c) black-headed gull (*Larus ridibundus*), UK; (d) great white egret (*Egretta alba*), Kenya. These species may also be found in other wetlands.

3.3 Swamps

3.3.1 Properties and development of swamps

Properties of swamps

Swamps are wetland communities that are dominated by trees, with waterlogged but generally non-peaty soils (Figure 3.17). They are often flooded, sometimes deeply and for long periods, although most swamps show alternating periods of flooded and non-flooded conditions. Since most swamp plants need oxygen for germination, they require the water-table to occasionally fall below the soil surface. Swamps are distributed widely, especially in tropical and subtropical parts of the world, and may be dominated by fresh water or salt water.

There are about $1.3 \times 10^6 \, km^2$ of freshwater swamps, predominantly found in Central and South America, southeastern North America, sub-Saharan Africa and Southeast Asia (Figure 3.17). Many of these areas are also floodplain wetlands as they are influenced by river flooding. Freshwater swamp trees include cypress (*Taxodium* spp.), white cedar (*Thuja occidentalis*), and tupelo (*Nyssa* spp.)

Figure 3.17 A tropical freshwater swamp in Indonesia.

Freshwater swamps are often densely forested and dark, with little understorey growth. Instead of ground vegetation, the swamp trees themselves often support a vigorous and diverse community of other plants and animals (Figure 3.18). These include **epiphytes** (plants which grow on other plants, generally on tree branches), such as orchids, ferns and bromeliads. A wide variety of specialized animals from invertebrates to primates are associated with the canopy and the canopy plant communities in different swamps (Figure 3.19). Many species of plants and animals are still being discovered in tropical and subtropical swamps.

The major type of saltwater swamp is the **mangrove swamp** (Figure 3.20). Around $240\,000 \, km^2$ of mangrove swamps are found throughout the world (Figure 3.21). They are restricted to the tropical and subtropical climate zones of

Figure 3.18 (a) Spanish moss growing on cypress trees, and (b) bromeliads and other epiphytes clinging to a tree in a tropical rainforest.

Figure 3.19 Chamaeleon (*Chamaeleo parsoni*) in the canopy of a freshwater swamp, Madagascar.

the world (about 30° N–30° S), grading into saltmarshes at higher latitudes. Mangrove swamps are composed not just of mangrove trees, but form a characteristic association of salt-tolerant plants. Their eponymous tree encompasses a group of over 60 species distributed among 12 genera and 18 families; most of them occurring in the Old World (only 10 species are in the New World). Mangroves, and the species associated with them, show numerous adaptations to the dual stresses of salinity and waterlogging. Like saltmarshes, many mangrove swamps show a clear succession among species, from submerged to intertidal to infrequently inundated zones of the swamp, probably due to the specific physiological adaptations required in these different environments.

Topic 3 Wetlands and the Carbon Cycle

Figure 3.20 A mangrove swamp along coastal waters, Cuba.

areas with abundant mangrove swamps

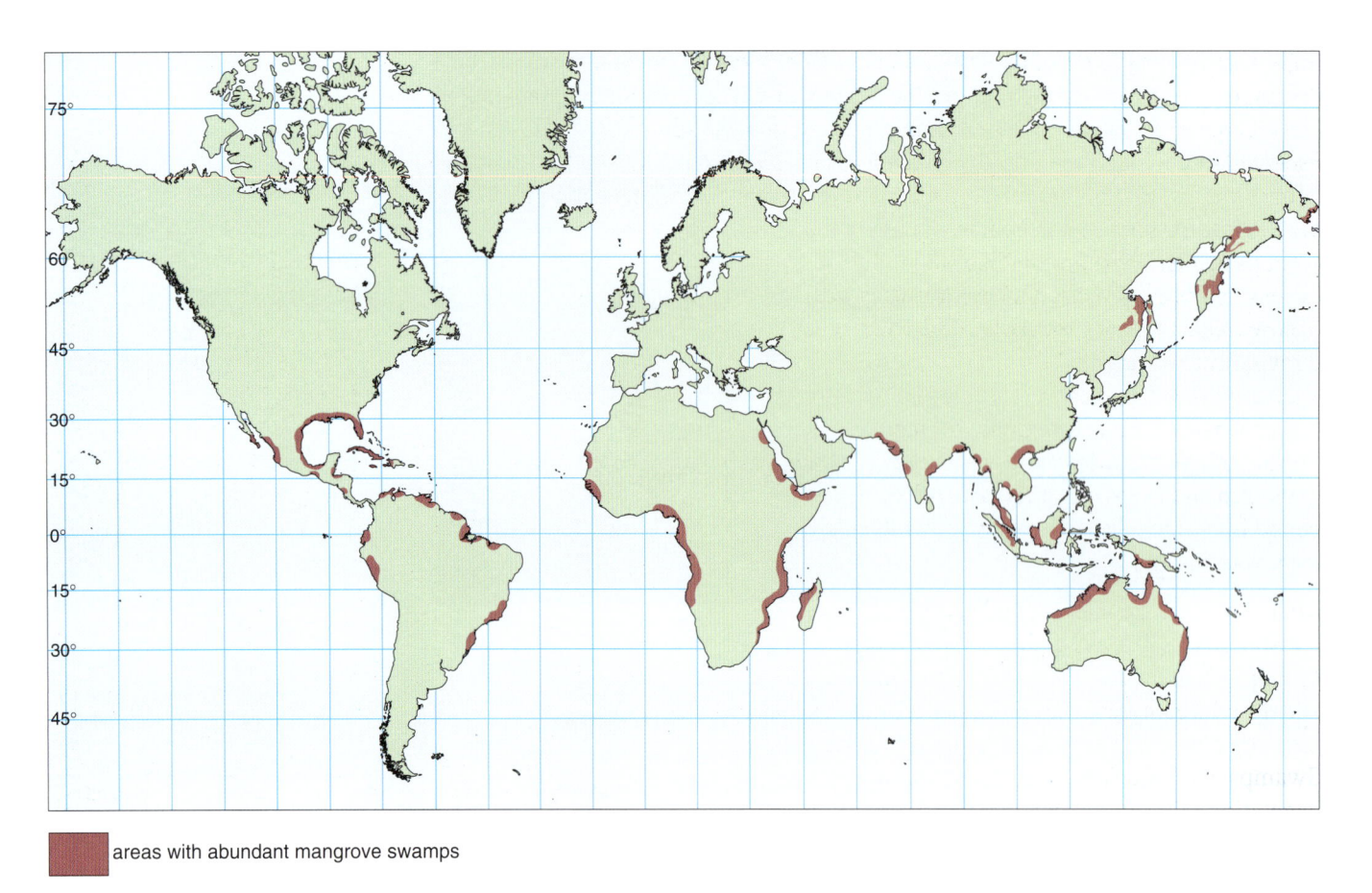

Figure 3.21 General distribution of mangrove swamps in the world.

172

Mangrove soils are often acidic due to the presence of metallic sulfides produced by microbial activity in a sulfate-rich environment. When mangrove swamps are drained or otherwise aerated the sulfides are oxidized, resulting in the formation of sulfuric acid:

$$4FeS_2(s) + 15O_2(g) + 8H_2O(l) = 2Fe_2O_3(s) + 8H_2SO_4(aq) \quad (3.1)$$

For this reason, drained mangrove swamps make poor agricultural soils unless they are heavily limed to neutralize the acid.

Development of swamps

Mangrove distribution is limited by the extent and severity of frost, as the trees cannot survive temperatures below freezing. In the absence of frost stress, mangroves are competitively superior to saltmarsh vegetation because they can grow rapidly and shade out ground vegetation. The year-round warm climate of tropical and subtropical regions, and the absence of major physical disturbances such as ice, allows these trees to replace non-woody vegetation in coastal areas. Mangrove swamps can only develop where there is protection from wave action. Although salt water is a physiological stress, the sea brings in nutrients, oxygen and organic sediments, and flushes accumulated salts. Salt water also prevents the establishment of freshwater species that may outcompete the mangroves.

Freshwater swamps can develop in permanently wet depressions underlain by an impermeable clay layer or by organic matter that impedes the downward drainage of water. Swamps span a wide variety of nutrient conditions, ranging between precipitation-dominated through to groundwater-dominated. This is reflected in a wide range of biogeochemical conditions from acidic (e.g. cypress dome swamps) to near-neutral (e.g. alluvial river swamps). They tend to be highly productive; the highest productivity occurs in swamps that have shallow surface water (thus increasing the chance that a temporary drought exposes soil to the air), regularly receive oxygenated throughflow water, or have a distinct 'pulsing' hydroperiod, with a wet season followed by a dry season (Figure 3.22), each lasting for months. Pulsing brings in nutrients, flushes out toxins, brings in oxygen and keeps out terrestrial species. If the dry period is too long, however, the plants will suffer from water stress.

Many swamps are nutrient sinks, especially for phosphorus. This fact has led some to suggest that swamps could be used to process, to some degree, water polluted by high levels of phosphorus and nitrogen from wastewater and agricultural runoff. Swamps can sequester nutrients in soil and vegetation, but this capacity has limits, and too-high levels of nutrients can change the swamp from a nutrient sink to a nutrient source.

Freshwater swamps are distributed more widely than mangrove swamps but still occur

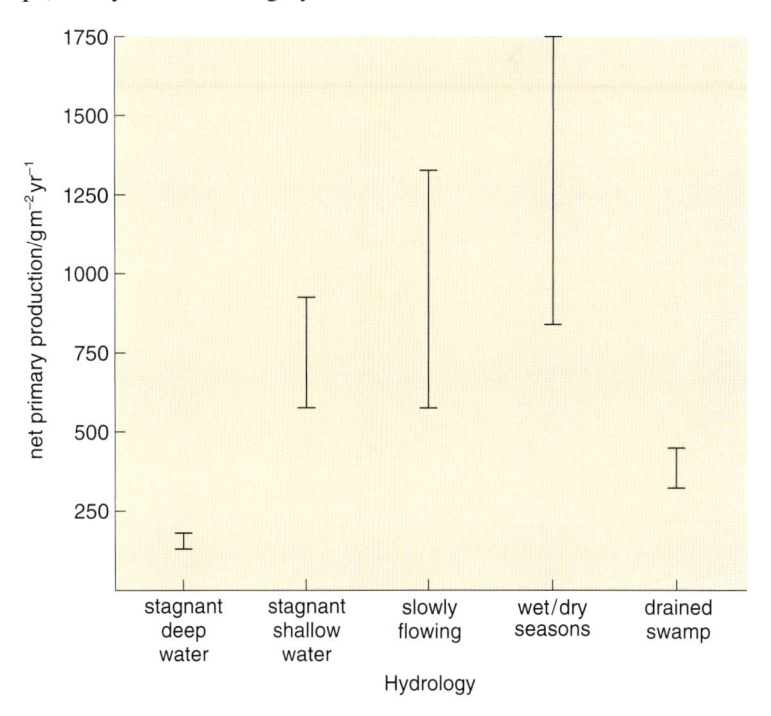

Figure 3.22 Swamp NPP as a function of hydrology. Bars denote the range of measured values in different swamps.

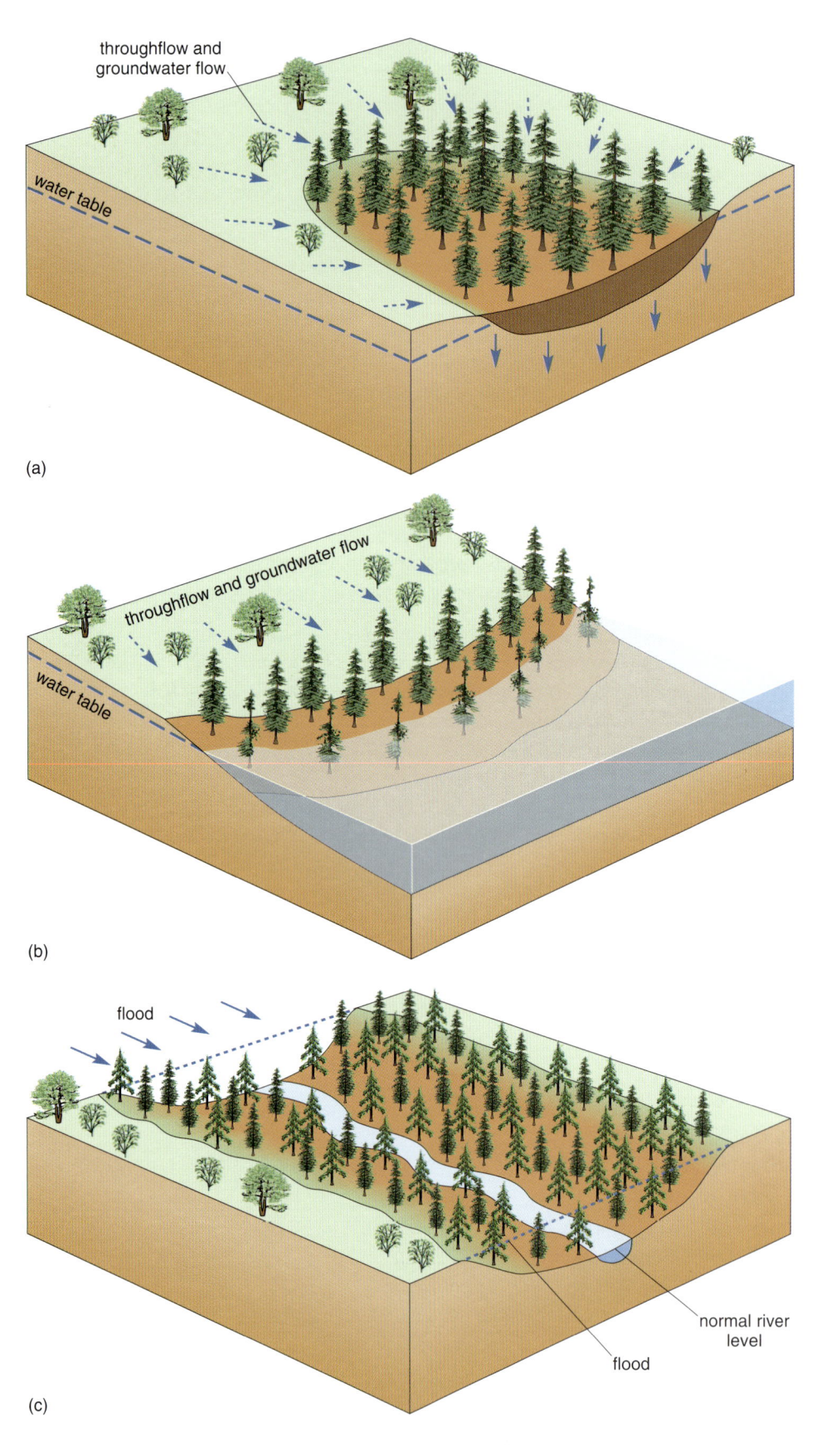

predominantly in warm, subtropical or tropical regions. Trees are, in general, less tolerant of very harsh conditions than are non-woody plants, since the latter can die back and then re-sprout from underground reserves when conditions improve. Thus, freshwater marshes tend to dominate over swamps in more extreme areas, such as where cold winters cause water to freeze deeply and over long periods, where inundation is deep or long-lasting, or in areas that are regularly disturbed by storms.

Some swamps, particularly those formed in depressions, may show a characteristic 'domed' shape, in which the trees in the middle are larger than those at the edges of the swamp (Figure 3.23a). Although no one is certain, there are several explanations for this shape, including deeper organic soils at the centre of the depression allowing better tree growth, or frequency of fire being higher along the drier edges. Swamps may also form along the edges of protected lakes (Figure 3.23b) or in floodplains (Figure 3.23c).

Figure 3.23 Schematic of (a) a domed swamp, (b) a lake-edge swamp, and (c) a floodplain swamp.

3.3.2 Biota

As with other wetland vegetation, swamp trees must be flood tolerant, and have developed adaptations to high water levels. These adaptations include structures that bring oxygen directly to the roots from the atmosphere, including pneumatophores and prop roots. **Pneumatophores** extend from the root system below ground to well above the level of the water (Figure 3.24a). A variation of these are thickened 'knees', seen in species such as cypress and tupelo, which also protrude out of the water. **Prop roots** (Figure 3.24b) are roots connected to the main stem of the tree, but which arch out over the surface of the water before descending underground. They are common in many mangrove species, and can provide both physical support and oxygen to the tree. Other wetland trees have a dense network of shallow roots that sit close to the surface, where oxygen diffusion occurs.

Figure 3.24 Root adaptations in swamp trees: (a) pneumatophores in a mangrove swamp in Australia; (b) prop roots in a mangrove swamp in Australia; (c) buttress roots in a freshwater swamp in Trinidad.

Another requirement of swamp trees is physical stability, often in semi-solid muds or soft sediments. In some species the main function of the prop roots (or 'buttress roots', Figure 3.24c) is support and stabilization. Some species, such as cypress and tupelo, show a pronounced swelling at the base, again for support.

Many swamps, both freshwater and saltwater, are thickly forested and therefore quite dark. Since the low amount of sunlight penetration inhibits the regeneration of seedlings, swamps generally show little understorey vegetation. Significant regeneration of seedlings usually occurs only when gaps open up; succession in swamps is therefore often 'disaster-led', following events such as hurricanes which may wipe out huge swathes of trees.

Freshwater swamps (Figure 3.25) can support many animal species, particularly the larger invertebrates such as worms, crayfish, snails and insect larvae. Partially decomposed wood in freshwater swamps is an excellent substrate for

Topic 3 Wetlands and the Carbon Cycle

(a)

(b)

Figure 3.25 Freshwater swamp animals. (a) American alligator (*Alligator mississippiensis*), Florida, USA; (b) Green pygmy goose (*Nettapus pulchellus*), northern Australia. Both may be found in other warm-climate wetlands.

(a)

(b)

(c)

(d)

Figure 3.26 Mangrove swamp animals: (a) fiddler crab, Australia; (b) oysters, New Zealand; (c) green heron (*Butorides virescens*), Florida, USA; (d) vervet monkey (*Cercopithecus aethiops*) eating a ghost crab. These species may also be found in other wetlands.

invertebrates to colonize. Freshwater shrimp and fish can be common in the surface water of swamps that have a close connection to regional hydrology. Since most swamps periodically dry out, reptiles and amphibians that can survive the changing water levels are more common than fish. Snakes and alligators (Figure 3.25a), frogs and salamanders are all common in various freshwater swamps.

A wide variety of animals is associated with mangrove swamps, including crabs, barnacles, oysters, snails and worms (Figure 3.26). Vertebrates can include wading birds such as egrets, herons and ibis, numerous species of fish, alligators, wildcats and other mammals. Crabs are particularly important consumers in saltwater swamps, as they remove detritus and are a food source for wading birds and fish. Crabs may also maintain open water areas in mangrove swamps by eating small roots, shoots and seedlings.

3.4 Linking peatlands, marshes and swamps through hydrology

The major differences between peatlands, marshes and swamps in productivity, soils, nutrients and vegetation come down to differences in climate and hydrology. Although there are exceptions to all generalizations, these hydrologic differences can be described broadly in terms of differences in the depth of the water-table and the duration of flooding.

The water-table of a peatland is generally near the surface and relatively constant over the year. For example, the water-table in a hypothetical fen, shown in Figure 3.27a, remains at or just below the surface all year round. In contrast, swamps and marshes are characterized by more fluctuating water-tables (Figure 3.27b and c), often linked to periods of drought and flooding of major rivers, or rainy and dry seasons. A high water input can remove organic matter, reducing the amount that accumulates in the wetland. Both swamps and marshes can potentially be flooded to great depths. An extreme example of this is Amazon rainforest swamps, which can have seasonal fluctuations in water level of 5–10 m.

Swamps often have extended dry periods; swamp trees can only tolerate extremes of flooding over long periods if they have specialized adaptations, such as pneumatophores. Marshes can tolerate longer periods of inundation; in fact, if dry periods are too long, trees may take root. Keep in mind, though, that these are generalizations—you may find some swamps that are flooded nearly all year, or some marshes that have long, dry periods. Other constraints such as climate and nutrients also determine the community observed at any time, as do factors that are unique to each ecosystem.

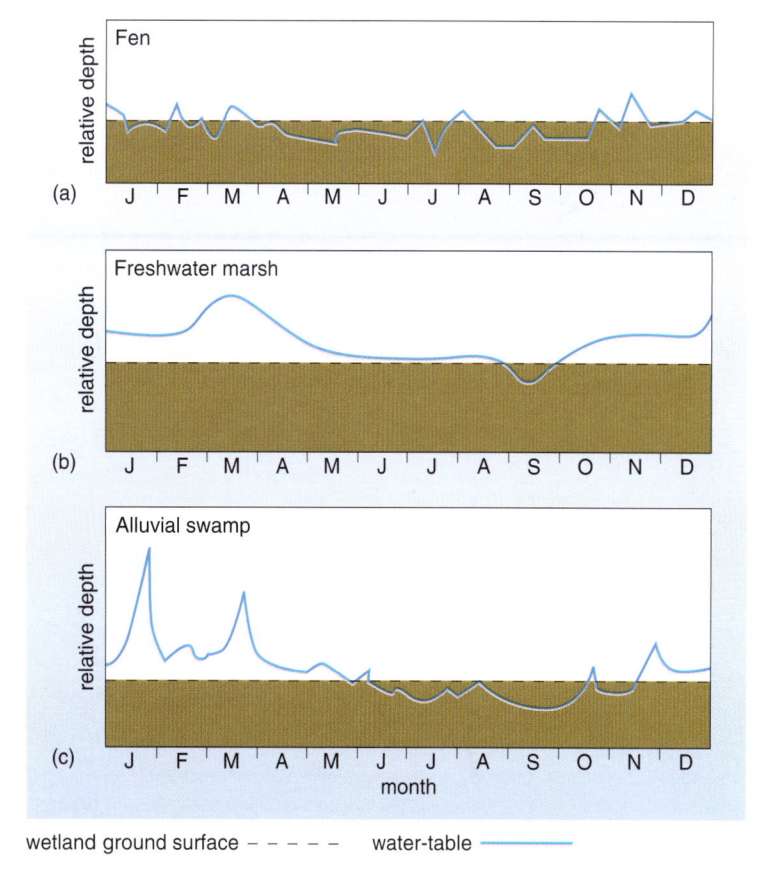

Figure 3.27 Annual water-table fluctuations for three different hypothetical wetlands: (a) a fen; (b) a freshwater marsh; (c) an alluvial swamp.

Table 3.1 Consequences of wetland hydrology, and adapted plant species.

Wetland type	Mean water level	Duration of dry periods	Ecosystem characteristics	Species adaptations
peatland	near surface	low	infertility	evergreenness carnivory
marsh	above surface	medium	regular disturbance herbivory fire	buried rhizomes annual shoots seed banks
swamp	above surface	high	shade occasional major disturbance	shade tolerance gap colonization

Table 3.1 summarizes the primary hydrologic, ecosystem and species characteristics we associate with peatlands, marshes and swamps. For instance, since there is little water flowing through peatlands to carry away decomposed organic matter, it accumulates, forming peat. This leads to isolation from the water-table and infertility, which in turn favours species that can conserve nutrients, such as evergreens or carnivorous plants. Marsh plants are subject to long periods of anoxia, short periods of dryness and exposure to herbivory, winds, waves, and possibly ice and fire. These conditions favour species with buried rhizomes that anchor shoots during disturbance, provide storage when conditions are poor, and allow rapid regeneration when conditions are poor enough or disturbance is high enough to kill off the top growth of the plant. Swamp trees, which can grow extremely densely, must be shade tolerant and develop seedlings that can germinate rapidly during dry periods, or when gaps in the canopy open up.

3.5 Wetland transitional communities

It is probably clear from the above discussion that, whereas many wetlands can be clearly placed in a certain category (e.g. peat bog, mangrove swamp), other wetlands may share characteristics of two or more groups. In addition, there are some plant communities that are transitional between wetlands and uplands; and that do not sit easily in either camp. These are communities that are influenced by periodic wet conditions or flooding, but not to an extent that a persistent community of swamp or marsh vegetation is supported. We may call them 'wetland transitional' communities, although it is important to note that many of these communities may be highly stable. They are often found along occasionally inundated floodplains.

Wet meadows (Figure 3.28), for instance, are grassland communities in which soils are occasionally waterlogged. However, they are without standing water for most of the year, and are not dominated by typical marsh plants. Wet meadows may occur along river floodplains, or along the shores of lakes, and may grade into marshes in wetter areas or grasslands in drier areas. The forested equivalent of a wet meadow is a **bottomland forest** (Figure 3.29) — a floodplain forest that is only intermittently flooded. Such forests can grade into swamps or marshes if inundation increases. Common moisture-loving tree species in bottomland forests include willow (*Salix* spp.), alder (*Alnus* spp.) and poplar (*Populus* spp.).

Figure 3.28 Wet meadow in flood stage, USA.

Figure 3.29 Bottomland forest in a floodplain, UK.

A temporally transitional wetland community is a **vernal pool** (Figure 3.30). Vernal pools are small depressions (about 50–5000 m^2), found particularly in northern latitude forests, that only exist as pools for a few weeks each year. They form when a large snowpack melts and saturates the soil in early spring. With low rates of evapotranspiration because of cool temperatures, and restricted runoff, the meltwater accumulates in small ponds dotted through the landscape. Vernal pools disappear by late spring when increasing temperatures and growing vegetation increase evapotranspiration, or the water percolates into the soil and drains away.

A number of animals and plants live at least some of their lifespan in these unique habitats. These organisms are a 'rapid response' group: waterlogging in spring often leads to a bloom of plant and animal life within days. A deafening chorus of spring frogs in northern woodlands signals the onset of reproduction, with tadpoles growing to adulthood in a few weeks. The eggs of some insects, such as many species of mosquito, also hatch in vernal pools in spring and live out all their larval stages there, emerging as adults before the ponds dry.

Since vernal pools may last for only a few weeks each year, conditions are not suitable for the establishment of a wetland vegetation community. However, vernal pools can rapidly take on the biogeochemical characteristics of wetlands, such as low oxygen levels, low decomposition rates and high organic matter accumulation, even in this short time period.

Figure 3.30 Vernal pools in woodland, with marsh marigold (*Caltha palustris*) in bloom.

3.6 Wetland classification in Britain

The British classification scheme for wetland plant communities has been recently standardized with the advent of the National Vegetation Classification system (NVC). As with all regional classifications, it emphasizes those wetlands that are common, and de-emphasizes wetlands that are rare or non-existent. Thus the British classification distinguishes three different kinds of non-forested freshwater wetlands based on their vegetation, because these types are much more frequent in Britain than are wooded wetlands. Confusingly, the British scheme uses the same terms as other classifications, such as 'swamp', to mean something quite different. In the North American system a 'swamp' denotes a woodland, whereas in Britain it is a marsh community without trees!

In brief, the classification system recognizes the following categories.

Mires. This category corresponds in part to the 'bogs' of the North American system. They are nutrient poor, with the water source usually dominated by precipitation, and the vegetation often dominated by *Sphagnum* mosses. They are peat-forming and generally have very acidic soils. They range from the moss-rich communities of bog pools (M1 in the NVC) to the iris-dominated stands at the upper edges of Scottish saltmarshes (M28).

Swamps. This category corresponds in part to both the 'fens' and the 'marshes' of the North American system. They have some water from throughflow or groundwater and so are richer in nutrients than are mires. Vegetation is dominated by grasses and sedges. They can be peat-forming, with a slightly acid to near-neutral pH. The vegetation may be species poor and dominated by a single species such as the common reed (*Phragmites* sp.). This type of swamp is the 'reedswamp' or 'reedbed' (S4, Figure 3.31). Alternatively they may be quite species rich with a mixture of grasses, sedges and herbs. Such communities are called rich fens in Britain (e.g. S24).

Figure 3.31 Reedswamp dominated by a single species (*Phragmites communis*), Surrey, UK.

Mesotrophic grasslands. Seven of the grassland communities within this category rely on inundation by floodwaters or consistently high water-tables from groundwater, and are often generically referred to as wet grasslands. They are similar to swamps in terms of their nutrient status and pH, but differ in their management. They are usually grazed by cattle or sheep and may be mown in summer for hay, and correspond in part to the North American wet meadow category. Regular management prevents tall swamp species from colonizing, and favours those low-growing grasses that can tolerate grazing. Wet grasslands include some of Britain's most diverse plant communities, such as floodplain meadows (MG4) and water meadows (MG8), which are the target of much conservation effort.

Saltmarshes. These are similar to saltmarshes in the North American description. The category is subdivided into 28 separate communities, which range from permanently submerged sea grass beds (SM1) to associations of shrubby species along the drift line at the high-tide mark (e.g. SM25).

Woodlands. The NVC lists just seven plant communities that could be described as wet woodland, because forested wetlands have all but disappeared from the British countryside following forest clearance. Those stands that do remain are dominated by trees such as alder (*Alnus glutinosa*, e.g. W5), willow (*Salix* spp., e.g. W1) and Birch (*Betula* spp., e.g. W4). W5 is found on river floodplains and is the closest British equivalent of North American bottomland forest. W1 occurs around the margins of water bodies, colonizing newly exposed silt, whilst W4 can invade the peaty soils of some upland mires when conditions allow, for example when grazing pressure is relaxed.

Table 3.2 summarizes the distinction between the North American terminology and the British terminology.

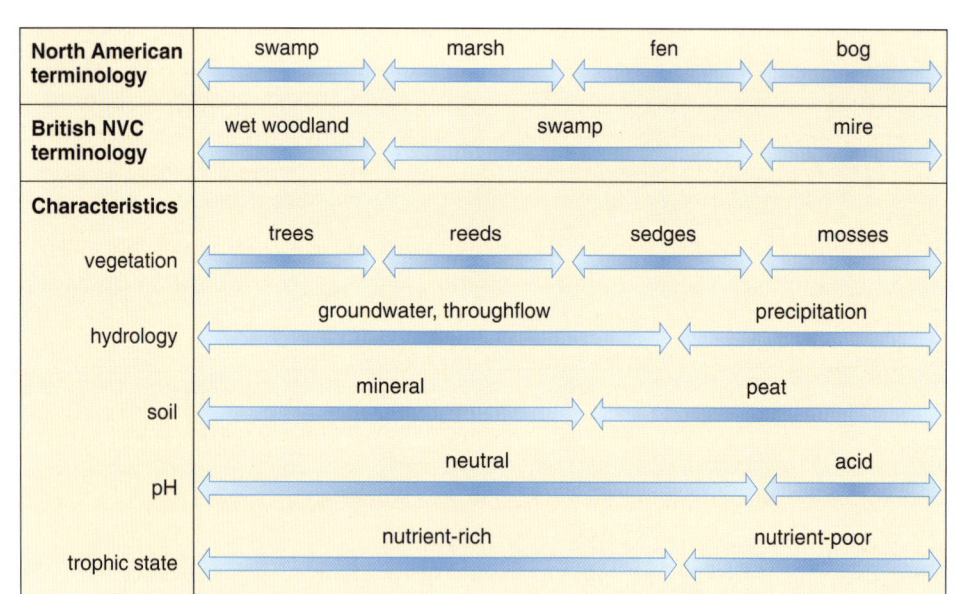

Table 3.2 Comparison of terms used to describe similar inland freshwater wetlands in North America and Britain.

North American terminology	swamp	marsh	fen	bog
British NVC terminology	wet woodland	swamp		mire
Characteristics				
vegetation	trees	reeds	sedges	mosses
hydrology	groundwater, throughflow			precipitation
soil	mineral		peat	
pH		neutral		acid
trophic state		nutrient-rich		nutrient-poor

3.7 Summary of Section 3

1. Based on their vegetation, hydrology and soils, wetlands can be described as peatlands, marshes or swamps.

2. Peatlands are wetlands that accumulate significant amounts of partially decayed plant material, or peat. They are the dominant wetland in cool, northern latitudes. Peat formation requires organic matter production in excess of decomposition, and restricted drainage.

3. Peatlands can be further divided into bogs or fens, depending upon their hydrology. Bogs are isolated from the groundwater, with their main source of water being from precipitation, so they have a low pH and are nutrient poor. Fens receive some groundwater or throughflow water, so are more nutrient rich.

4. Peatlands may form through either terrestrialization (the gradual filling in of shallow lakes), or paludification (the flooding of terrestrial vegetation).

5. Peatlands are generally less productive than other vegetation communities in the same climatic zone, and less productive than other wetlands. Because of this, and other features such as high acidity and unpalatable vegetation, peatlands support relatively low levels of animal populations.

6. Marshes are wetland communities that are dominated by non-woody, vascular plants, without a deep accumulation of peat. Marshes are characterized by predominance of throughflow water or groundwater that has been in contact with mineral soils, rather than water from precipitation. This often leads to a rich nutrient status.

7. Marshes can be divided into freshwater marshes and saltmarshes. They are some of the most productive natural ecosystems on Earth and often support a highly diverse biota.

8. Marsh plants are exposed to perhaps the widest variety of conditions of all wetland plants. Since marshes are often alternately flooded and dry, plants must be able to tolerate deficits in oxygen, but must also tolerate oxidation and drying of soil, herbivory, competition and fire when water levels are low, and freezing. Saltmarsh plants must also tolerate extremes in salinity, waves, tides, tidal floods, etc.

9. Marsh plants have adapted to these conditions by strategies including producing thick roots to act as food stores, or dense roots to anchor the plant and deter the establishment of other plants.

10. Swamps are wetland communities that are dominated by trees, with waterlogged, but generally non-peaty soils. They are often flooded, sometimes deeply and are often densely forested and dark, with little understorey growth. Swamp trees have developed numerous adaptations to high water levels, including pneumatophores and prop roots.

11. Swamps may be dominated by fresh water or salt water. The major type of saltwater swamp is the mangrove swamp. Both freshwater and saltwater swamps are distributed widely in tropical and subtropical parts of the world.

12. Swamps span a wide variety of nutrient conditions, from precipitation-dominated to groundwater-dominated. Like marshes, they can be highly productive and support a rich and diverse biota.

13. Transitional between marshes and swamps are wet meadows and bottomland forests; a vernal pool is a highly ephemeral wetland.

14. Wetland classification in Britain has been recently standardized. The British classification scheme distinguishes three major kinds of non-forested freshwater wetlands — mires, swamps, and the wet grassland communities within mesotrophic grasslands.

Question 3.1

Using information found in this section, estimate the proportion of the world's natural (non-rice-paddy) wetlands that are: (a) peatlands, (b) freshwater marshes, (c) saltwater marshes, (d) freshwater swamps and (e) saltwater (mangrove) swamps. Then choose an effective way of plotting these proportions and plot them.

Question 3.2

Describe the changes to the air and soil temperature, the net carbon accumulation, the hydrology, the concentration of nutrients in the soil solution and the vegetation that would be expected to occur if you travelled from a landscape dominated by prairie potholes to one dominated by kettlehole bogs. In which dominant direction would you be travelling?

Question 3.3

Why do you think we have saltmarshes and saltwater swamps, but no 'salt peatlands' (i.e. no extensive peatlands in areas influenced by salt water)?

Question 3.4

Why do you think vernal pools are rare in warm climates?

Carbon cycling processes in a wetland

If you put on your boots and take a stroll through your local wetland you are likely to notice one thing above all others — the black squelchy soils that ooze up around your feet. Many wetland soils are rich in organic matter; dig down in a peat bog and you may well continue digging for 5 or 10 m without reaching mineral soil. And this organic matter is about 50% carbon.

Not only do wetlands store more carbon than other terrestrial biomes, they circulate it more actively through oxidized and reduced forms, and release carbon in a wider range of different compounds into the atmosphere than does any other biome. Awareness of the global carbon cycle is important in order to gain an overview of the major sources and sinks of carbon in the biosphere, and to appreciate how future increases in atmospheric CO_2 will probably be partitioned among these sources and sinks. In the next three sections we examine these crucial biogeochemical properties of wetlands in carbon cycling from three perspectives: the local perspective (Section 4), the global perspective (Section 5), and the impact of people on wetland carbon cycling (Section 6).

4.1 Carbon gains and losses in a 'model' wetland

In the global carbon cycle, the sizes of the major carbon reservoirs, such as carbonate rock or coal, are enormous, on the scale of 10^{19} kg C. But these major fluxes and transformations of carbon on a global scale are all ultimately due to processes occurring at a local scale. To appreciate the global implications of the carbon cycle, it is essential to understand these local processes. Thus, in this section, we bring the global carbon cycle 'down to earth' by exploring in some detail the gains, losses and transformations of carbon in one important reservoir: wetland soils.

Although bogs, fens, marshes and swamps differ in factors such as the levels of nutrients they receive, their hydrology and their dominant vegetation, their carbon biogeochemistry is broadly similar.

4.1.1 Carbon enters the wetland ecosystem

Consider a 'model' wetland, for example a northern peat bog, a mid-latitude marsh or a subtropical swamp (Figure 4.1). The surface of the wetland is covered in vegetation, perhaps *Sphagnum* moss, sedges and grasses, reeds and rushes, trees such as black spruce, cypress or tamarack, or some combination of these wetland plants. The surface is very gently undulating and the water-table varies throughout the year from about 5 cm above the surface to about 10 cm below the surface.

In the spring and summer, photosynthesis by vegetation causes CO_2 to be drawn out of the local atmosphere (Figure 4.1 **A**) and converted to organic carbon according to the following overall reaction:

$$\text{light energy} + 6CO_2(g) + 6H_2O(l) = 6O_2(g) + C_6H_{12}O_6(s) \qquad (4.1)$$
$$\text{carbohydrates}$$

Topic 3 Wetlands and the Carbon Cycle

The total carbon fixation by photosynthesis is known as gross primary production (GPP).

Note that Equation 4.1 describes the *overall* reaction. As with all the biological reactions described in the rest of this section, the actual conversion from 'reactants' to 'products' occurs in many stages, often involving different organic molecules that act as enzymes, intermediate compounds and removal agents for intermediate products. We simplify this greatly by showing only the net result.

Figure 4.1 Major input fluxes of carbon into wetland soils. The whole area of the figure represents a model wetland, including aerobic and anaerobic soils. **A**, photosynthesis (GPP); **B**, plant respiration releasing CO_2 into (**B₁**) the atmosphere, and (**B₂**) the soil solution (becoming DIC); **C**, release of dissolved organic carbon (DOC) through roots into both aerobic and anaerobic soil solution; **D**, input of DOC and dissolved inorganic carbon (DIC) from surrounding catchments; **E**, shed leaves releasing particulate organic carbon (POC) to the soil solution; **F**, input of POC from surrounding catchments.

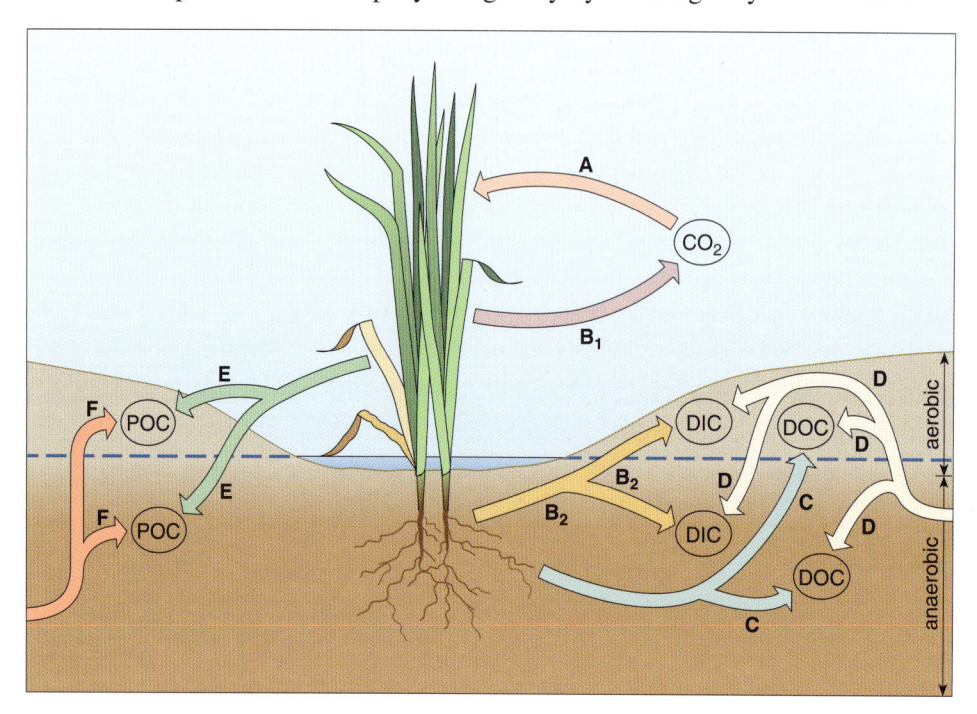

About half the total amount of carbon fixed in photosynthesis becomes plant tissue.

○ What happens to the rest of the fixed carbon, and how is it represented in Figure 4.1?

● It is oxidized by the plant itself for energy (through respiration), and is released as CO_2 through the leaves (Figure 4.1 **B₁**) and roots (Figure 4.1 **B₂**).

○ What is the overall equation for aerobic respiration, and how is it related to the overall equation for photosynthesis?

● $C_6H_{12}O_6(s) + 6O_2(g) = 6CO_2(g) + 6H_2O(l)$ (+ energy) (4.2)
carbohydrates

The overall reaction is the reverse of photosynthesis, and instead of *requiring* solar energy, it *releases* chemical energy.

The carbon dioxide released by respiration can go into two major pools, only one of which continues cycling in the wetland. Much of the CO_2 respired by plants is released by the leaves and goes directly back into the atmosphere (Figure 4.1 **B₁**). Some of the carbon is used to provide energy for the roots, releasing carbon dioxide in the process through root respiration (Figure 4.1 **B₂**). In waterlogged wetland soils this carbon dioxide dissolves into the soil solution forming $CO_2(aq)$

(often represented as carbonic acid H_2CO_3), which can, depending on the pH, further dissociate into HCO_3^- and CO_3^{2-}:

$$CO_2(aq) + H_2O(l) = H_2CO_3(aq) = HCO_3^-(aq) + H^+(aq) = CO_3^{2-}(aq) + 2H^+(aq) \quad (4.3)$$

$CO_2(aq)$ and HCO_3^- dominate in most wetland soil solutions. At higher pH levels, some of the HCO_3^- dissociates to CO_3^{2-} and H^+.

These various forms of dissolved CO_2 ($CO_2(aq)$, HCO_3^- and CO_3^{2-}) are known collectively as **dissolved inorganic carbon**, or **DIC**.

Many wetlands are highly productive. The roots of wetland plants, through respiration, release large amounts of CO_2, leading to high concentrations of dissolved CO_2 in the soil solution, often many times higher than atmospheric concentrations. As CO_2 dissolved in water forms an acidic solution (Equation 4.3), the more CO_2 that is released, the more acid the soil solution can become. Therefore high productivity in a wetland can increase the acidity of the soils and sediments.

○ In light of the above paragraph, why do you think we don't show an arrow from atmospheric CO_2 into the soil solution DIC in Figure 4.1?

● Because the concentration of CO_2 is often higher in the soil solution than in the atmosphere, CO_2 would not generally diffuse directly from the atmosphere into the soil solution. Rather, to achieve net equilibrium, the CO_2 would usually move the other way.

The carbon that is *not* used for energy and re-released as CO_2 is built into the complex molecules, tissues and organs that compose the plant. This *net primary production* (NPP) is the nutritional and energy basis for nearly all the rest of life on Earth.

Plant roots also release carbon compounds other than CO_2. Most commonly these are simple organic molecules or ions, such as the ethanoate ion (CH_3COO^-, informally known as acetate), collectively grouped under the category of **dissolved organic carbon** or **DOC** (Figure 4.1 **C**). These organic species are often waste products of plant metabolism and are also known as root **exudates**. The amount and type of carbon leached through plant roots varies with the type of vegetation and the season. In general, vascular plants such as sedges and rushes leach more DOC than do mosses, and more DOC is leached in summer than winter. DOC can also be released from dying roots.

DIC and DOC may also enter the wetland from the surrounding catchments, in groundwater, throughflow or floodwater (Figure 4.1 **D**). Although most wetlands are net exporters of DOC, groundwater can often bring in large amounts of dissolved inorganic carbon as dissolved CO_2, HCO_3^- or even CO_3^{2-}.

Both dissolved inorganic carbon and DOC play vital roles in the carbon cycling of wetlands, particularly in the production of gases. However, neither is the most important source of carbon to wetland soils.

○ What do you think is the most important source of carbon to wetland soils?

● It is the fixed carbon that is stored in plant tissue and which becomes incorporated into the soil when the plant dies or sheds its leaves.

Senescence of leaves or death of plants, especially in autumn, releases large amounts of carbon to the sediments as **particulate organic carbon** or **POC** (Figure 4.1 **E**). Particulate organic carbon may be as small as a bacterium or as large as an entire tree, but in general we think of new inputs of particulate organic carbon as parts of leaves or dead roots from the vegetation. In the case of annual plants (plants which die within a year), nearly all the carbon that has been fixed over the summer goes into the soil at the end of the growing season. Perennial plants release fixed carbon to the soil when leaves die or drop onto the surface, or when the plants themselves eventually die.

POC can also be transported into wetlands from other areas (Figure 4.1 **F**). Floodplain wetlands, for example, can be a large sink for particulate carbon as sediment or bits of vegetation when the rivers flood. In some major river systems floodplain deposits during flooding can significantly reduce the supply of organic carbon and nutrients transported to downstream ecosystems.

To summarize, the major source of carbon to the wetland ecosystem is the fixation of atmospheric CO_2 by the vegetation (Figure 4.1 **A**). Through root respiration (Figure 4.1 B_2), leaching from roots (Figure 4.1 **C**) and, most importantly, death and senescence of the vegetation (Figure 4.1 **E**), carbon is transferred into wetland soils as DIC, DOC and POC. Carbon can also enter the wetland in all these forms from the surrounding environment (Figure 4.1 **D** and **F**). The input of carbon into wetland soils in temperate regions peaks in the autumn.

4.1.2 Carbon transformation in wetland soils

The major 'theme' of carbon transformation in wetland soils is that the energy of POC and DOC (and, to some extent, DIC) is exploited by organisms for growth and maintenance. (There are also organisms that can oxidize reduced inorganic ions such as NH_4^+ and Fe^{2+} for energy. Since we are following the chemistry of *carbon* in wetlands, however, we will not describe these in any more detail.) The pulse of organic matter added to soils in the autumn provides a carbon bonanza for soil microbes.

The reduced carbon compounds composing plant tissue provide a source of energy, primarily for a rich and varied population of microbes. But releasing this energy requires an oxidizing agent: in aerobic environments this is oxygen; in anaerobic environments it can be a number of other chemical species. Wetlands have both types of environment and thus have many types of carbon transformations. Here we describe the major ones.

If you would like to review concepts of oxidation and reduction, take a look at Box 4.1.

Aerobic respiration of POC and DOC

It is perhaps surprising that aerobic respiration can be highly important in wetlands, or even the dominant form of respiration. One major reason for this is that many wetlands are not completely waterlogged over the course of a year, or the water-table lies close to the surface so that some oxygen diffuses to the upper soil. The latter is especially so in peatlands (Figure 3.27a). Peatlands also often show great variability in surface topography, with small hillocks or **hummocks** raised above the surface and often drying out, alternating with lower areas or **hollows** that are saturated for most of the time.

Box 4.1 Oxidation and reduction: the currency of life

As described earlier in the Course, a simple definition of oxidation that works satisfactorily for many situations is 'the addition of oxygen to a chemical compound'. Examples of this are the combustion of methane (the same overall reaction as oxidation of methane by methanotrophic bacteria — see Section 4.1.3), and aerobic respiration:

$$CH_4(g) + 2O_2(g) = CO_2(g) + 2H_2O(l) \; (+ \text{energy})$$

$$C_6H_{12}O_6(s) + 6O_2(g) = 6CO_2(g) + 6H_2O(l) \; (+ \text{energy}) \qquad (4.2)$$

Similarly, reduction may be considered the opposite of oxidation: the removal of oxygen from a compound during a reaction.

However, under *anaerobic* conditions, the oxidizing agent is a species other than molecular oxygen, such as nitrate (NO_3^-) or sulfate (SO_4^{2-}). For the case of oxidation of organic matter (C_{org}) under either aerobic or anaerobic conditions, a more general definition of oxidation is the reaction of C_{org} with an oxidizing agent (X_{ox}), leading to a more oxygen-rich compound of carbon (C_{ox}, e.g. CO_2) and a reduced form of X (X_{red}):

$$C_{org} + X_{ox} = C_{ox} + X_{red} \; (+ \text{energy})$$

Other species, such as H^+ or H_2O, may also feature in the equation (in aerobic respiration (Equation 4.2), H_2O is the reduced (X_{red}) form of O_2).

Using this generalization, we can write the aerobic respiration of organic matter as:

$$\underset{C_{org}}{C_6H_{12}O_6(s)} + \underset{X_{ox}}{6O_2(g)} = \underset{C_{ox}}{6CO_2(g)} + \underset{X_{red}}{6H_2O(l)} \; (+ \text{energy})$$

An example of respiration of organic matter using nitrate (NO_3^-) as the oxidizing agent is denitrification:

$$\underset{C_{org}}{5C_6H_{12}O_6(s)} + \underset{X_{ox}}{24NO_3^-(aq)} + 24H^+ = \underset{C_{ox}}{30CO_2(aq)} + \underset{X_{red}}{12N_2(g)} + 42H_2O(l) \; (+ \text{energy})$$

Swamps and marshes usually have higher water-tables, often well above the surface. Their sediments can come into contact with the air, not so much through local topography, but over time, as the wetlands periodically dry out, exposing sediments to oxygen (Figure 3.27b and c). Peatlands may also show water-table fluctuations, but they are generally less dramatic. In addition to periodic lowering of the water-table, even seemingly saturated wetlands can have numerous aerated microsites, for example, around the roots of vascular plants (Figure 2.9). Another source of oxygen, especially to fens, marshes and swamps, is water flowing through the wetland as throughflow or groundwater, or over the wetland during floods. Spring meltwater may also release a large pulse of oxygen-rich water to high-latitude wetlands.

Returning to our model wetland, let us assume it has an aerobic area of a few centimetres depth corresponding roughly to the unsaturated zone above the water-table (Figure 4.2). This aerobic area supports a vigorous suite of detritivore and decomposer organisms, including insects, nematodes, protozoa, fungi and bacteria, which decay plant material through aerobic respiration (Figure 4.2 **G**).

These organisms use the energy released by aerobic respiration (Equation 4.2) to power their metabolism.

The energy yield per unit of carbon oxidized is higher from aerobic respiration than any other respiration process. Because of this high energy yield, aerobic organisms are very efficient at breaking down organic carbon and can outcompete other organisms for the carbon present. In addition, oxygen is a toxin for many anaerobic organisms. Thus when oxygen is present, decomposition proceeds through aerobic respiration via its associated organisms.

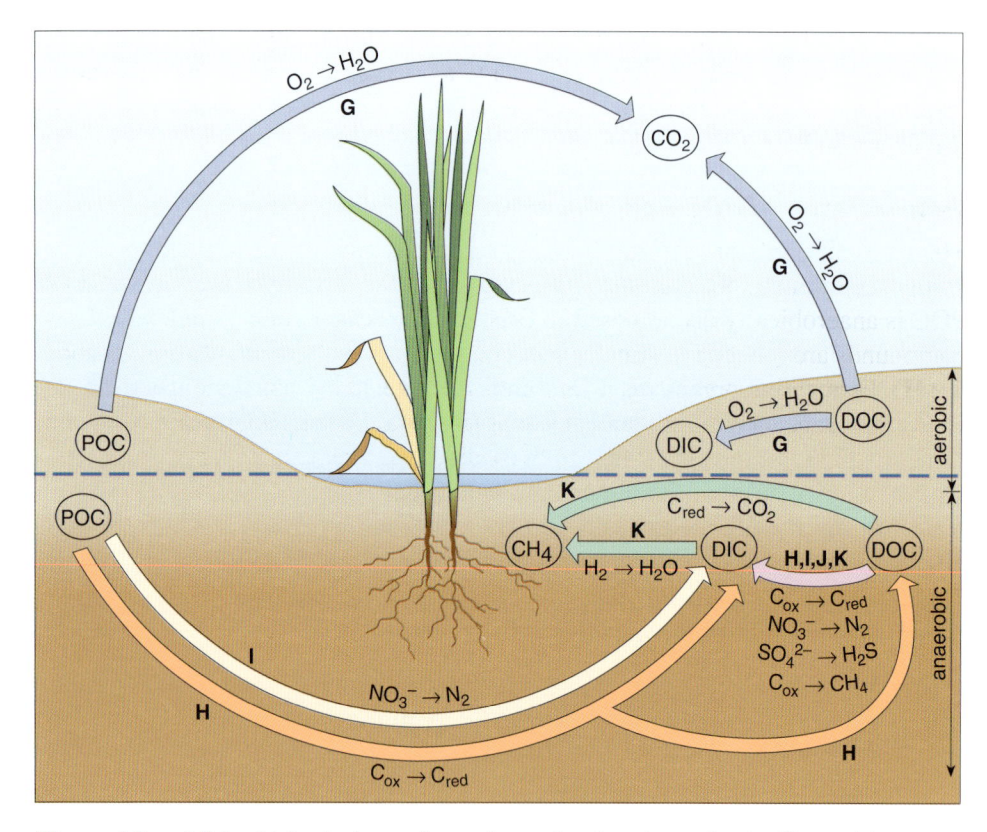

Figure 4.2 Major biological transformations of carbon in wetlands: **G**, aerobic respiration; **H**, fermentation reactions, producing DOC and DIC; **I**, denitrification; **J**, sulfate reduction; **K**, methanogenesis via ethanoate fermentation and CO_2 reduction. Written along each arrow is the corresponding reduction reaction except for two **K** arrows leading to CH_4; here is written the corresponding oxidation reaction.

The proportion of carbon from NPP that is released through aerobic respiration can vary greatly depending upon the type of wetland and the local climatic conditions, but typically more than half of NPP in wetlands is decomposed aerobically. In peatlands it typically ranges around 70–80%, and in swamps or marshes it can be as low as 5% or as high as 95% depending upon factors such as the amount of oxygen-rich water flowing through the wetland, and the depth and duration of flooding. In our 'model' wetland, we allow 75% of the fixed carbon to be decomposed aerobically, with most of the CO_2 resulting from respiration released into the atmosphere (Figure 4.2 **G**).

The large supply of organic carbon, and relatively limited aerated zone, in wetlands means that the oxygen used by aerobic respiration is eventually depleted, with organic matter as POC and DOC still remaining. Both the concentration of oxygen and the activity of aerobic organisms decrease at or just below the position of the water-table. Below this point, **anaerobic respiration** dominates.

All exclusively anaerobically respiring organisms are single-celled (although clearly not all single-celled organisms are anaerobic!). This is because the energy yield from anaerobic respiration is too low to meet the high energy requirements of multicellular organisms. The energy yield of oxidizing one mole of carbohydrate with oxygen is about six times higher than that from oxidizing one mole of carbohydrate with sulfate. But for most of the history of the Earth the atmosphere was essentially oxygen free and anaerobic respiration the only form of respiration. Anaerobic bacteria therefore include members of some of the earliest life forms on Earth.

Fermentation of POC and DOC into simpler compounds

The particulate organic carbon (POC) in the anaerobic zone of wetlands is composed of numerous complex organic compounds. An important way in which POC is anaerobically decomposed to DOC, and in which more complex DOC compounds are reduced to simpler compounds, is through **fermentation** (Figure 4.2 **H**). Fermenting organisms power their metabolism without recourse to an oxidizing agent such as oxygen, nitrate or sulphate. Instead, an organic compound (glucose, $C_6H_{12}O_6$, say) provides both oxidizing and reducing agent — or, in the terminology of Box 4.1, the same compound is both C_{org} and X_{ox}. In the fermentation process, more oxidized forms of carbon in the same compound are effectively used to oxidize more reduced forms of carbon in the same compound. The upshot is that relatively complex organic molecules are broken into simpler components. For example, sugars may be broken down into a variety of organic acids:

$$C_6H_{12}O_6(s) = 2CH_3\text{–}CHOH\text{–}COOH(aq)\ (+\ energy) \tag{4.4}$$
$$\text{lactic acid}$$

or

$$C_6H_{12}O_6(s) = CH_3COOH(aq) + C_2H_5COOH(aq) + CO_2(aq) + H_2(g)\ (+\ energy) \tag{4.5}$$
$$\text{ethanoic acid} \quad \text{propionic acid}$$

Note again that Equations 4.4 and 4.5 represent the *net* result of complex multi-step reactions which we do not attempt to reproduce here.

There are many fermentation reactions, all of which produce simpler organic compounds, including CO_2, from more complex organic compounds. As anyone who has brewed beer knows, oxygen must be excluded for significant fermentation of organic carbon to occur. In wetlands, fermentation reactions can operate across a broad spectrum of conditions from borderline aerobic to strongly reducing.

Respiration through fermentation may be as old as life itself. Even today, in all organisms the initial stages of metabolism are through anaerobic fermentation pathways. Aerobic organisms can also temporarily switch to fermentation to gain energy when oxygen is limiting. For example, the muscle spasms that can occur during extreme exercise are due to a build-up of lactic acid from anaerobic metabolism under an 'oxygen deficit'. However, organisms that rely exclusively on fermentation for energy are either bacteria or fungi (e.g. yeasts).

The carbon compounds that are broken into simpler organic molecules through fermentation are available for other anaerobic bacteria to act upon. Among the organisms that make up this community are the **denitrifying bacteria**, **sulfate-reducing bacteria** and **methanogens.** These bacteria gain energy by using the nitrate ion (NO_3^-), sulfate ion (SO_4^{2-}) and CO_2, respectively, as oxidizing agents, instead of O_2.

Just as aerobic organisms exclude anaerobic organisms by their far greater efficiency, there is a competitive hierarchy among anaerobic bacteria broadly determined by differences in the energy yield of the metabolic pathways employed. In general, denitrifying bacteria are more efficient than sulfate-reducing bacteria, and both outcompete methanogenic bacteria.

○ Why should a higher energy yield determine the winner of a competition over resources?

● The 'winners', organisms that can achieve higher energy from compounds, can grow and reproduce more rapidly and efficiently than the 'losers'.

POC and DOC oxidation through reduction of NO_3^-

If there is a high enough concentration of NO_3^- in the soil solution, carbon can be oxidized by **denitrification** (Figure 4.2 I). Denitrification is the process by which bacteria use dissolved NO_3^- to oxidize organic carbon to CO_2, releasing energy in the process (recall Box 4.1). The NO_3^- is reduced to N_2O or N_2:

$$5C_6H_{12}O_6(s) + 24NO_3^-(aq) + 24H^+ = 30CO_2(aq) + 42H_2O(l) + 12N_2(g) (+energy) \quad (4.6)$$

All denitrifying organisms are bacteria. Denitrifying bacteria can use a wide variety of carbon compounds, but require a source of NO_3^- and a soil oxidation–reduction (redox) status that is on the border between oxic and anoxic. Some denitrifying bacteria can respire aerobically when oxygen is present, switching to NO_3^- when oxygen is depleted.

Denitrification is a highly important global process as it is the main mechanism by which NO_3^-, which can be used by plants, is released back to the atmosphere as N_2. For people, this can be an undesirable process if it removes nutrients in areas where nitrogen is limited. On the other hand, denitrification can be a highly effective way of removing nitrogen in areas that have been polluted by nitrogen-rich runoff.

○ Which terrestrial ecosystems often have particularly nitrogen-rich runoff?

● Agricultural soil solutions are often rich in nitrate (NO_3^-) and ammonium ions (NH_4^+) from fertilizer input.

In the presence of oxygen, ammonium ions are readily oxidized to nitrate by nitrifying bacteria, which do not require carbon for energy:

$$NH_4^+(aq) + 2O_2(g) = NO_3^-(aq) + H_2O(l) + 2H^+(aq) (+energy) \quad (4.7)$$

Thus much of the NH_4^+ in agricultural soil solution that is not taken up by plants is oxidized to NO_3^- by the time that solution has drained to any wetlands downslope.

Unfortunately, it is notoriously difficult to get an accurate estimate of rates of denitrification in the biosphere. One reason is that a major product of denitrification is N_2. Since the atmosphere is already composed of 80% N_2, it is very difficult to measure any small 'additional' N_2 released through denitrification and added to that enormous pool.

Equation 4.6 shows the net reaction for denitrification. However, denitrification often does not proceed to completion, producing nitric oxide (NO) or nitrous oxide (N_2O) instead of N_2. This can occur under suboptimal conditions for the reaction, such as low temperature, organic carbon limitation or the presence of some oxygen.

○ What is important about N_2O in the atmosphere?

● N_2O is a powerful greenhouse gas.

As with rates of denitrification, it is difficult to get an accurate estimate of the total N_2O emission from land such as a wetland or a riparian zone receiving NO_3^- in runoff water. Measurements in the field show that N_2O emission may occur sporadically, often in brief but intense bursts. It is not clear what conditions trigger the pulses of N_2O, but factors such as rapid changes in the water-table height or air pressure, or heavy rainfall displacing N_2O in porewater, are all likely contenders. Because N_2O emission is so variable, the estimate of N_2O emission from wetlands globally is subject to high uncertainty, although wetlands are definitely a major source.

Despite the sparse data on both denitrification rates and N_2O emission, one thing is fairly clear. Except in wetlands receiving runoff that is relatively high in NO_3^-, such as wetlands draining agricultural land, denitrification is not usually a major pathway of carbon breakdown in wetlands.

○ Why do you think this is so? (Hint: Why do farmers add nitrogen to crops?)

● The growth of many terrestrial plants, including many wetland plants, is limited by nitrogen. Because plants scavenge the available NO_3^- in natural systems, there is usually insufficient NO_3^- for denitrification to take place.

Denitrification can be an important pathway for microbial carbon metabolism in wetlands that drain areas where high levels of fertilizer are applied, or in areas that receive high levels of nitrogen in rainfall deposition, such as in the Netherlands or northern Germany (where rainfall is greatly enriched in nitrogen by volatilized ammonia from intensive animal husbandry). These ecosystems are so enriched in nitrogen that the vegetation is 'saturated', and surplus nitrate remains in the soil solution.

Denitrification in floodplain wetlands can be an important process by which nitrate in agricultural runoff is removed from the soil solution before it reaches a stream or river. This has led to the suggestion that 'buffer strips' of wetlands should be constructed along rivers and streams that drain agricultural land (Figure 4.3). These buffer strips would remove NO_3^- by both plant assimilation and denitrification. As long as the buffer areas are wet and have a sufficient supply of available carbon to support denitrifying bacteria (Equation 4.6), denitrification could remove a significant proportion of the nitrogen in agricultural runoff. The 'catch', of course, is that under some conditions the buffer strips could be a significant source of N_2O to the atmosphere.

Figure 4.3 Buffer strips for removing NO_3^- from agricultural runoff through both biological uptake and denitrification.

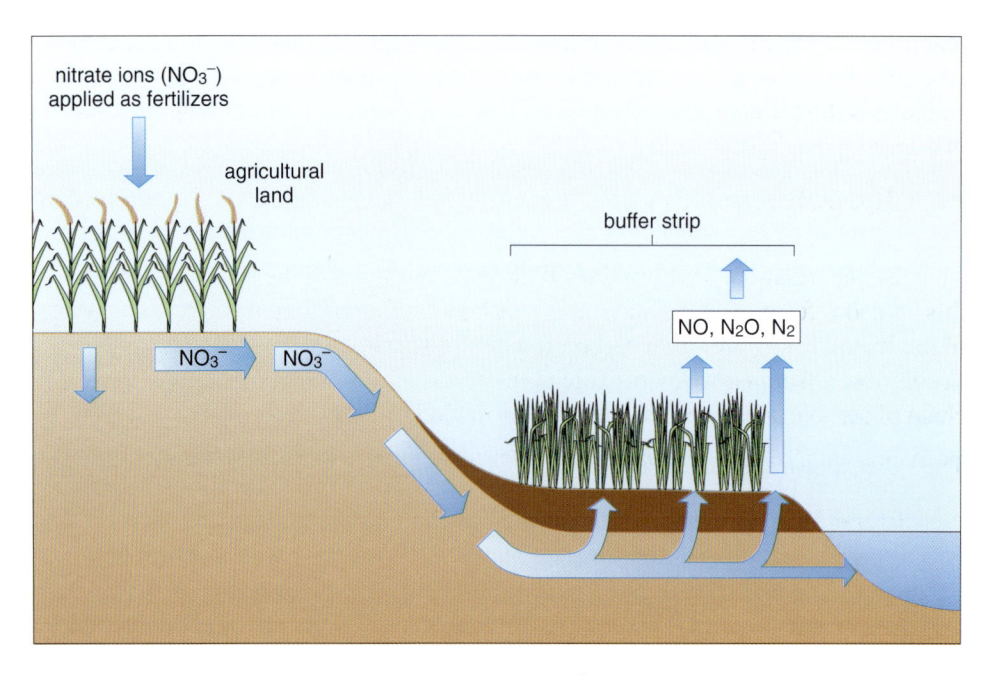

DOC oxidation through reduction of SO_4^{2-}

If there is little O_2 or NO_3^- in the environment, microbes can oxidize organic carbon by reducing sulfate, SO_4^{2-}, in the process of **sulfate reduction** (Figure 4.2 **J**).

$$\underset{\text{ethanoate}}{CH_3COO^-(aq)} + SO_4^{2-}(aq) + 3H^+(aq) = 2CO_2(aq) + 2H_2O(l) + H_2S(g) (+ \text{energy}) \quad (4.8)$$

Sulfate reduction is a strictly anaerobic process carried out by a few groups of specialist bacteria, which can use only a limited number of simple dissolved organic compounds such as ethanoate (acetate), lactate, or alcohols as carbon sources. The energy yield from sulfate reduction is far less than from aerobic respiration or denitrification, or many forms of fermentation. Despite this, in estuaries, saltmarshes and mangrove swamps, anaerobic carbon metabolism is often dominated by sulfate-reducing bacteria, the reason being that seawater contains a relatively high and constant supply of SO_4^{2-}.

Most freshwater wetlands, however, show relatively low concentrations of SO_4^{2-} in soil solution, and consequently low levels of sulfate reduction. The reason is somewhat different than the reason for low nitrate concentrations, since sulfate is not usually a limiting ion for plant metabolism. Instead, sulfate concentrations are low in wetlands because the sources of sulfate are usually minor. In freshwater wetlands the major source of sulfate is either atmospheric deposition (sulfate in rainfall and snowmelt) or geological sources such as iron pyrite or gypsum. Where these sources are low or absent (and sometimes even if they are present, if a thick layer of organic matter isolates the biologically active zone from the underlying geology), anaerobic metabolism moves down another energy level.

DOC oxidation through the reduction of CO_2

If oxidizing agents such as oxygen, nitrate or sulfate are not present, carbon metabolism can proceed through **methanogenesis** (Figure 4.2 **K**). As its name describes, methanogenesis (*methano*, methane; *genesis*, creation) is a chemical reaction that creates methane, CH_4. There are two major types of methanogenic reactions in wetlands that are carried out by two different groups of bacteria. One of these is the reduction of ethanoic acid or a limited number of ethanoate-like dissolved organic substrates:

$$CH_3COOH(aq) = \underset{\text{methane}}{CH_4} + CO_2 \; (+ \text{ energy}) \qquad (4.9)$$

Because the ethanoic acid provides both the reducing and the oxidizing agent, this is also a form of fermentation. As with sulfate reduction, the substrates used in methanogenesis are often carbon molecules that have been progressively decomposed from complex organic compounds to simpler compounds down a chain of aerobic and anaerobic metabolism, including other fermentation reactions.

The other major methanogenic reaction is reduction of CO_2 using hydrogen:

$$CO_2 + 4H_2 = CH_4 + 2H_2O \; (+ \text{ energy}) \qquad (4.10)$$

Here, under the most reducing conditions, CO_2 itself is reduced. The hydrogen is a by-product of the fermentation of more complex organic molecules (e.g. Equation 4.5) that could be harmful to cells if not removed.

If it is present, sulfate can also be used as an oxidizing agent for H_2, by a group of sulfate-reducing bacteria:

$$SO_4^{2-} + 4H_2 + 2H^+ = H_2S + 4H_2O \; (+ \text{ energy}) \qquad (4.11)$$

These two reactions (Equations 4.10 and 4.11) are the final stage of the chain of organic carbon breakdown in wetland environments in which electrons are passed from the components being oxidized to those being reduced, releasing progressively less energy.

Thus the organic carbon from vegetation that is incorporated in wetland soils as POC and DOC is decomposed by aerobic organisms (Figure 4.2 **G**), fermenting bacteria and fungi (**H**), denitrifying bacteria (**I**), sulfate-reducing bacteria (**J**) and methanogens (**K**). In the process, more complex carbon compounds are broken into simpler compounds, and the soil solution is progressively depleted of O_2, NO_3^- and SO_4^{2-}.

The loss of these compounds can be measured over time if an aerobic soil becomes waterlogged (Figure 4.4). In some cases they can also be measured vertically in the soil profile of a wetland, as more oxidized surface soils give way to more reduced, deeper soils. Figure 4.4 shows the changes in concentration of different ions in one soil after it was fully saturated with water.

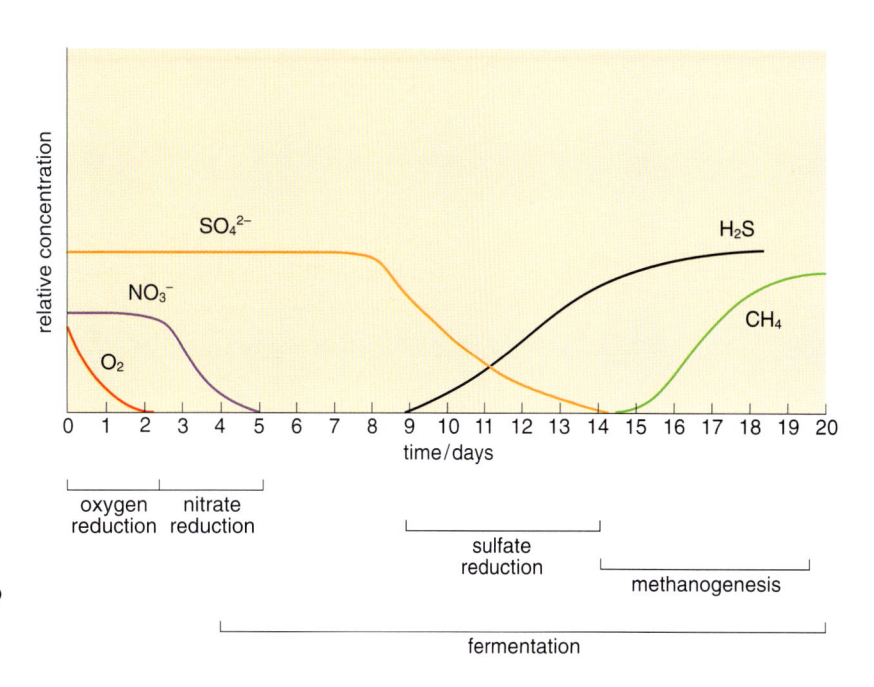

Figure 4.4 Changes in relative concentrations of O_2, NO_3^-, SO_4^{2-}, H_2S and CH_4 over time in a soil after it was saturated with water.

4.1.3 Carbon release from wetlands

Carbon released into the atmosphere as CO_2 and CH_4

The CO_2 produced in the respiration reactions described in the previous section is soluble in water and is rapidly hydrated (Equation 4.3). This occurs both in aerobic and anaerobic soils, since even the driest soils contain some soil water. Since the concentration of CO_2 in soil solution is higher than in the atmosphere, most of the CO_2 ultimately diffuses out of the soil solution into the atmosphere (Figure 4.5 **L**, arrows going into the aerobic zone and into the atmosphere from the soil).

Methane is only sparingly soluble in water. In anaerobic soils, methane produced by methanogenesis diffuses upward toward the surface and can be released by diffusion from either saturated or unsaturated sediments (Figure 4.5 **M**, arrow directly to the atmosphere from the soil). If the rate of CH_4 production is high enough, tiny bubbles of gas form that grow as they rise through the soil and collect more gas, including CO_2, to eventually break the surface in a process called **ebullition**. Sometimes large quantities of bubbles collect just below the surface in the saturated soil of a wetland, and stirring up the sediment of shallow anaerobic ponds, marshes or waterlogged depressions can cause the surface to appear to boil. These bubbles are usually a mix of gases, primarily CO_2 and CH_4, but may be nearly pure methane gas (and would explode if collected and touched with a lit match, but we don't recommend this experiment!).

The aerenchyma tissue of stems and roots of vascular wetland plants can provide a shortcut for CH_4 and CO_2 to bypass the soil and pass directly into the atmosphere (Figure 4.5 **L**, **M**, arrows into and out of the plant). Emergent plants may act as passive conduits for gas diffusing from the soil, like straws stuck in the mud, or may actively transport gases formed around the rooting zone as part of the reverse flow of pressurized air moving through the roots (Figure 2.6b). In wetlands that are dominated by vascular plants like sedges, reeds and cattails, or in rice paddies, transport through stems and leaves can be the dominant way by which CH_4 is released to the surface. Up to 90% of the methane emitted from rice fields may be released through the plant stems.

In highly productive saturated wetlands such as subtropical swamps, the CH_4 produced may escape rapidly to the surface, especially through ebullition and plant transport. However, in wetlands where there is some soil in contact with oxygen, methane may become re-oxidized to CO_2 (Figure 4.5 **N**) by a variety of microbes — the **methanotrophic** (*methano*, methane; *trophic*, feeding), or 'methane consuming' bacteria.

Methanotrophic bacteria use the energy released by the oxidation of CH_4 to CO_2 to power their metabolism:

$$CH_4 + 2O_2 = CO_2 + 2H_2O \; (+ \text{energy}) \tag{4.12}$$

Because this reaction releases a considerable amount of energy, it can also occur without the intervention of organisms (as shown by the exploding bubbles above). Indeed, controlled burning of methane gas is used to heat homes and cook food for much of the population in the industrial world. But there is one important requirement: air. Methanotrophic organisms can only thrive in the presence of oxygen.

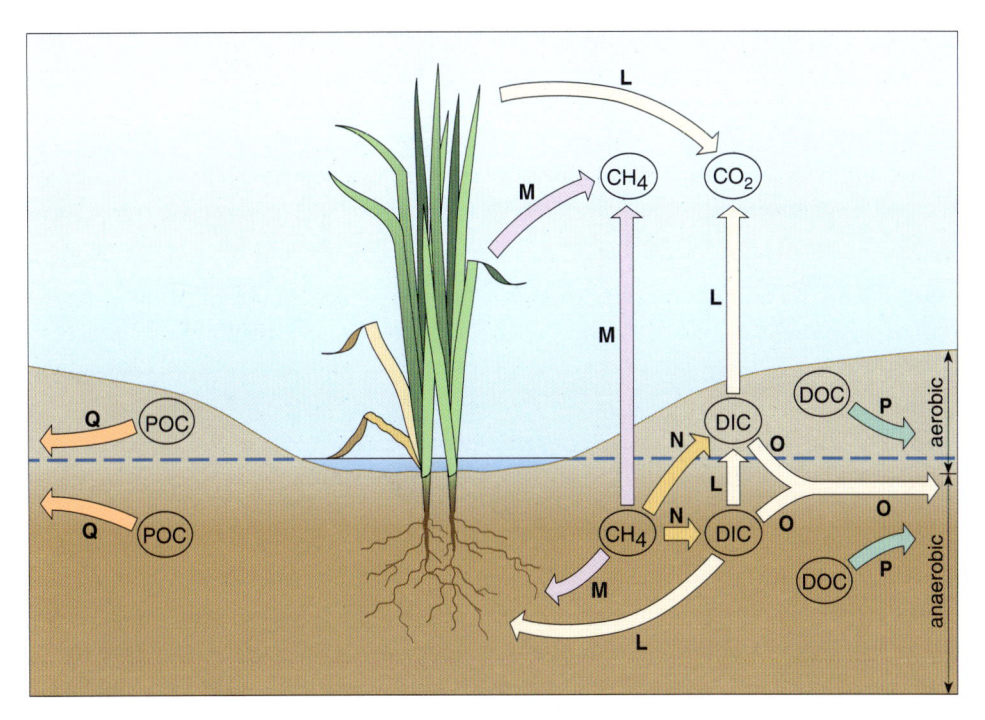

Figure 4.5 Carbon removal through **L**, CO_2 diffusion, ebullition and plant transport; **M**, CH_4 diffusion, ebullition and plant transport; **N**, CH_4 oxidation to CO_2; **O**, DIC leaching; **P**, DOC leaching; **Q**, POC leaching.

Methanotrophic bacteria in soils can oxidize methane wherever they come into contact with oxygen. Where the water-table is below the surface, methanotrophic bacteria colonize the aerobic zone and oxidize CH_4 diffusing through it on its way to the surface. The proportion of CH_4 removed by oxidation in wetlands is thus related to the depth of the water-table. If the water-table is well below the surface, much of the methane diffusing to the surface is oxidized to CO_2 and little CH_4 is released. If the water-table is close to the surface, much of the methane escapes oxidation and is released to the atmosphere.

Methanotrophic activity also occurs around the roots of vascular plants. Recall that some of the oxygen transported by plants down to waterlogged roots leaches from roots into the surrounding area (Figure 2.9). This O_2 can be used by methanotrophic bacteria to oxidize methane. In addition, even in seemingly fully water-saturated soil, tiny air pockets or 'microsites' can exist. Rates of CH_4 oxidation can potentially reach an order of magnitude greater than rates of methanogenesis, so these aerobic zones do not have to be large to remove significant amounts of CH_4. Globally, oxidation through methanotrophic bacteria is thought to remove about 20% of the CH_4 produced in natural wetlands.

Since rates of anaerobic decomposition are related to the initial amount of carbon available, and since most of this carbon was initially fixed by the wetland vegetation, it is not surprising that the amount of reduced gases like CH_4 and H_2S emitted from a wetland is related to the productivity of the wetland. Indeed, in freshwater wetlands, mean seasonal CH_4 emission from ecosystems as disparate as tundra, boreal peatlands, subtropical marshes and rice paddies is remarkably consistent at about 2–5% of NPP.

- ○ Why might this relationship between wetland productivity and reduced gas emission be less strong for N₂O release from wetlands?

- ● Highly productive wetlands may remove large quantities of NO_3^- through assimilation by plants. In addition, denitrification can produce significant amounts of NO or N_2 as well as N_2O.

Recall (Section 4.1.2, p. 190), that we have allowed 75% of the fixed carbon in our model wetland to be decomposed aerobically. A reasonable partition of the remaining 25% of carbon that ends up in the anaerobic layer is to allow 10% to be released back to the atmosphere as CO_2 from anaerobic respiration (via fermentation, denitrification, sulfate reduction and/or methanogenesis, depending upon the oxidizing agents present), and a further 3% of NPP released to the atmosphere as CH_4 through methanogenesis.

Carbon leached from the wetland as DOC, DIC and POC

The carbon fixed by NPP may also be removed from wetlands in leachate water, mainly as dissolved inorganic carbon (DIC), dissolved organic carbon (DOC), or particulate organic carbon (POC). The leached DIC (Figure 4.5 **O**) is mainly dissolved CO_2 from root respiration or microbial respiration, or, depending on the pH, HCO_3^- or CO_3^{2-} (Equation 4.2). At the typically low pH of bogs and nutrient-poor fens it is mostly dissolved CO_2; wetlands on alkaline soils or receiving runoff with a high pH have higher proportions of hydrogen carbonate (HCO_3^-) and carbonate (CO_3^{2-}) ions. When the leached water reaches the surface in streamflow or in a lake, some of the DIC may be released back to the atmosphere as CO_2, and some may remain dissolved, potentially available for photosynthesis by aquatic vegetation.

Dissolved organic carbon compounds (DOC) that escape microbial metabolism can also be leached from the wetland into runoff water and subsequently into terrestrial or aquatic ecosystems downstream (Figure 4.5 **P**). The flux of this DOC from shallow subsurface water flowing through wetlands is particularly important, because it is a source of relatively easily degraded carbon for use by a wide variety of microbes, which in turn support organisms further up the food chain, such as aquatic invertebrates and fish. Wetlands can be a major source of DOC to streams and lakes. Indeed, several studies have shown that the proportion of wetland area in a catchment is a good predictor of the concentration of DOC in surface waters draining the catchment.

Unlike DOC, the net loss of CH_4 in leachate from wetlands is generally low — it is only poorly dissolved and most of that is oxidized rapidly upon contact with throughflow or groundwater.

More DOC is leached from wetlands that have significant subsurface flow or a connection to regional groundwater, since flowing water can continually remove the DOC that is produced in the wetland. Therefore, in general, swamps and marshes release more DOC than do peatlands. Among peatlands, fens, with some throughflow water and relatively high productivity, release more DOC than do bogs. Typical fluxes of dissolved organic carbon for streams draining peatlands are around $10 \text{ g C m}^{-2} \text{ yr}^{-1}$ or less; rivers draining swamps or marshes may carry 1–2 orders of magnitude more DOC. Floods can remove a high proportion of the annual flux of DOC from a floodplain wetland over a brief time period.

Carbon may also be lost from a wetland in particulate form (POC) through erosion or surface flow (Figure 4.5 **Q**). This is commonly in the form of sediment washed away from floodplain wetlands during floods. Bogs and fens, which have a more tenuous connection to regional hydrology than swamps or marshes, lose less carbon as POC.

Of the 12% of NPP left in our model wetland after aerobic and anaerobic respiration, let us say the wetland loses in leachate an additional 2% as DIC, 2% as DOC and 1% as POC. The remaining 7% of NPP that is neither released to the atmosphere nor exported in leachate remains in the ecosystem. This carbon is added at the interface between the aerobic and anaerobic zones, or roughly around the mean location of the water-table, and the wetland grows. *This is net carbon accumulation.*

4.2 Processes of carbon storage and accumulation in wetlands

If more carbon enters a wetland through NPP or from upstream catchments than is lost through decomposition or leaching, the wetland accumulates carbon. Wetlands form some of the most important carbon-accumulating ecosystems on Earth. In aerobic ecosystems there is a rough equilibrium between the rate of carbon entering the soil and the rate at which it is decomposed or otherwise removed, but on wetland soils deposited carbon may remain undecomposed, even over relatively long time scales. The factors controlling carbon accumulation in wetlands are of intense interest, because carbon locked up in wetlands is carbon that is not in the atmosphere as the greenhouse gases CO_2 or CH_4.

Net carbon accumulation in wetlands is the difference between the carbon *gained* through primary production and in water entering the wetland, and carbon *lost* as CO_2 and CH_4 through decomposition and emission to the atmosphere, or simply physical removal of carbon in water flowing out of the wetland. Thus:

$$C_{acc} = (NPP + DIC_{in} + DOC_{in} + POC_{in}) - (CO_{2out} + CH_{4out} + DIC_{out} + DOC_{out} + POC_{out}) \quad (4.13)$$

Note that in Equation 4.13 we are referring only to the *carbon* in each pool.

The larger the difference between the rate of total carbon input and the rate of total carbon output, the higher the carbon accumulation rate. Thus, wetlands with high carbon accumulation rates have relatively high carbon gains or low carbon losses. Each of the major types of wetland — bogs, fens, swamps and marshes — shows a different balance among the various inputs and outputs in Figure 4.13.

- ○ Which type of wetland generally has the highest carbon input through NPP?
- ● Because they are richer in nutrients and are generally found in warmer climates, NPP in swamps and marshes is usually greater than in fens, which in turn is greater than in bogs.

Apart from NPP, the other major carbon input is dissolved or particulate carbon in throughflow or leachate water. Here also, swamps and marshes, with a stronger connection to the regional groundwater or throughflow water, have a higher carbon input than peatlands. Among peatlands, fens receive more throughflow water than bogs and thus receive more carbon in throughflow water. Overall then, carbon input as DIC, DOC or POC follows the same order as carbon input in NPP: swamps, marshes > fens > bogs.

The major loss of carbon is through partial or complete decomposition and removal as either gaseous carbon (CO_2/CH_4), dissolved carbon (DIC and DOC) or carbon as particles (POC). The rate of decomposition is partly a function of the chemical quality of the plant material, and the pH and nutrient status of the wetland. Wetlands with non-woody vegetation, high levels of nutrients or neutral to slightly alkaline pH in porewater generally decompose carbon at a faster rate than nutrient-poor or acid wetlands. Decomposition may also be enhanced from more hydrologically active wetlands, for example if a prolonged water-table drawdown or exposure to oxygenated groundwater brings previously anaerobic sediments into contact with oxygen.

○ Given the above, in which major wetland types are carbon outputs through decomposition likely to be highest?

● Again, swamps, marshes > fens > bogs.

Carbon output through leakage of DOC, DIC and POC can be a major loss from wetlands; indeed outputs of DOC and DIC from wetlands generally exceed inputs in that form. Of the major wetland types, swamps and marshes, with their connectivity to groundwater and surface water, generally show the highest output of dissolved carbon.

Thus both peatlands and swamps/marshes accumulate carbon, but due to a different balance of processes. In general, swamps and marshes accumulate carbon because they have very high rates of carbon input (and high, but not *as high*, rates of carbon output). Bogs and fens, on the other hand, generally accumulate carbon because they have very low rates of carbon output.

Changes in the rates of any of the factors identified in Equation 4.13 can potentially affect the carbon balance of any wetland. Let's start to examine this by considering the first component of the carbon balance: NPP.

○ What are the main environmental controls on NPP in wetlands?

● Your answer should include temperature, precipitation, nutrients and hours of sunlight as the major factors, and may also include the degree of waterlogging.

Temperature, precipitation, nutrients and sunlight play a major role in NPP in all ecosystems. Wetlands are unique in that *too much* water can also inhibit productivity, and dry periods can enhance productivity. A declining water-table exposes more of the sediment to oxygen. For many wetland plants, additional exposure to oxygen increases their productivity. Increased growth and productivity means that more carbon is removed from the atmosphere.

The implication of this may be that to increase the storage of carbon by wetlands we should simply drain them, thus increasing plant productivity. But not so fast!

○ What happens to the organic carbon compounds in saturated soil when that soil is exposed to oxygen, and what are the consequences for the atmospheric CO_2 balance?

● They become subject to aerobic microbial respiration, removing carbon from the soil and releasing it to the atmosphere as CO_2.

So, lowering the water-table in a wetland increases carbon storage in wetland plants, but decreases carbon storage in the soil. Again, it is the *balance* of carbon gain or loss from the entire system that must be considered.

Table 4.1 summarizes the major processes involved in carbon storage and release in wetlands, and the major environmental factors influencing those processes.

Table 4.1 Major processes involved in carbon storage and release in wetlands, and major environmental factors influencing those processes. Note that each environmental factor influences more than one (and in some cases, all) of the processes.

Processes	Environmental factors
net primary production (NPP)	temperature
aerobic respiration	precipitation
anaerobic respiration	atmospheric CO_2
DOC/DIC/POC leaching	nutrients
non-biological oxidation (primarily through fire)	dissolved oxygen
	hydrology
	fire

The environmental factors listed can affect the processes both directly and indirectly. For example, increasing temperature could increase NPP, potentially leading to an increase in carbon storage. On the other hand, higher temperatures may increase both aerobic and anaerobic respiration, *decreasing* carbon storage. Higher temperatures would also enhance evapotranspiration, which may then lower the water-table and expose more soils to aerobic decomposition. In this case, carbon loss through aerobic respiration would increase, but carbon loss through anaerobic respiration would decrease.

We can only conclude that the net effect on the storage of carbon in wetlands of a change in external conditions such as temperature or precipitation can be extremely difficult to predict! This unfortunate fact faces those scientists involved in modelling the effects of climate change on global carbon storage.

In theory at least, the total annual loss of carbon from a wetland will eventually more or less equal the total input of carbon. At that point, the wetland neither gains nor loses appreciable amounts of carbon. Wetlands differ from other terrestrial ecosystems in that the time taken to reach this 'steady state', where carbon accumulation balances carbon release, is generally very long. A newly planted forest accumulates carbon at a rapid rate for several decades, then at a slower rate for several more decades. Wetlands, on the other hand, can continue to accumulate carbon for thousands of years. Some of the peatlands that were formed after the last glaciation are over 10 000 years old; many marshes and swamps are several thousand years old.

Even in wetlands that seem highly stable, changes occur. Because factors such as the weather change from year to year, many wetlands show an annual net gain or loss of carbon. But, over the course of several years, in these old wetlands the carbon fluxes roughly balance. It then takes a more persistent change, such as a long-term change in the climate or vegetation, or human alteration such as drainage, to change the carbon balance again.

4.3 Summary of Section 4

1. The major source of carbon to wetland ecosystems is fixation of atmospheric CO_2 by the vegetation. Carbon can also enter the wetland from the surrounding environment as DIC, DOC and POC.

2. Through root respiration, leaching from roots and, most importantly, death and senescence, carbon is transferred into wetland soils from the vegetation. The input of carbon into wetland soils peaks in the autumn.

3. The organic carbon from vegetation that is incorporated in wetland soils as POC and DOC is decomposed by aerobic organisms, fermenting bacteria and fungi, denitrifying bacteria, sulfate-reducing bacteria and methanogenic bacteria. In the process, more complex carbon compounds are broken down into simpler compounds, and the soil solution is progressively depleted of O_2, NO_3^- and SO_4^{2-}. This may happen either vertically down a soil profile, or over time, as the oxidation sources for aerobic organisms, denitrifying bacteria and sulfate reducing bacteria, respectively, are used up.

4. Where the water-table is below the surface, methanotrophic bacteria can colonize the aerobic zone and oxidize CH_4 diffusing upward through it. CH_4 can also be removed in saturated soils by oxidation around the aerated root zone of vascular wetland plants, or in small oxidized microsites within the soil.

5. Carbon can be removed to the atmosphere as CO_2 and CH_4 through diffusion, ebullition and plant transport. It can also be removed via leaching of DIC and POC or transport out of the catchment of POC.

6. If net carbon input is greater than net carbon loss, the wetland accumulates carbon. This new carbon is generally added at the interface between the aerobic and anaerobic zones, or roughly around the mean location of the water-table.

7. Both peatlands and swamps/marshes accumulate carbon, but due to a different balance of processes. Peatlands accumulate carbon primarily because decomposition rates and leaching losses are low despite the relatively low NPP; swamps/marshes accumulate carbon primarily because NPP is exceptionally high despite relatively high carbon losses through leaching.

8. The major processes involved in carbon storage and release in wetlands are NPP, aerobic respiration, anaerobic respiration, leaching of dissolved carbon and transport of particulate carbon. These processes are influenced by numerous interconnecting environmental factors, including precipitation, temperature, levels of atmospheric CO_2, nutrients, dissolved oxygen, hydrology and fire.

9. Wetlands can continue to accumulate carbon for thousands of years.

Question 4.1

Describe, using equations from Section 4, in particular Box 4.1, the similarities between the *net* effects of sulfate reduction and eating a sandwich.

Question 4.2

We have determined in this section that wetlands are particularly important for carbon accumulation in the biosphere. Overall, what would you say is the single most important reason for this?

Question 4.3

A wildfire sweeps through a peat bog in Canada, incinerating most of the vegetation and some of the surface soils. Describe, using one equation from Section 4, the major effect of the fire in terms of carbon gain or loss to the wetland.

Topic 3 Wetlands and the Carbon Cycle

5 Carbon cycling in wetlands and the global picture

In this section you take some of the local properties of wetlands we have been discussing and use your knowledge of those properties to consider questions of global importance. We first ask the question: How much carbon is stored globally by wetlands? Is it a significant amount in terms of global carbon storage overall? Is it significant to the atmospheric balance of CO_2?

5.1 Global carbon storage by wetlands

Any decisions we ultimately make about managing the global carbon storage potential of wetlands must be based upon an understanding of how much carbon is already stored, what the potential is for more storage, and how stable the storage is in the face of environmental change. If you were given this task, how would you go about it?

The first step would be to write a general equation. To estimate the amount of carbon stored in the world's wetlands (let's call this 'C'), we need to first estimate the area of the world's wetlands ('A'), and then multiply this area by the average amount of carbon stored in the vegetation and the soil below every unit area of a wetland ('c'). In equation form:

$$C = A \times c \quad (5.1)$$

○ What are the SI units for C, A and c?

● C = units of mass, or kg, A, units of area, or m², so c must be in units of kg m^{-2}.

A, the area of the world's wetlands, is fairly straightforward to understand, if not estimate. A bit more thought is needed for c: it tells us how many kilograms of carbon are contained in every square metre of an 'average' wetland, if that square metre were projected in an imaginary column from the top of the highest plants down to the mineral soil or bedrock.

Let us ignore the vegetation for the time being, and concentrate on the soil. To estimate the average amount of carbon in a 1 m² column of wetland soil projected down to the bottom of the wetland, we could first take a representative cubic metre of soil and calculate the amount of carbon in the cube (Figure 5.1). To do this we calculate the mass of solid particles in that volume (i.e. not air or water) and the proportion of carbon in the solid particles. Once we know the amount of carbon in the cube we then multiply that value by the depth in metres of the column.

Thus, c or the carbon contained in a unit area of wetland soil = mass of solid particles × proportion of carbon × depth.

○ What is another word for the first term in this equation, the mass of solid particles in a volume of soil?

● The soil bulk density

So, c = bulk density × proportion of carbon × depth (5.2)

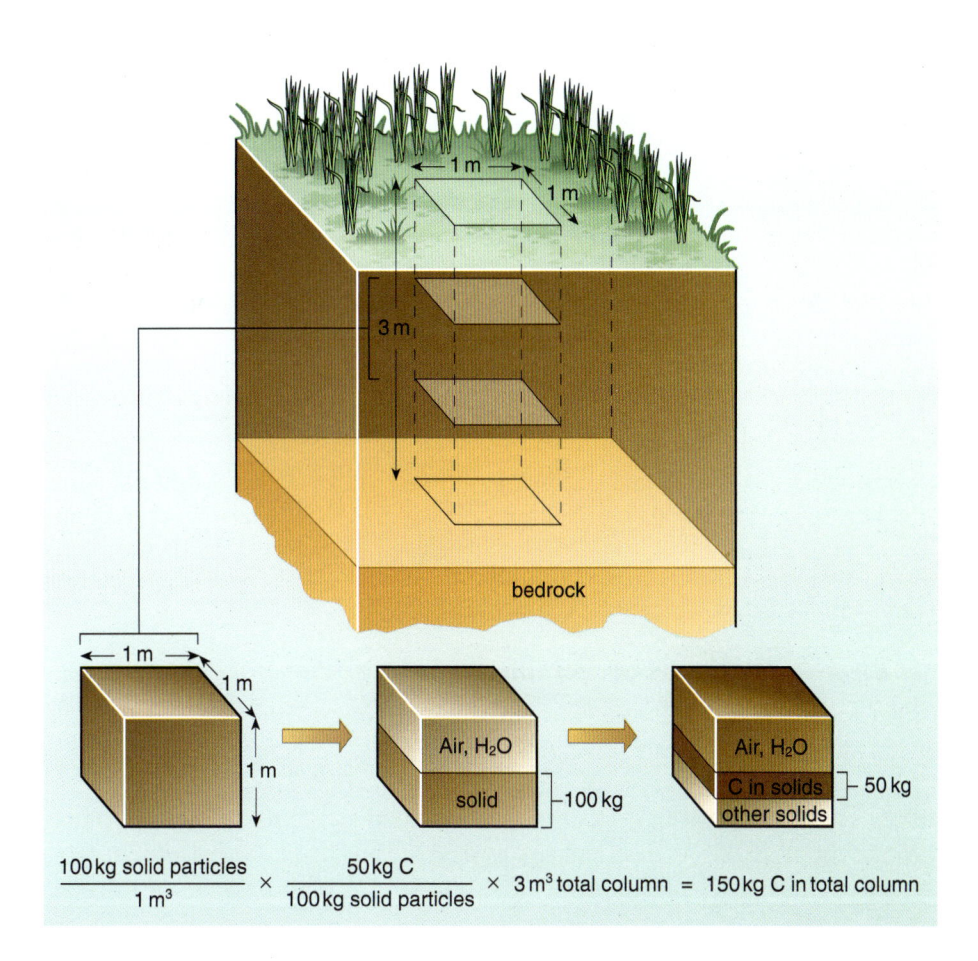

Figure 5.1 Estimating the amount of soil carbon in a 1 m² unit area of a wetland.

Note that you can take *any* representative volume of soil to calculate its bulk density and proportion of carbon. In practice, the bulk density and proportion of carbon of a wetland's organic soil is usually estimated by averaging the values of multiple samples, whereas the mean depth of a wetland's organic soil may be estimated with as few as one or two deep cores.

After understanding the logic, an important check of an equation you set up is to check that the units in the equation balance. For this you can re-write both Equations 5.1 and 5.2, adding in the SI units for each term:

c = bulk density (kg m^{-3}) × proportion of carbon (no units) × depth (m)

Therefore units of c are kg m^{-2}.

(All this checking and re-checking may seem a bit tedious, but it is essential when making complicated calculations of this kind to continually verify that your equations make sense!)

The calculation for global carbon storage in wetland soils assumes many things — among them that, of the huge variety in the world's wetlands, we can come up with an average amount of carbon stored by them all. A better approach would be to take mean values of soil carbon stored for each of the wetland types separately, multiply them by the estimated global area of each wetland type, and sum the estimates. But even with this refinement, the uncertainty in soil carbon storage and wetland distribution is very large.

Having set up an equation that makes sense, and verified that the units also make sense, it 'only' remains for us to put numbers onto the equation. We follow this calculation for peatlands, where data are the best known.

Area of the world's wetlands

Perhaps surprisingly for the 21st century, estimates of the global amount and distribution of the world's natural wetlands continue to be highly uncertain. Major discrepancies still exist in estimates of the area of peatlands in Russia, in the global extent and importance of floodplains, and as we discussed earlier, in the definitions of different wetland types used by different countries. However, as a rough estimate, the world's natural wetlands are distributed mainly between northern latitude peatlands and tropical–subtropical latitude marshes, swamps and floodplain wetlands. You have already collated estimates of the areas of the world's major wetlands in Question 3.1. However, this is only one set of estimates; there are several others based on different methods, and the choice of which to use is fairly arbitrary. Uncertainties in the estimated global area of different wetland types can reach as high as 30%!

Mass of carbon per area of wetland

In a 1991 paper, Professor Eville Gorham of the University of Minnesota, USA, compiled all the known data to calculate the value of c for boreal and subarctic peatland soil. He used a number of studies to estimate the mean depth of peatlands as 2.3 m, the mean bulk density of peatland soil to be 112 kg m^{-3}, and the proportion of carbon in peatland soil as 0.517. Thus, from Equation 5.2:

$$c = 112 \text{ kg m}^{-3} \times 0.517 \times 2.3 \text{ m}$$
$$= 133 \text{ kg C m}^{-2}$$

This estimate says that a square metre of 'average' peatland, if projected downward from the soil surface to the mineral soil or bedrock, would contain an average of about 130 kg of carbon. Some shallow peatlands would of course contain less carbon, and very deep peatlands would contain more.

Multiplying this by the total area of subarctic and boreal peatlands (3.42×10^{12} m^2 in Gorham's paper) gives us (Equation 5.1):

$$C = A \times c$$
$$C = (3.42 \times 10^{12} \text{ m}^2) \times (133 \text{ kg C m}^{-2})$$
$$= 455 \times 10^{12} \text{ kg C}$$

(Note that the value of A, the global area of subarctic/boreal peatlands, is the same as the estimate we previously gave for *all* peatlands (Section 3.1). This is simply because there are several different estimates for the area of subarctic/boreal peatlands, and no one estimate is clearly superior.)

Gorham also estimated that the total amount of carbon stored in peatland vegetation is only about 2 kg C m^{-2}, as opposed to 133 kg C m^{-2} for peatland soil. Thus about 98.5% of the total carbon stored in boreal and subarctic peatlands is in the peat.

Table 5.1 summarizes estimates calculated in similar ways (albeit with fewer data, and thus estimated with less confidence) for all natural wetlands.

Table 5.1 Estimated global storage of carbon in natural wetland soil (does not include rice paddies).

Wetland type	Carbon stored/10^{12} kg
boreal + subarctic peatlands	455
Arctic peatlands	43
tropical/low latitude peatlands	22
marshes/swamps	220
total	740

There are more detailed estimates of carbon storage for peatlands than for marshes or swamps, which are lumped together in Table 5.1. Partly this is because of the major importance of peatlands in global carbon storage — due to their greater age and longer persistence in the landscape, peatlands store much more carbon than do marshes or swamps, despite often lower annual rates of carbon accumulation. But another reason for the focus on peatlands is more mundane. Most wetland researchers work in the temperate climate zone, where peatlands are fairly easily accessible. In contrast, a large area of the world's swamps and marshes are in tropical and subtropical areas that have been only infrequently studied by scientists.

Table 5.2 compares the estimated global carbon storage in peatlands with that of other global carbon pools.

Table 5.2 Carbon pools in various global reservoirs.

Reservoir	C storage/10^{12} kg
carbonate rock	40 000 000
coal and oil	10 000 000
deep ocean	38 000
surface ocean	1000
non-wetland soil	760
boreal/subarctic peatland soil	455
other wetland soil	285
atmosphere	760
biota	560

○ What conclusions can be drawn from Table 4 as regards the relative amount of carbon stored in wetland soil?

● The amount of carbon stored in wetland soils is far less than that stored in carbonate rock, coal or oil, but it is about as much as in the atmosphere and in non-wetland soil.

There is indeed much more carbon in coal, oil and carbonate rock than in wetland soil. These are our major carbon reservoirs. But although the sizes of these reservoirs are enormous, the transfers into and out of them are small. The amount of carbon stored in rock is over 30 000 times more than the amount stored in all soil, but the amount of carbon that moves into or out of soil each year is about 240 times greater.

Thus carbon stored in rocks and oil can be considered essentially inert to modification through climate change, without the direct input of humans. We may burn more or less coal, but a warmer or wetter climate over a hundred years will not enhance the rate at which carbon is stored in coal or carbonate rock deposits significantly enough to make a difference to atmospheric CO_2. In contrast, the rate at which carbon is stored in wetland soils is extremely responsive to climate. For example, if summers are persistently warmer and longer in the boreal and subarctic zone, evapotranspiration is enhanced and more peat is exposed to aerobic decomposition. (At the end of this section you will calculate for yourself just how dramatic the effect could be.)

Because it can respond so rapidly to changes in climate, the carbon stored in wetland soils is a globally important reservoir. Indeed, the total amount stored is about equal to the amount of carbon in the atmosphere, and although natural wetland soils account for only 5% of all soils (with an extra 1% for rice paddy soils), they store about half the world pool of soil carbon. Rice paddy soils store considerably less carbon than natural wetlands because they are constantly harvested, although many rice soils store more carbon than other soil types.

The carbon in wetlands is extremely vulnerable to increasing or decreasing storage through climate change, in ways that are very difficult to predict. Carbon storage may be altered by any of the environmental factors in Table 4.1, especially temperature, precipitation, nutrients, hydrology and fire.

5.2 Global carbon accumulation in wetlands

We can use the same approach as in Section 5.1 to estimate if wetlands are currently accumulating a globally significant amount of carbon. A value would be considered globally significant if, for example, the accumulation of carbon were of the same order as the release of carbon by human activity. In this exercise we consider carbon accumulation of freshwater wetlands only — primarily peatlands in northern latitudes and tropical/subtropical swamps and marshes. The global carbon budget of mangrove swamps and saltmarshes is assumed here to be in a rough balance due to the regular action of waves and tides removing carbon, or a slight negative, due to destruction of tidal wetlands by human development.

Wetlands can accumulate carbon globally by either expanding into new areas or through accretion at existing sites. Although some peatlands, such as in western Siberia, are still increasing in area globally, the lateral spread of peatlands is relatively small and is probably at least balanced by the loss of peatlands through mining and draining. However, many wetlands accumulate carbon vertically, as we saw at the end of Section 4.1 when working through our 'model' wetland. Is that amount globally significant?

○ Using as an example your calculations in Section 5.1 on the global storage of carbon in wetlands, what do we need to know to calculate the global accumulation rate of carbon in wetlands?

● We need to know the global area of major wetland types and the carbon accumulation rate of those types, per unit area.

In equation form we could express this as:

$$U = A \times u \tag{5.3}$$

where U = the global accumulation of carbon in wetlands, A is the area of the world's wetlands, and u is the mean carbon accumulation per unit area of wetlands. The SI units for U are in kg C yr^{-1} and A in m^2, and so the units for u would be kg C m^{-2} yr^{-1}.

Current rates of carbon accumulation have been measured in a number of peatlands, including in Canada, the USA, Russia and Finland. Average values for bogs and fens are of the order of 20 g C m^{-2} yr^{-1} or 0.02 kg C m^{-2} yr^{-1}. Although only a small proportion of the initially fixed carbon may remain, the extremely high productivity of marshes and swamps (of the order of ten times higher than bogs and nutrient-poor fens) means that carbon accumulation from these ecosystems can still be substantial. It is, however, highly variable and unfortunately there have been few studies made, so data are sparse. As a very crude estimate, based on a few studies from Northern Hemisphere swamps, we can assume carbon accumulation of (primarily low-latitude) swamps, marshes and floodplain wetlands to be about 50 g C m^{-2} yr^{-1} (0.05 kg C m^{-2} yr^{-1}).

Scaling these values to their respective global areas (see Sections 3.1, 3.2 and 3.3.1, or the answer to Question 3.1) gives us:

Peatlands: (0.02 kg C m^{-2} yr^{-1}) × (3.4 × 10^{12} m^2) = 68 × 10^9 kg C yr^{-1}

Other wetlands: (0.05 kg C m^{-2} yr^{-1}) × {(0.95 + 1.3) × 10^{12} m^2} = 113 × 10^9 kg C yr^{-1}

We thus arrive at a total global carbon accumulation for natural wetlands of 181 × 10^9 kg C yr^{-1}, or 1.8 × 10^{11} kg C yr^{-1}, with slightly more carbon being accumulated in low-latitude swamps and marshes than in northern latitude peatlands. (Although swamps and marshes often accumulate carbon at a higher rate than peatlands, globally they store less carbon because they are usually shorter-lived features in the landscape, converting to peatlands, grasslands or woodlands over time.)

○ The annual rate of carbon released to the atmosphere by fossil fuel production of CO$_2$ is fairly accurately known, at about 5.5 × 10^{12} kg C yr^{-1}. The release of CO$_2$ due to clearing and burning tropical forests is less well known, but 1–2 × 10^{12} kg C yr^{-1} is a good approximation. Thus about 7 × 10^{12} kg C yr^{-1} are released to the atmosphere through human activity. What proportion of this released carbon can be offset by global carbon accumulation in natural wetlands?

● (1.8 × 10^{11} kg C yr^{-1}) / (7 × 10^{12} kg C yr^{-1}) = 0.026 = 2.6%

Even accounting for large errors in the back-of-the-envelope calculation described above, it is clear that current levels of carbon sequestration by wetlands are at most only a few per cent of the rate at which CO_2 is being released by human activities. The conclusion is unavoidable: we cannot count on carbon accumulation by the world's wetlands to counteract the carbon released to the atmosphere by human activity. Wetlands may perform many useful functions, but they will not save us from the human-accelerated greenhouse effect.

5.3 Global carbon release by wetlands

We have determined that wetlands currently hold a significant amount of carbon, approximately as much as all the carbon held as ($CO_2 + CH_4$) in the atmosphere. We have also determined that current rates of carbon accumulation in wetlands are not sufficient to counteract the amount of carbon that is being released into the atmosphere each year by the combustion of fossil fuels. We now turn to practically the opposite question: could wetlands be releasing the greenhouse gases CO_2 or CH_4 in globally significant amounts? If so, are there any appropriate mitigation strategies to reduce this rate of release?

The CO_2 emitted during decomposition in wetlands originated from fixed carbon that was *itself* originally atmospheric CO_2. Thus CO_2 released through decomposition can be thought of as part of a loop from atmospheric CO_2 through plants and back to atmospheric CO_2. However, through methanogenesis, wetlands also convert some atmospheric CO_2 into another greenhouse gas, CH_4.

CH_4 is a highly effective greenhouse gas and, although its concentration in the atmosphere is much lower than CO_2, molecule for molecule it is a much more potent greenhouse gas. It accounts for about 16% of the enhanced greenhouse effect due to post-industrial changes in concentrations of greenhouse gases, although it contributes only 0.5% of atmospheric carbon. Clearly, the extent to which CO_2 is converted to CH_4 in wetlands is of major importance in understanding and limiting greenhouse warming.

The global methane release for wetlands is estimated by using data from long-term studies of individual wetlands (Figure 5.2) and summing them over the estimated global area of that particular wetland type. We leave this calculation for you to make in Question 5.1.

The major sources of methane include ruminant animals, termites, landfills, biomass burning, venting of gas in production and wetlands. Most estimates agree that wetlands are *the* major global source of CH_4, with natural wetlands and rice agriculture together contributing about 33% of the global budget. Current calculations suggest that tropical and subtropical regions (mostly swamps and marshes) contribute slightly more methane than do northern and temperate region peatlands. Methane emission from rice agriculture is concentrated in two latitude bands: 0–10° S and 20–30° N.

Figure 5.2 Methane emission from a peat bog in Minnesota is measured with a closed chamber. Samples are removed with a syringe and the concentration of methane is measured over time as the gas accumulates in the chamber.

This brings us to the end of Section 5. We have determined that wetlands store an appreciable part of the world's active pool of carbon, that this pool is vulnerable to climate change, and that emission of methane from this pool contributes significantly to methane in the atmosphere. We have also determined that current rates of carbon accumulation by wetlands will not remove enough atmospheric CO_2 to make a significant impact on human CO_2 release.

Extrapolating carbon storage, accumulation and release from wetlands to a global scale is a risky business, involving large uncertainties. But, given the importance of these questions to humankind, such extrapolations must be done. Today some of the most advanced computers in science are used to provide answers, such as those approximated by your calculations, to questions about the impact of climate change on current and future generations.

5.4 Summary of Section 5

1. Wetlands are a major storage reservoir for carbon. The amount of carbon stored in natural wetland soils is about as much as in the atmosphere. It is also about as much carbon as is stored in non-wetland soils, despite only covering 5% of the area.

2. Estimating the importance of any global carbon reservoir in terms of climate change needs to take into account the amount of carbon stored and how reactive that reservoir is to environmental change. The amount of carbon stored in wetland soils is potentially highly sensitive to changes in climate.

3. Current levels of carbon sequestration by wetlands are at most only a few per cent of the rate of CO_2 being released by human activities.

4. Wetlands are the major global source of atmospheric CH_4, with natural wetlands and rice agriculture together contributing about 33% of the global total.

Question 5.1

By compiling databases for vegetation classification, soil classification and inundation, NASA scientists Elaine Matthews and Inez Fung in 1987 created a global map of the world's wetlands, divided into major groups. Using measurements reported in the scientific literature, they then estimated the mean methane emission rate from each of these groups and a 'production season' during which methane is emitted (limited by cold in middle to high latitudes and dry periods in the tropics/subtropics; Table 5.3). They used these values to produce the first major systematic estimate of the global methane flux from natural wetlands.

(*Note*: the estimates of area are slightly different from those in Section 3 due to slightly different data and categories.)

Topic 3 Wetlands and the Carbon Cycle

Table 5.3 Summary of data on methane emissions collected by Matthews and Fung (1987).

Wetland group	Area/10^9 m^2	Mean emission rate/ g CH$_4$ m^{-2} d^{-1}	Production season/days
peatlands	2974	0.20	127
freshwater marshes	1007	0.12	171
freshwater swamps	1088	0.07	180
floodplain wetlands	195	0.03	171
mangrove swamps		not considered	
saltmarshes		not considered	

(a) Using the information in Table 5.3 (without the arduous work of collecting the data yourself!), estimate the total annual methane emission for all natural wetlands.

(b) The estimated global emission of methane from *all* sources is about 5.4×10^{11} kg CH$_4$ yr^{-1}. What proportion of this global emission is accounted for by natural wetlands?

(c) Why do you think CH$_4$ emission from saltmarshes and mangrove swamps was not considered?

Question 5.2

(a) In the highly unlikely event that all the carbon stored in wetland soils was oxidized and released to the atmosphere, and none of this 'new' carbon was removed by factors such as enhanced NPP or dissolution in surface ocean water, by how much would atmospheric CO$_2$ levels increase?

(b) (i) Somewhat less dramatically, imagine that declining water-tables and rising temperature due to climate change were to halt carbon accumulation and instead decompose an annual average of 0.5 cm of peat per year from boreal and subarctic peatlands. Assuming the mean depth of all of these peatlands to be 2.3 m and the total amount of carbon stored in them to be 455×10^{12} kg, approximately how much carbon per year would be added to the atmosphere by this process?

(ii) The annual rate of CO$_2$ release by fossil fuel burning is 5.5×10^{12} kg C yr^{-1}. What would be the percentage increase in the rate of CO$_2$ release if 0.5 cm of peat per year were decomposed from boreal and subarctic peatlands and the carbon released as CO$_2$?

Impacts of wetland management and climate change on carbon cycling

6

Wetlands have been directly exploited by humans for centuries for food, energy, fertilizer and forestry but, as with most other human influences on the environment, our impact has accelerated enormously in the last 50 years. Human management is now a significant force in the global wetland carbon cycle. In this final section we shall briefly explore the implications of some major management practices on the carbon balance of wetlands. We shall also touch upon the implications of what may be the major 'management practice' of humans today—climate change.

6.1 Drainage and carbon storage

To date, large-scale drainage is probably the human activity that has had the most significant impact on the carbon cycle in wetlands. Drained peatlands are used primarily for forestry, pasture and agriculture. The total area of drained tropical and subtropical wetlands is not accurately known.

Wetland drainage can severely impact wildlife and reduce the carbon export to downstream ecosystems. In terms of carbon storage, the effects of drainage vary, but the major impact is a rapid, usually deep, and permanent drop in the water-table. This exposes wetland soils to aerobic decomposition.

○ What is the consequence of a drop in the water-table in terms of the release of CO_2 from the soil, the release of CH_4 from the soil and the overall carbon balance of the wetland?

● Because it would expose more soil to air, drainage would enhance the emission of CO_2 and decrease the emission of CH_4. Overall, because of greatly increased rates of aerobic decomposition, the wetland would lose carbon to the atmosphere.

CO_2 release has been measured from many drained wetlands. Soil carbon loss as CO_2 due to drainage tends to be higher from the more nutrient-rich sites, probably due to higher rates of microbial activity. Thus, organic matter in nutrient-rich soil is decomposed faster and more completely than in nutrient-poor soil. Carbon release is highest in the months to years immediately after drainage, and decreases over time as more of the available carbon is decomposed and a greater proportion of the remaining carbon is refractory.

The *net* effect of drainage on the long-term carbon balance of a wetland ecosystem depends partly upon the climate, the soil chemical characteristics and the depth of drainage, and critically on the use to which the drained land is put. If the drained wetland is planted with trees, the carbon lost through oxidation may be compensated for by the extra tree biomass. The net carbon balance of the ecosystem is then dependent upon the ratio between the rate of accumulation of carbon in woody biomass and litter, and the rate of decay of carbon in the previously accumulated wetland soil.

Trees will not accumulate carbon for nearly as long as the original wetland (hundreds of years at best, versus thousands of years), but the short-term net effect of drainage and planting for forestry could be a continuation or even enhancement of the carbon sink in most circumstances. Draining wetlands for pasture or agricultural use results in a net carbon loss to the atmosphere, since the amount of carbon that can be accumulated (in increased NPP and litter production) is not enough to compensate for the increased decomposition due to drainage.

6.2 Rice and greenhouse gases

About 1.5×10^6 km^2 of land on Earth is under cultivation for rice (Figure 6.1). Much of this land is natural wetland that has been converted to paddy agriculture, but some is artificially flooded former 'dry land'. Rice agriculture is a major source of CH_4, adding about 60×10^9 kg C yr^{-1}, or some 11%, to the global CH_4 budget, and mitigation of CH_4 release from rice paddies is a major research area. Rice paddies cover an area only about 25% the size that natural wetlands cover, but even so, they release in total about half as much methane. Methane emission *per unit area* from a rice paddy is thus about two times higher than the mean methane emission from a natural wetland.

Figure 6.1 Rice terraces in 2000 year-old paddy fields, Philippines.

Factors controlling CH_4 emission from rice paddies are the same as those controlling CH_4 emission from natural wetlands: temperature, depth to the water-table, carbon supply (and therefore NPP), and competition between anaerobic bacteria. Rice under cultivation in warmer climates has higher NPP and releases more carbon (as both CO_2 and CH_4), than does rice in cooler regions. Permanently flooded rice paddies release more CH_4 than seasonally flooded rice, or sites where the water-table is periodically lowered. Fields planted with different cultivars of rice emit different amounts of CH_4, related to their final yield, the length of the growing season and the mineral fertilizers added to them.

Managing rice paddies to reduce CH_4 emission must also be of economic benefit, or it will not be done. Fortunately, as with many conservation strategies, managing rice plantations for economic reasons often has environmental benefits. For instance, reducing the amount of water irrigating a rice field (which increases aeration) can save money as well as decrease CH_4 emission. Similarly, periodic draining increases CH_4 oxidation levels and also reduces the concentration of intermediate products of anaerobic decomposition, which can be toxic to plants. Lower intensity ploughing can reduce soil compaction and puddling, which, by increasing soil aeration, can both increase yields and decrease CH_4 emission. Finally, addition of sulfate fertilizer as part of the management to boost grain yields can also reduce CH_4 emission.

○ Why would adding sulfate fertilizer reduce methane emissions?

● Adding sulfate fertilizer could stimulate sulfate-reducing bacteria, which can outcompete methanogenic bacteria.

6.3 Wetland restoration and carbon balance

Restoring wetlands that have previously been degraded through drainage or river channelization can have many desirable benefits. Among these are increasing wildlife diversity, enhancing downstream productivity and, in many cases, increasing carbon storage. A restored wetland may, however, release CH_4, an undesirable effect from the perspective of global warming. However, since aerobic decomposition has already removed much of the easily decomposable carbon as CO_2, CH_4 emission due to anaerobic decomposition from restored wetlands may, at least for several years, be significantly lower than emission at similar pristine sites.

Restoring floodplains and riparian zones along watercourses that have been channelled can increase their hydrologic connectivity, enhancing the amount of time that the river or stream receives seepage water from wetlands. As a result, the nutrient input to these surface waters would increase. This could greatly enhance the productivity of aquatic ecosystems downstream. Current river restoration projects, such as a major project on the Danube, aim to increase seepage time by restoring the connectivity between the floodplain and channel, and thus create a more balanced nutrient export to enhance downstream productivity.

6.4 Climate change

Climate change could affect wetlands in many different ways, some of them in opposition to each other. Higher CO_2 itself could increase NPP, since plants grown in a CO_2-rich atmosphere often show enhanced photosynthesis rates, at least over the short term. Warmer temperatures could clearly increase NPP in many areas. Changes in precipitation, nutrient release or species composition would all profoundly affect the carbon balance of a wetland. Estimates of the *net* effect of these different factors on wetland carbon cycling must be made on a region by region basis, using information on the projected changes in temperature and precipitation for each region and the type of wetlands that occur there.

Northern and boreal peatlands

For peatlands, much warmer air temperatures (up to a 9 °C increase at high-latitude is predicted to occur with a doubling in atmospheric CO_2) would probably lead to much higher levels of evapotranspiration and declining water-tables. This in turn would enhance CO_2 release from soil (due to decomposition) and decrease CH_4 emission (due to CH_4 oxidation and a decreased carbon supply to methanogenic bacteria). The effect on the total carbon storage of the ecosystem would depend upon how the vegetation community changes. If drying peatlands were colonized by trees, net carbon storage could be enhanced, at least temporarily.

Trees need nutrients, however, and increasing carbon storage through building tree biomass would only happen for relatively nutrient-rich fens. Since trees may grow only poorly on exposed bog soil, due to nutrient limitations, many bogs may shift from a carbon sink to a net carbon source if the climate warms significantly.

Persistent changes in hydrology would greatly affect the flux of DOC from northern wetlands. If the wetland areas in a region shrink under drier climatic conditions, their hydrologic connection to drainage waters would be reduced. As a result, the export of DOC to aquatic ecosystems downstream would decrease. This could have profound consequences for downstream biota.

A drier landscape is also more prone to fire. Should severe droughts become more common in northern peatlands, the frequency of fires may well increase. This could cause rapid oxidation of peat and may potentially be a significant source to the atmosphere of carbon previously stored for hundreds or thousands of years in peat deposits.

The response to climate change of Arctic and subarctic peatlands may be dramatic, as thawing of permafrost may change the landscape in ways that are a challenge to predict. In low-lying areas, earlier snowmelts, melting permafrost and thawed slumped peat could cause an increase in the number of shallow lakes (enhancing the carbon sink due to reduced rates of decomposition), whereas in higher areas, melting permafrost may enhance runoff and erosion, lowering regional water-tables and creating a local carbon source.

Beyond the effect on existing peatlands, global warming may be expected to expand the zone of peat accumulation northward (wherever relief allows) and contract it southward (due to more frequent summer droughts and fires). It is likely that the degradation of southern peatlands will more than compensate for the area of new peatland colonized in polar regions, and the net effect will be a contraction of the overall area of boreal/subarctic peatland.

Tropical and subtropical wetlands

The increase in mean surface temperature in tropical and subtropical wetlands is expected to be less than in northern wetlands, but still substantial at a predicted 2–4 °C in response to a doubling of atmospheric CO_2. Major changes in precipitation may occur over tropical wetlands, especially if there is a change in the pattern or frequency of monsoons, tropical storms and El Niño events. Indeed, the carbon balance of equatorial and subtropical wetlands (many of which are already flooded well above the surface) may be more affected by changes in climate variability or extreme weather events than by fluctuations in mean water-table due to climate change.

In wetlands where the water-table is close to the surface, a reduced water supply due to higher evapotranspiration rates may result in a drawdown of the water-table below the surface, and oxidation of formerly waterlogged soils. Wetlands where the water-table is well above the surface may still remain flooded even with lowered water levels. In these cases, the soil remains anaerobic but becomes warmer. Higher temperatures increase the rates of most microbial processes, including both methane production and methane oxidation. Higher temperatures without new sites of oxidation would favour methanogenesis, however, and increase the rate of CH_4 emission. More methane in the atmosphere could lead to yet warmer temperatures — an example of a positive feedback. An increase in temperature would also enhance methanogenesis through increasing NPP.

Thus the overall effect of climate change will mainly depend upon the extent of warming and the natural hydrology of the wetland. Wetlands in which the water-table is close to or below the surface may begin to dry out if the water-table is reduced further, releasing large amounts of previously stored carbon to the atmosphere as CO_2. If these wetlands are nutrient poor, the carbon lost through oxidation may not be regained through growing trees. Warming of saturated swamps and marshes may not drop water-tables enough to cause soil oxidation, but may greatly enhance the emission of methane from these wetlands.

6.5 Summary of Section 6

1. Some of the major human influences on wetlands are drainage, rice agriculture and channelization. Climate change may potentially have a major effect on wetlands.

2. The net effect of drainage on the long-term carbon balance of a wetland ecosystem depends on the depth to which the water-table is lowered and the use to which the drained land is put. If drainage is shallow and the drained wetland is planted with trees, the carbon lost through oxidation may in the short term be compensated for by the extra tree biomass.

3. Much of the 1.5×10^6 km² of land under cultivation for rice is natural wetland that has been converted to paddy agriculture; some is artificially flooded former 'dry land'. Rice agriculture is a major source of CH_4, accounting for about 11% of the global CH_4 emissions budget.

4. Managing rice plantations to mitigate CH_4 release can have economic benefits. Advantageous management strategies include reducing the amount of irrigation water, periodic draining, reducing the intensity of compaction and adding sulfate fertilizer.

5. Restoring floodplains and riparian zones along watercourses that have been channelized can increase their hydrologic connectivity, increasing the amount of time that the river or stream receives seepage water from wetlands. This can greatly increase the supply of dissolved organic carbon to the biota of the river or stream.

6   The overall effect of climate change on the carbon balance of a wetland would depend upon the amount and seasonal timing of warming, the effect of warming on storm and precipitation patterns, and the natural hydrology of the wetland. Wetlands where the water-table is close to or below the surface may begin to lose stored carbon, releasing large amounts to the atmosphere as CO_2. Warming of saturated swamps and marshes may not drop water-tables enough to cause soil oxidation, but may greatly enhance the emission of methane from these wetlands.

Question 6.1

Many streams that drain agricultural land in the UK have been channelized, resulting in very steep banks. Describe some of the major benefits of restoring gradual slopes into such streams, thus increasing the area of floodplain wetlands. What may be some of the drawbacks?

Learning outcomes for Topic 3

Ater working through this topic you should be able to:

1. Using your own words, define 'wetland'. (*All questions*)

2. Outline how inundation leads to the unique characteristics of wetland soils. (*Question 2.1*)

3. Describe the overall adaptations of wetland biota to the major stresses of oxygen limitation, toxic compounds, and nutrient limitation. (*Questions 2.1 and 2.2*)

4. Explain how (a) the amount of water (especially the height of the water-table above or below the surface), (b) the rate of flow of water, and (c) the water quality each affect the chemistry and biology of wetlands. (*Question 2.2*)

5. Give the three major classifications of wetlands and describe the differences between them based on their hydrology, vegetation and soils. (*Questions 2.2, 3.1, 3.2, 3.3 and 4.2*)

6. Outline the major vegetation adaptations to the nutrient-poor peatland environment. (*Question 3.1*)

7. Describe the two major ways in which peatlands form. (*Questions 3.2–3.4*)

8. List some of the environmental stresses that face marsh vegetation, and describe the major plant adaptations to those stresses.

9. Describe the adaptations shown by swamp trees to a waterlogged environment. (*Question 2.1*)

10. Outline how carbon enters wetland ecosystems, and how carbon is transferred from wetland vegetation into wetland soils. (*Question 4.2*)

11. Describe five major pathways of microbial respiration of organic carbon in wetland soils, and indicate the circumstances under which each pathway occurs. (*Questions 4.1, 5.1 and 6.1*)

12. Outline how carbon is removed from wetland ecosystems. (*Questions 4.2, 4.3 and 6.1*)

13. Describe the major similarities and the differences between carbon accumulation in (a) peatlands and (b) marshes and swamps. (*Question 4.2*)

14. Outline some of the major environmental factors that influence the carbon balance in wetlands. (*Questions 2.2, 3.2, 3.3, 4.2 and 4.3*)

15. Describe the importance of wetlands in producing the greenhouse gas methane and explain the conditions in a wetland that would favour the conversion of methane to CO_2 by methanotrophic bacteria. (*Question 5.1*)

16. Compare (without quoting actual values) the amount of carbon stored in wetland soils to that stored in other major reservoirs of the global carbon cycle, such as rocks, non-wetland soils, and the atmosphere. (*Question 5.2*)

17. Explain why the carbon stored in wetland soil is highly important in the global carbon cycle. (*Question 5.2*)

18. Given appropriate information, perform basic calculations such as unit conversions and 'scaling-up' from a local scale to a regional or global scale. (*Questions 3.1, 5.1 and 5.2*)

Answers to questions

Question 2.1

(a) In contrast to non-wetland soils, wetland soils are often anaerobic due to the fact that they are saturated with water during at least part of the year. In mineral soils this causes gleying and/or mottling due to the loss of oxidized forms of iron and manganese and retention of reduced forms. Wetland soils also tend to produce, accumulate and release the products of anaerobic metabolism, including sulfides, oxides of nitrogen, methane, or organic matter in various stages of decomposition, leading to the characteristic odour of some wetland soils. Inundation by water also restricts rates of decomposition, leaving many wetland soils highly rich in organic matter. In some, the organic layer can be many tens of metres thick. The organic matter in wetland soils is often less decomposed than in non-wetland soils — original plant parts may still be visible deep in the sediment.

(b) Inundation by water has led to adaptations by vascular wetland plants to bring oxygen to the roots. These include development of aerenchyma to allow diffusion of air into roots, building up pressure gradients to pump air into roots and development of root structures that protrude into the air. Since seedlings can be especially vulnerable to oxygen deficiency, some wetland plants also time seed production or germination to coincide with low water-tables. Others produce persistent seeds, living offspring, or seeds that float.

Question 2.2

Being hydrologically dominated by precipitation, Wetland A is probably nutrient poor. The soil fertility would be correspondingly low, although the soils may be rich in organic matter since there is little throughflow or groundwater to remove accumulated carbon. Since the soil solution would not have reacted with soils or rocks in mineral soil from upstream catchments to any great extent, the concentration of basic cations and the pH would also be low. Therefore, in addition to being adapted to waterlogging, the vegetation in Wetland A would probably be adapted to low nutrient levels. Wetland A would probably be dominated by mosses, with few species of vascular plants. Vascular plant adaptations may include thickened leaves, evergreen habits, slow rates of growth and deep rooting.

The hydrology of Wetland B suggests that it is a downstream wetland receiving runoff from surrounding catchments. Being hydrologically dominated by throughflow, with a major groundwater component, Wetland B most likely receives higher levels of nutrients than Wetland A. Soil fertility would therefore be relatively high, with a higher proportion of Ca^{2+} and Mg^{2+} due to ions weathered from bedrock and released from soils in upstream catchments. Depending upon the proportion of agriculture (and possibly disturbance) in the upstream catchments, Wetland B may or may not be rich in dissolved nitrogen and phosphorus. Soil pH would largely depend upon the bedrock of the upstream catchment, but a good guess is that it would be higher than Wetland A, especially given the high input of groundwater in Wetland B. Vegetation in Wetland B would probably have a greater proportion of vascular plants and be more diverse than Wetland A. The plants would show adaptations to waterlogging, such as aerenchyma tissue, but would not necessarily have the adaptations to low nutrient levels shown in Wetland A.

Question 3.1

Section 3.1 tells us there are approximately $3.4 \times 10^6 \, km^2$ of peatlands in the world; Section 3.2 that there are about $950\,000 \, km^2$ of freshwater marshes and $380\,000 \, km^2$ of saltmarshes. Mangrove swamps cover approximately $240\,000 \, km^2$ and freshwater swamps approximately $1.3 \times 10^6 \, km^2$ (Section 3.3.1).

First we sum the individual numbers to arrive at the total area of wetlands from these estimates:

$(3.4 \times 10^6 \, km^2) + (0.95 \times 10^6 \, km^2) + (0.38 \times 10^6 \, km^2) + (0.24 \times 10^6 \, km^2) + (1.3 \times 10^6 \, km^2) = 6.27 \times 10^6 \, km^2$

(In Section 2.4, the area of the world's natural wetlands is given as approximately $6 \times 10^6 \, km^2$.)

Then we divide the area of each individual wetland by the total:

Area of peatlands = $3.4 \times 10^6 \, km^2 / (6.27 \times 10^6 \, km^2) = 54\%$

Area of freshwater marshes
 = $0.95 \times 10^{\,6} \, km^2 / (6.27 \times 10^6 \, km^2) = 15\%$

Area of saltmarshes = $0.38 \times 10^6 \, km^2 / (6.27 \times 10^6 \, km^2) = 6\%$

Answers to questions

Area of mangrove swamps
$= 0.24 \times 10^6 \text{ km}^2/(6.27 \times 10^6 \text{ km}^2) = 4\%$

Area of freshwater swamps
$= 1.3 \times 10^6 \text{ km}^2/(6.27 \times 10^6 \text{ km}^2) = 21\%$

Summing the individual percentages gives us 100%.

A pie chart is the most effective way of showing these proportions (Figure 3.32). You may also have chosen a bar chart.

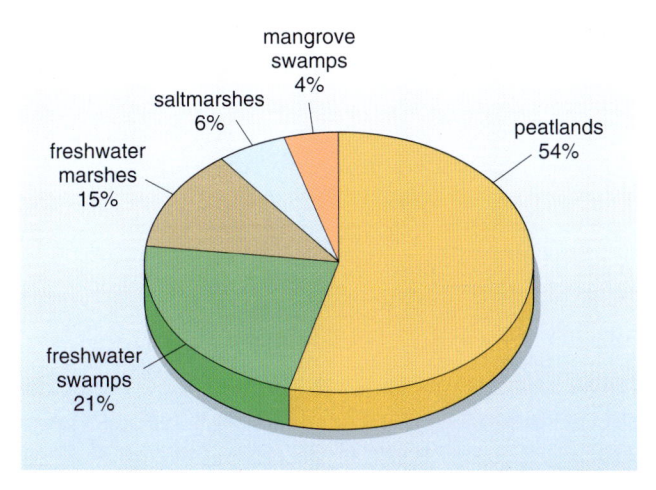

Figure 3.32 Distribution of the major wetland types in the world represented as a pie chart.

Question 3.2

For a prairie pothole landscape to grade into a kettlehole bog landscape we would expect air and soil temperature to decline and more organic matter to accumulate as peat. As peat builds up, connectivity to the groundwater and/or throughflow water decreases and nutrient levels would decline. The density and number of vascular plants would decrease and mosses such as *Sphagnum* would increase. A marsh or fen type of wetland would become more boglike.

As prairie potholes are most commonly found in North America, you would be travelling north.

Question 3.3

The major source of water to tidally influenced wetlands (saltmarshes, mangrove swamps) is throughflow water or surface flow water from the sea. This water constantly removes organic matter that is fixed in the wetland in dissolved and particulate form. The removal leads to low net rates of organic matter accumulation (i.e. little peat can accumulate).

Question 3.4

Warm climates do not allow a large snowpack to build up in winter, therefore there is no large flush of water in the spring to saturate the soil and create a pool. Higher evapotranspiration rates in warm climates also mean less water accumulates in the soil. These factors reduce the likelihood of spring 'pools' forming in a warm landscape.

Question 4.1

The net effect of both processes is breakdown of organic matter (C_{org}) using an oxidizing agent (X_{ox}), leading to a more oxidized form of carbon (C_{ox}, in both cases CO_2) and a reduced form of X (X_{red}), and energy:

$$C_{org} + X_{ox} \longrightarrow C_{ox} + X_{red} \; (+ \text{ energy})$$

In sulfate reduction (Equation 4.8), the form of organic carbon (C_{org}) is simple organic species like the ethanoate ion (CH_3COO^-), and the oxidizing agent (X_{ox}) is sulfate, which is reduced to H_2S (X_{red}):

$$CH_3COO^-(aq) + SO_4^{2-}(aq) + 3H^+(aq) =$$
$$\quad C_{org} \quad\quad\quad\quad X_{ox}$$
$$2CO_2(aq) + H_2S(g) + 2H_2O(l) \; (+ \text{ energy}) \quad (4.8)$$
$$\;\; C_{ox} \quad\quad\;\; X_{red}$$

The energy that is liberated is used by sulfate reducing bacteria for growth, maintenance, metabolism and reproduction.

The net result of eating a sandwich is also a gain in energy, this time through aerobic respiration (Equation 4.2). A simple net equation for the fate of a sandwich is:

$$C_6H_{12}O_6(s) + 6O_2(g) = 6CO_2(g) + 6H_2O(l) \; (+ \text{ energy}) \quad (4.2)$$
$$\quad\; C_{org} \quad\quad\;\; X_{ox} \quad\quad\; C_{ox} \quad\quad\;\; X_{red}$$

The carbohydrate in the sandwich (C_{org}) is more complex than the carbon compounds used in sulfate reduction — here we represent it as $C_6H_{12}O_6$. The oxidizing agent (X_{ox}) is oxygen and it comes from the air (which is why we need to breathe). The reaction between the carbohydrate and the oxygen produces CO_2 and water (X_{red}), liberating energy in the process. That energy is what you use to run your body.

Like the sulfate-reducing bacteria, you use that energy for growth, maintenance, metabolism and reproduction.

Topic 3 Wetlands and the Carbon Cycle

Question 4.2

The main reason that wetlands accumulate more carbon than other ecosystems is that rates of aerobic decomposition are limited due to inundation by water. Since anaerobic decomposition is a much slower process overall, not all of the carbon that is fixed each year can therefore be oxidized and removed as CO_2.

Question 4.3

The most important effect of the fire is the oxidation of large amounts of organic matter, both in the vegetation and in the carbon-rich peat soils. Although a non-biological process, chemically the effect of fire is the same as that of aerobic respiration (Equation 4.2):

$$C_6H_{12}O_6(s) + 6O_2(g) = 6CO_2(g) + 6H_2O(l) \text{ (+ energy)}$$

This represents a major loss of fixed carbon from the wetland.

Question 5.1

(a) To estimate the global emission of methane from each group, we first need to multiply the mean daily emission per square metre by the mean number of days in a year in which methane is emitted. This gives us the total annual methane emission per square metre of wetland. Multiplying that value by the total area of that wetland group gives us the total annual global CH_4 emission for each group. As a word equation:

total annual emission = emission per area and day × number of days × total area

We then repeat the calculation for the other wetland groups and sum the values to find the annual global emission from all wetlands.

We'll first set up the equation for peatlands, and check that the equation gives an answer in the correct units (in this case, grams):

$$0.2 \text{ g } CH_4 \text{ m}^{-2} \text{ d}^{-1} \times 127 \text{ d} \times (2974 \times 10^9) \text{ m}^2 =$$

Does the equation produce an answer in the right units?

$$0.2 \text{ g } CH_4 \, \cancel{\text{m}^{-2}} \, \cancel{\text{d}^{-1}} \times 127 \, \cancel{\text{d}} \times 2974 \times 10^9 \, \cancel{\text{m}^2} = \text{g}$$

All units cancel out except grams, so yes.

Then we can proceed. Multiplying the equation through gives us the total annual methane emission for peatlands as $75\,540 \times 10^9$ g CH_4.

Following the same procedure for the other wetland types gives us:

Freshwater marshes, $20\,660 \times 10^9$ g CH_4; freshwater swamps, $13\,710 \times 10^9$ g CH_4; and floodplain wetlands, 1000×10^9 g CH_4. Summing all four wetland groups together gives us: $110\,910 \times 10^9$ g CH_4 emitted annually by natural wetlands, or 1.12×10^{14} g CH_4 yr^{-1} = 1.12×10^{11} kg CH_4 yr^{-1} (to 2 significant figures).

(b) Natural wetlands:
1.12×10^{11} kg CH_4 yr^{-1} / (5.4×10^{11} kg CH_4 yr^{-1})
= 20%.

(c) The mean CH_4 emission from saltmarshes and mangrove swamps was neglected because these wetlands are rich in sulfate (SO_4^{2-}). The presence of sulfate allows microbial carbon cycling to be dominated by sulfate-reducing bacteria, which outcompete methanogenic bacteria (Section 4.1.2), so production of methane is suppressed.

Question 5.2

(a) From Table 5.1, the total amount of carbon stored in natural wetlands is approximately 740×10^{12} kg. This is about equal to the total amount of carbon in the atmosphere (760×10^{12} kg). If all the carbon stored in wetlands were oxidized and released to the atmosphere, atmospheric CO_2 levels would double.

(b) (i) A 0.5 cm 'slice' of peat is 1/460th of the total amount of peat in a peatland of depth 2.3 m (0.5 cm / 230 cm). Since we assume this is the mean depth of *all* subarctic and boreal peatlands, oxidizing and removing 0.5 cm of peat removes 1/460th of the total carbon stored in these peatlands.
(455×10^{12} kg) / 460 = 1.0×10^{12} kg C oxidized per year.

(ii) 1.0×10^{12} kg C yr^{-1}/(5.5×10^{12} kg C yr^{-1}) = 18% increase to the rate of CO_2 release.

Question 6.1

Some of the benefits of encouraging development of floodplain wetlands are:

- Enhancing export of DOC to downstream ecosystems
- Removing NO_3^- from agricultural runoff
- Providing habitat for various species
- Removal of toxins such as pesticides and heavy metals
- Increasing carbon storage.

Some of the drawbacks are:

- Removing some land from agricultural development
- Potentially increasing the emission of CH_4 and/or N_2O due to flooded anaerobic conditions.

You may have thought of other benefits or drawbacks.

Acknowledgements for Topic 3 *Wetlands and the Carbon Cycle*

Grateful acknowledgement is made to the following sources for permission to reproduce material in this book:

Figures

Figures 1.1, 3.6a, 3.7a, 3.7b, 3.8a, 3.8c, 3.15a, 3.15b, 3.26b, 3.28, 3.29: Mike Dodd; *Figures 2.3a, 2.4, 2.5, 3.1a, 3.4, 3.8b, 3.30*: Geoscience Picture Library; *Figure 2.3b*: John Holford; *Figures 2.7, 3.26d*: Mark Deeble and Victoria Stone/Oxford Scientific Films; *Figure 2.8*: Kathie Atkinson/Oxford Scientific Films; *Figures 2.10. 3.9, 3.21, 3.22, 3.23, 3.27, 4.4*: Mitsch, W. J. and Grosslink, J. G. (2000) *Wetlands*, © 2000 by John Wiley & Sons, Inc. This material is used by permission of John Wiley & Sons, Inc.; *Figure 2.11*: The Nature Conservancy (US); *Figure 2.12*: Sue Cunningham; *Figure 3.1b*: David Nunuk, Science Photo Library; *Figure 3.2*: Gordon Maclean/Oxford Scientific Films; *Figure 3.5*: Silkeborg Museum, Denmark/Munoz-Yague/Science Photo Library; *Figure 3.6b*: Professor David Firmage; *Figure 3.6c*: Niall Benvie/Oxford Scientific Films; *Figure 3.10*: Environment Protection Agency; *Figure 3.11*: Sally Morgan/Ecoscene; *Figure 3.12*: Richard Herrman/Oxford Scientific Films; *Figure 3.14a*: Bob Gibbons/Oxford Scientific Films; *Figure 3.14b*: David Fox/Oxford Scientific Films; *Figure 3.15c*: Denis Crawford/Graphic Science; *Figure 3.15d*: Roger and Donna Aitkenhead/Oxford Scientific Films; *Figure 3.16a*: Papilio/Bryan Knox/Ecoscene; *Figure 3.16b*: Ecoscene; *Figure 3.16c*: Papilio/Dennis Jordan/Ecoscene; *Figure 3.16d*: Richard Packwood/Oxford Scientific Films; *Figures 3.17, 3.26a*: Wayne Lawler/Ecoscene; *Figure 3.18a*: Gregory Dimijian/Science Photo Library; *Figures 3.18b, 3.24a*: Andrew Brown/Ecoscene; *Figure 3.19*: Doug Allan/Oxford Scientific Films; *Figure 3.20*: Robert Cinti/Science Photo Library; *Figure 3.24b*: Simon Grove/Ecoscene; *Figure 3.24c*: Dr Morley Read/Science Photo Library; *Figure 3.25a*: David Tipling/Oxford Scientific Films; *Figure 3.25b*: Scott Pearson; *Figure 3.26c*: Daybreak Imagery/Oxford Scientific Films; *Figure 3.31*: David Gowing; *Figure 5.2*: Nancy Dise; *Figure 6.1*: Michael Cuthbert/Ecoscene.

Tables

Table 2.1: Mitsch, W. J. and Grosslink, J. G. (2000) *Wetlands*, © 2000 by John Wiley & Sons, Inc. This material is used by permission of John Wiley & Sons, Inc.

Every effort has been made to trace all the copyright owners, but if any has been inadvertently overlooked, the publishers will be pleased to make the necessary arrangements at the first opportunity.

TOPIC 4
CRYOSPHERE

Richard Hodgkins

1	**Ice on Earth**	**227**
1.1	Defining the cryosphere	227
1.2	Mass balance: credit and debit in the crysophere's deposit account	231
1.3	Summary of Section 1	236
	Learning outcomes for Section 1	236
2	**Ice on land**	**237**
2.1	Energy balance: credit and debit in a catchment's current account	237
2.2	The cryosphere and hydrology	238
2.3	Creeping and sliding: the flow of glaciers and ice sheets	243
2.4	Glaciers and the landscape	249
2.5	Summary of Section 2	253
	Learning outcomes for Section 2	254
3	**Ice underground**	**255**
3.1	The periglacial environment	255
3.2	Permafrost	256
3.3	Processes and features of periglacial environments	260
3.4	Summary of Section 3	265
	Learning outcomes for Section 3	265
4	**Ice in the oceans**	**266**
4.1	Why the Arctic Ocean can be crossed on foot	266
4.2	Icebergs	266
4.3	Sea ice	270
4.4	Summary of Section 4	275
	Learning outcomes for Section 4	275

5	**Ice and people**	**276**
5.1	For better or for worse	276
5.2	Cryospheric hazards	276
5.3	Cryospheric resources	283
5.4	Summary of Section 5	286
Learning outcomes for Section 5		**286**
6	**Ice and environmental change**	**287**
6.1	Global warming	287
6.2	The cryosphere and climate change	287
6.3	Water resources	297
6.4	The West Antarctic Ice Sheet: 'Threat of disaster'	298
6.5	Summary of Section 6	302
Learning outcomes for Section 6		**303**
Answers to questions		**304**
Acknowledgements		**305**
Index		**306**

Ice on Earth

1.1 Defining the cryosphere

The cryosphere is that part of the hydrosphere that concerns water in its solid, or frozen, form (i.e. snow and ice in their many forms). In the UK, the only significant component of the cryosphere we come across is the seasonal snow cover of the higher mountains. However, large parts of the world are strongly influenced by features such as seasonal and perennial snow, glaciers and ice sheets, sea ice and icebergs, and permanently frozen ground (Figure 1.1). These naturally tend to occur in the Earth's colder environments, particularly high mountains, high latitudes (toward the North and South Poles) and continental interiors (beyond the moderating influence of the oceans). Like many features of the environment and landscape, parts of the cryosphere are heavily influenced by past climates and are still adjusting to the relatively warm conditions of the early 21st century: this is particularly true of glaciers and frozen ground. Furthermore, the cryosphere is crucial to human life and economic activity in many parts of the globe, most importantly in providing a reliable water supply during the dry summer season. Nevertheless, hazards associated with the cryosphere have also been responsible for great loss of life, for instance through flooding and avalanching. At the global scale, atmospheric warming threatens historically unprecedented rates of sea-level rise from the melting of snow and ice.

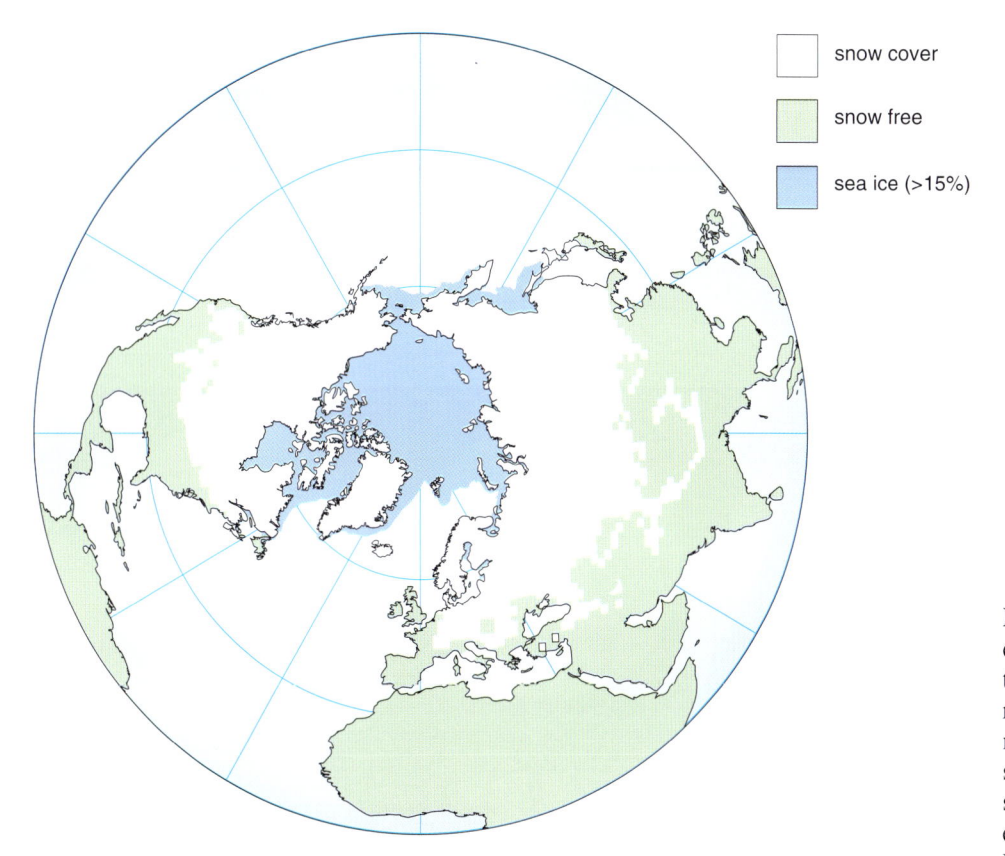

Figure 1.1 Living in a mild environment like the UK, it's easy to forget how cold much of the rest of the globe is, and how much of it experiences significant snowfall and freezing. This map shows snow cover and sea ice extent for the Northern Hemisphere during January 1995.

Topic 4 Cryosphere

1.1.1 Sources of frozen water on Earth

Frozen water accounts for the overwhelming majority — up to 75% — of all fresh water on Earth; some 27.5×10^6 km^3 is locked in glaciers and ice sheets! The different components of the cryosphere are enormously varied: snow, for example, is highly seasonal, whereas glaciers are perennial. These different components can be important in different ways. For example, the Greenland Ice Sheet has an area of 1.8×10^6 km^2 and a volume of 3.0×10^6 km^3 (mean thickness 1600 m), whereas Arctic sea ice (frozen seawater, discussed in Section 4) has an area of 4×10^6 km^2 and a volume of 6000 km^3 (mean thickness 1.5 m). The ice sheet would raise sea-level by about 8 m were it to melt completely, which would be catastrophic in many parts of the world. While the sea ice has no significance in terms of sea-level — as it is floating, it already displaces its own volume of water (Archimedes' Principle) — its relatively bright surface reflects a significant amount of incoming solar radiation back into space, which has important consequences for Earth's climate system.

On average, snow covers about a quarter of the Earth's surface (Figures 1.2 and 1.3), falling at sea-level poleward of latitudes 35° N and S, and at higher altitudes closer to the Equator; for example, Mount Kenya (0° 09' S, 5199 m in altitude) supports a perennial snow and ice cover. Snowflakes are formed from crystals of ice in the atmosphere, the size and shape of the crystals depending mainly on the temperature and amount of water vapour available as they develop. At temperatures above about -40 °C ice crystals condense around atmospheric aerosols, but at lower temperatures crystals form directly from water vapour. In warmer, humid air, the crystals are able to grow rapidly and clump together to form snowflakes. In colder, drier air, the particles remain small and compact. The texture and density of fallen snow undergo continual change (Section 2.2). Snow covers tend to become increasingly dense as further covers are deposited over them, and, where they survive annual melting, may form glacier ice (see below). There is a cliché which states that Eskimos have some extravagantly large number of different words for snow. This is also true of the English language: there are detailed classification schemes for snow crystals both deposited from the atmosphere and produced by subsequent changes within the deposited snow.

Figure 1.2 Seasonal distribution of Northern Hemisphere snow cover. In a typical year, virtually all Europe, Asia and North America above 40 °N have a seasonal snow cover of significant duration, as do the Andean Cordillera and smaller mountain ranges in Australasia and Africa. Colours denote likelihood of snow cover at a location in any given month.

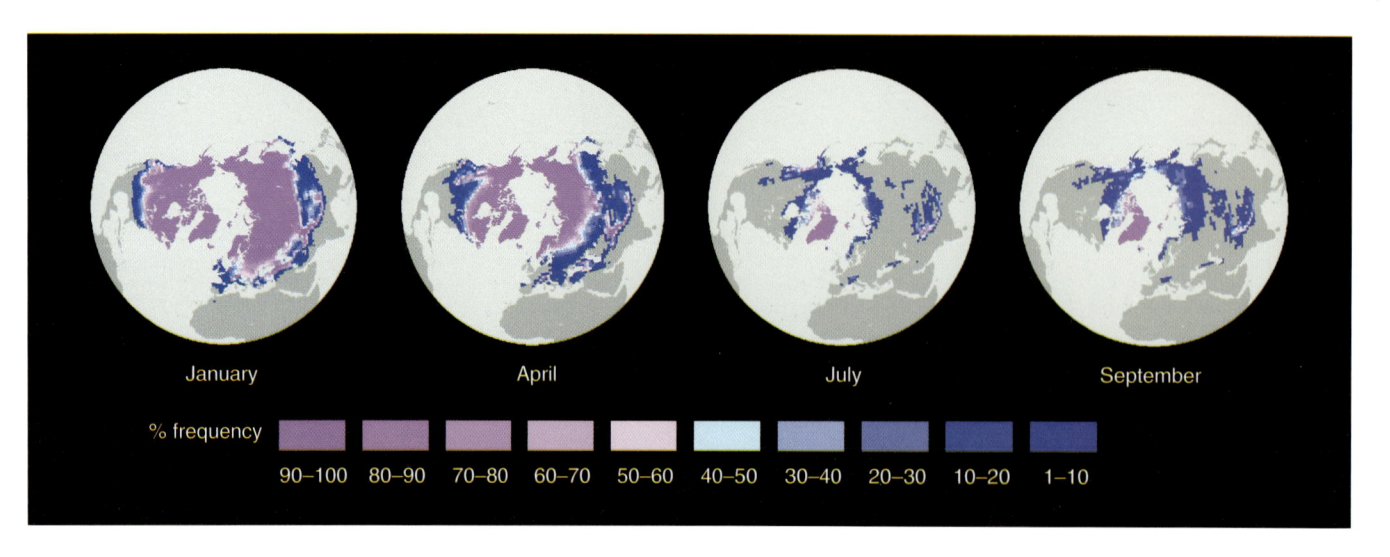

January · April · July · September

% frequency: 90–100, 80–90, 70–80, 60–70, 50–60, 40–50, 30–40, 20–30, 10–20, 1–10

Ice on land

Figure 1.3 Winter snow cover in the Sierra Nevada, southern Spain, January 2001.

1.1.2 Global glacierization: the distribution of glaciers

Of the many components of the cryosphere, the most permanent are glaciers and ice sheets, which will accordingly be the focus of the rest of this section. About 10% of the Earth's land surface is covered by glacial ice. The great majority of this ice is found in the two major ice sheets: Antarctica (85.7%) and Greenland (10.9%). The remaining ice consists of about two-thirds high-latitude ice caps in places such as the Canadian High Arctic islands, and one-third mountain glaciers in places like the Swiss Alps (Figure 1.4). Table 1.1 shows the global distribution of glacial ice by continent and region. Table 1.2 lists the principal characteristics of the world's major ice masses. Some of this is general and should make sense already, some is specific and probably won't: we'll be referring to this table as we go along, and fill in the meanings and significance of this information. There are alternative ways of viewing different regions of glacial cover. For instance, although tremendously important in global terms, Antarctic ice is very remote from permanent population and comparatively stable. Although trivial in global terms, the smaller ice cover of many mountain regions is critically important in terms of water supply, but rapidly diminishing: a combination of circumstances that poses serious problems for local populations. Differences in the nature and impact of glacial ice cover around the world are discussed in later sections.

Note that **glaciated** means *formerly* covered by glaciers (e.g. western Scotland), whereas **glacierized** means *currently* covered by glaciers (e.g. Greenland). This is not a semantic distinction, but helps us distinguish past from present glacial influences.

Figure 1.4 Global distribution of glaciers (dots) and ice sheets (grey). See also Table 1.1.

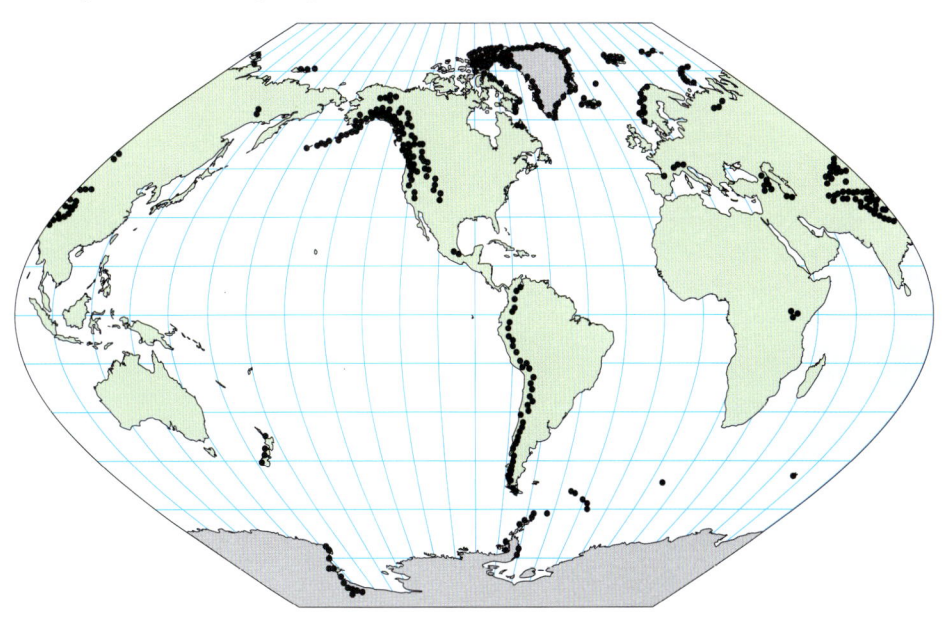

229

Table 1.1 Global distribution of glacier ice. The Antarctic (85.7%) and Greenland (10.9%) ice sheets together constitute 96.6% of the world's total glacierized area (13 586 310 km^2). Of the remaining 3.4% (about 550 000 km^2), high-latitude ice caps and ice fields comprise about two-thirds, and mountain glaciers one-third.

Continent	Region	Area/km^2	Totals/km^2
South America	Tierra del Fuego, Patagonia	21 200	
	Argentina north of 47.5° S	1385	
	Chile north of 46° S	743	
	Bolivia	566	
	Peru	1780	
	Ecuador	120	
	Columbia	111	
	Venezuela	3	25 908
North America	Mexico	11	
	USA (including Alaska)	75 283	
	Canada	200 806	
	Greenland	1 726 400	2 002 500
Africa		10	10
Europe	Iceland	11 260	
	Svalbard	36 612	
	Scandinavia (including Jan Mayen)	3174	
	Alps	2909	
	Pyrenees, Mediterranean mountains	12	53 967
Asia	Former USSR	77 223	
	Turkey, Iran, Afghanistan	4000	
	Pakistan, India	40 000	
	Nepal, Bhutan	7500	
	China	56 481	
	Indonesia	7	185 211
Australasia	New Zealand	860	860
Antarctica	Subantarctic islands	7000	
	Antarctic continent	13 586 310	13 593 310
Grand total			**15 861 766**

Table 1.2 General data on global glacier ice. Actual sea-level is smaller than equivalent sea-level because some ice is below present sea-level. Similarly, floating ice shelves have no direct effect on sea-level and are excluded. Large uncertainties are associated with most of these figures. Some of the variables in this table will be unfamiliar at the moment, but will be explained in subsequent sections.

Variable	West Antarctica	East Antarctica	Greenland	Other glaciers
Area/10^6 km^2	2.1	9.9	1.8	0.6
Volume/10^6 km^3	3.4	25.9	2.95	0.18
Mean accumulation/ m w.e. yr^{-1}	0.3	0.1	0.3	1.2
Equivalent sea-level/m	9.4	71.7	8.3	0.5
Actual sea-level/m	6.0	60.0	7.5	0.5
Major mass loss mechanism	calving	calving	calving and melting	melting
Meltwater discharge/ 10^6 m^3 yr^{-1}	53 (Antarctica total)		237	690
Iceberg discharge/ 10^6 m^3 yr^{-1}	2016 (Antarctica total)		316	50
Response time/yr	100–1000	10 000–100 000	1000–10 000	10–100

1.2 Mass balance: credit and debit in the crysophere's deposit account

Glaciers are good indicators of climate trends. Their growth or shrinkage integrates many features of the climate, such as total winter precipitation, its spatial distribution, and variations in the length and intensity of the summer melt season, over years, decades or longer.

1.2.1 Definitions of mass balance terms

Ice masses (glaciers and ice sheets) are fed largely by snowfall in winter. Successive snowfalls cause snow to accumulate, become compressed and made more dense by burial, and eventually turn to ice. In the summer, ice masses experience melting as temperatures rise. These competing processes, which respectively supply and remove snow and ice, determine the **mass balance**. 'Mass' in this sense means the quantity of snow or ice added or removed, and it is expressed as a **water-equivalent** (w.e.). The reason we use the w.e. unit is because snow and ice have very variable densities: snow tends to be in the range 150–450 kg m^{-3}, and ice around 900 kg m^{-3}. So if we melt a 0.5 m layer of 350 kg m^{-3} snow, we get significantly less water than if we melt a 0.5 m layer of 900 kg m^{-3} ice. To ensure we're comparing like for like, all mass balance quantities are standardized to their equivalent in water. Mass balance quantities are usually expressed as m w.e. yr^{-1}, this is metres water-equivalent per year.

In any one year, a certain mass is gained from snowfall, and a certain mass is lost by melting. The net result will be a mass balance increase if more mass is gained than is lost, and vice versa. Many (though not all) of the world's glaciers currently have negative mass balances: they are losing more mass, in the form of summer melting, than they are gaining in the form of winter snowfall, consequently they are shrinking. To use an analogy from finance, a glacier's mass balance is like a bank account which has credits and debits, and at the end of the month may be in the black or in the red. The glacier has income or **accumulation** (processes that add mass) and expenditure or **ablation** (processes that remove mass). When these incomings and outgoings are put together, the glacier's account can show a surplus (a positive mass balance) or an overdraft (a negative mass balance). These ideas are developed in Box 1.1.

Processes of accumulation include:

- snowfall, which is affected by air temperature, humidity and the movement of air masses relative to oceans and mountains;
- avalanching and snow drifting, which provide sometimes very important re-distribution of snowfall; and
- other processes such as condensation, freezing of rain, re-freezing of meltwater, and freezing of seawater onto the base of floating ice, which may be locally important.

Processes of ablation include:

- melting, principally resulting from solar radiation;
- the breaking off or **calving** of icebergs from ice masses which terminate in seawater, which is particularly important for the large ice sheets in Greenland and Antarctica; and
- minor or local processes such as snow blowing and evaporation.

○ If snow or ice melts then re-freezes before it runs off the glacier, is that still ablation?

● No it isn't: the meltwater has actually to leave the glacier in order for mass to be lost.

From this discussion, we can define some parts of an ice mass in terms of their mass balance behaviour: the **accumulation areas** and **ablation areas** are those parts of the glacier that experience net accumulation ($b_n > 0$)) and net ablation ($b_n < 0$), respectively. These lie either side of an **equilibrium line** ($b_n = 0$), at which there is no net change in mass balance over the course of a balance year. A typical mountain glacier will have its accumulation area at its higher elevations where snowfall survives from year to year, and an ablation area close to its foot or terminus. Mass balance variations over the course of a year and from place to place on a glacier are examined in Box 1.2.

Box 1.1 Quantitative definitions of mass balance

Quantifying mass balance is crucial to understanding the 'health' of a glacier. Most simply, the relationship is

$$b = c + a$$

where b is mass balance, c is accumulation, and a is ablation. Note that a is a negative quantity, because ablation is the loss of mass. Alternatively

$$b_n = b_w + b_s = c_t + a_t$$

where b_n is net mass balance, b_w is winter balance, b_s is summer balance, c_t is total accumulation and a_t is total ablation. Separate winter and summer balances can be defined because, in most cases, accumulation overwhelmingly takes place during the winter by snowfall and ablation during summer by melting.

○ If the accumulation c on a glacier is 2 m w.e. yr^{-1}, and the ablation a is 2.1 m w.e. yr^{-1}, what is the mass balance?

● The mass balance of the glacier is given by $b = c + a$, so

$b = 2 - 2.1$ m w.e. yr^{-1}

$b = -0.1$ m w.e. yr^{-1}, and so the glacier is losing mass at a rate of 0.1 m w.e. yr^{-1}.

It is relatively simple to measure the total amount of accumulation at the end of winter and the total amount of ablation at the end of summer. However, episodes of winter melting and summer snowstorms are not unknown (central Asian glaciers, for example, seem to experience simultaneous accumulation and ablation during the summer, with little snowfall during very dry winters), so an alternative is to consider total accumulation and total ablation, although this presents greater difficulties in terms of measurement. Although a mass balance can be determined for any interval of time, it is conventional (and logical) to measure mass balance over the period of approximately one year, between successive mass balance minima, that is, from the end of one summer to the end of the next summer, the period over which the glacier experiences a full cycle of accumulation and ablation (a balance year). Note, however, that because the seasons vary, this period does not necessarily correspond to 12 months.

Box 1.2 Mass balance in time and space

Figure 1.5a is an idealised cross-section across an ice sheet, showing seasonal variations in accumulation and ablation at three localities. Near the summit (Figure 1.5b) no melting occurs, and so the cumulative plot of accumulation shows a continuous rise, steepest in the winter and early spring. Figure 1.5c plots accumulation at a lower altitude, showing that some melting occurs in the summer but this does not remove all the winter snowfall. The third plot, in the ablation zone, (Figure 1.5c) shows a higher winter accumulation of snow, but that melting starts in the spring and continues into the autumn, removing all this precipitation and melting some of the ice that has flowed down from the accumulation zone.

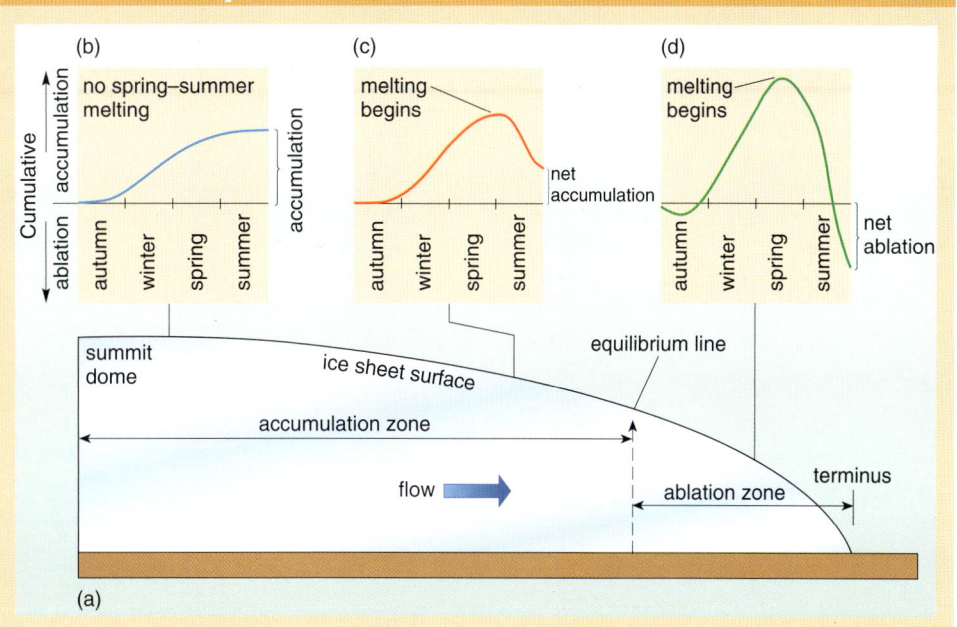

Figure 1.5 Idealized section through an ice sheet (a), showing its cumulative mass budget at three locations (b–d). In (b), the temperature never rises above freezing, so there is no ablation. In (c), there is some ablation during the summer, but not enough to melt all the snow that fell during the preceding winter and spring. However, in (d) late spring and summer ablation exceeds earlier accumulation.

1.2.2 Sensitivity of mass balance to climate change

Because climate is changing almost continuously, it is unlikely that a glacier or ice sheet is ever completely in a state of mass balance equilibrium (i.e. showing no net gain or loss of ice). However, because of concern over the environmental impact of glacier contribution to sea-level rise, it is important to define the sensitivity of mass balance to climate change: that is, if climate changes by a certain amount, how will the mass balance change? On the face of it, this is a simple task of relating climate to accumulation and ablation rates (something we look at in more detail in Section 2). However, numerous factors conspire to complicate this simple relationship. For instance, much of the mass lost from the large ice sheets is as icebergs rather than meltwater, while changes in the profile of a glacier or ice sheet resulting from mass balance change will provoke a response in the ice flow, which will attempt to compensate for those changes (something else we look at in more detail in Section 2).

Nevertheless, estimates of glacier sensitivity to climate change have been made, and a definition of glacier sensitivity has emerged: *the ratio of the rate of change in glacier volume or mass balance to a given change in some climatic parameter* (usually air temperature). Because this simply relates meteorology to surface change without taking into account other effects, particularly the feedback between mass balance and ice flow, this is sometimes referred to as the **static sensitivity**. An estimate of worldwide glacier sensitivity based on a study of 12 selected glaciers from a range of climates shows that for a uniform 1 °C atmospheric warming, glacier mass balance would decrease overall by 0.40 m yr^{-1}, corresponding to a sea-level rise of 0.58 mm yr^{-1}. These figures are not set in stone, however, and are subject to continued revision.

A final, very important, point is that glacier mass balance is essentially always attempting to catch up with climate. Although mass balance can vary enormously from year to year, overall changes in glacier configuration in response to changed climate take place very slowly. The flow of ice is a slow process and a single year's mass balance change is only a tiny fraction of the whole glacier volume. Current glacier changes may therefore result from ancient climate changes. Sometimes this is readily interpreted: the 20th century retreat of European glaciers (e.g. the Glacier d'Argentière shown in Figure 1.6) was a response to the end of a relatively cold period between 1350–1870 AD known as the 'Little Ice Age'. However, in the case of the large ice sheets, interpretation is not so straightforward. For example, the Greenland Ice Sheet (Figure 1.7) is at present generally thickening in the centre and thinning at the margins. Different researchers have suggested that these changes are due either to a reduced rate of outflow of old (> 10 000 yr), highly compressed ice at the base of the ice sheet, or to increased rates of accumulation associated with 20th century atmospheric warming (i.e. an effect of the last 100 yr). This uncertainty must be taken into account when we attempt to forecast change.

Figure 1.6 The cryosphere manifested at a local scale: the terminus of Glacier d'Argentière, above the Chamonix valley in the French Alps. Though trivial in size compared to the great ice sheets of Greenland and Antarctica, mountain glaciers such as this have had considerable impact on settlement and activity in many countries.

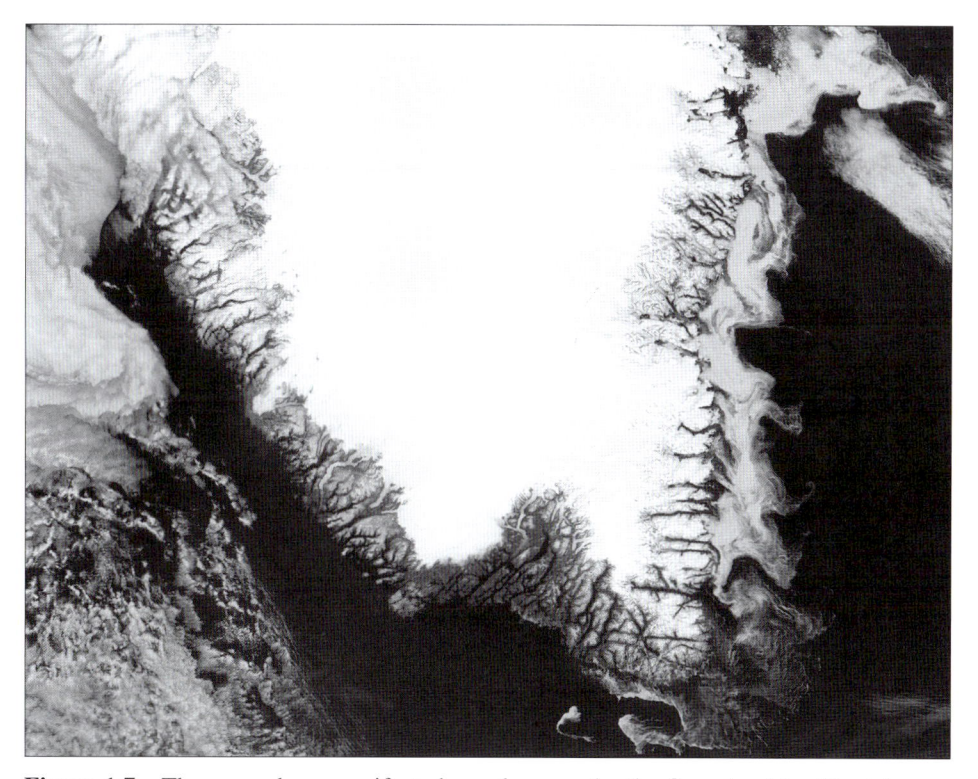

Figure 1.7 The cryosphere manifested on a large scale: the Greenland Ice Sheet from space.

1.3 Summary of Section 1

1. The great majority of fresh water on the surface of the Earth is in the form of snow and ice, and is known as the cryosphere. If frozen water melted then sea-level would rise by about 90 m.

2. The cryosphere is widely distributed in high latitudes and at high altitudes, on the land and over the ocean. Some components of the cryosphere are perennial, such as glaciers and ice sheets which currently cover about 10% of the Earth's land surface; other components exhibit great seasonal variability and can cover even larger areas, notably winter snow cover and sea ice.

3. The state of 'health' of an ice mass is expressed through its mass balance: the sum of mass gain, or accumulation (mainly by snowfall), and mass loss, or ablation (mainly by melting, and by calving of icebergs at the margins of the large ice sheets). Winter mass balance is usually positive (more mass gained by snowfall than lost by melting), and summer balance negative (more mass lost by melting than gained by snowfall). When an ice mass is in a state of equilibrium with climate, annual net mass balance is zero: this is rarely achieved in practice.

4. Zones of net accumulation and net ablation are separated by a conceptual line where there is no net change, known as the equilibrium line.

5. Defining the sensitivity of mass balance to climate change, in terms of the quantity of change for a given change in (usually) air temperature, is an important concern of cryospheric science. However, defining such a quantity is complicated by the dynamic response of ice masses, in which the slow flow of ice gradually adjusts to accommodate mass balance changes in an attempt to achieve equilibrium, giving significant lags between climate and mass balance change.

Question 1.1

If the rate of accumulation of snow on a glacier is 1.5 m w.e. yr^{-1} and the rate of ablation is 3 m w.e. yr^{-1} what is the mass balance?

Learning outcomes for Section 1

Now that you have completed your study of Section 1 you should be able to:

1.1 Describe the different forms of frozen water on the Earth.

1.2 Understand the meaning of the terms accumulation and ablation, mass balance and equilibrium line, the processes that influence them, and how they are related to each other and to climate.

1.3 Identify the difficulties in defining the sensitivity of ice masses to climate change.

Ice on land

2.1 Energy balance: credit and debit in a catchment's current account

Snow and ice are extremely important in the hydrology of the catchments where they occur, modifying the response of runoff (see Block 4, Part 1, Section 4.1) to meteorological processes and altering its seasonal distribution, in comparison with catchments not affected by snow or ice. For example, in the arid basin of the upper Indus River in Pakistan, 80% of runoff is derived from 20% of its area, consisting mainly of high-altitude snowfields and glaciers. Without inputs from high-altitude snow and ice, the flow of this river, of great importance in a densely populated region, would be enormously diminished.

As you saw in Block 2, Part 1, the surface of the Earth constantly gains and loses energy according to weather conditions and the properties of the surface. The energy balance can be determined via equations of greater or lesser complexity, given certain information about weather and snow or ice surface conditions, and is often used to calculate rates of melting of snow covers, glaciers and sea ice. The amount of energy available for melting of snow and ice, Q_M, is determined by the energy balance of the surface. In simple terms, this is given by

$$Q_M = Q_{NR} + Q_S + Q_L + Q_G$$

where Q_{NR} is net solar radiation, Q_S is sensible heat, Q_L is latent heat (of condensation or evaporation), and Q_G is heat derived from heat conduction through a snow cover. Q_{NR} is typically the most important component of the energy balance Q_S and Q_L are referred to as the 'turbulent heat fluxes', and are strongly influenced by conditions in the atmosphere close to the surface, particularly wind speed. Incoming solar radiation at any point can be predicted from a knowledge of the time of year and position on the surface of the Earth, but local factors such as cloudiness, slope aspect and the albedo of the surface greatly complicate the overall, net radiation budget. The turbulent fluxes tend to be responsible for most day-to-day variation in melting because they are influenced by factors such as air temperature and wind speed, which change very quickly; on the other hand, net radiation is influenced by incoming solar radiation and the surface albedo, which change only slowly over the course of a melt season.

The melt rate, M, can be determined from:

$$M = Q_M / S$$

where S is the latent heat of fusion, the quantity of energy required to turn a given mass of snow or ice into water.

From data generated in a study of Aletschgletscher (the largest glacier in Switzerland), *average* contributions to total energy available for melting were: Q_{NR} 71%, Q_S 21% and Q_L 8% (melting 47 mm d^{-1} at 2220 m above sea-level (a.s.l.), 12 mm d^{-1} at 3366 m a.s.l.); while *maximum* melt rates coincided with contributions in the ratio Q_{NR} 55%, Q_S 21%, Q_L 24% (melting 93 mm d^{-1} at 2220 m a.s.l., 17 mm d^{-1} at 3366 m a.s.l.).

○ Which variable is likely to give most day-to-day variation in melting?

● Q_L. The relative importance of this term will increase on windy days.

The relative contributions of the energy balance components also vary with altitude: while Q_{NR} increases, Q_S and Q_L tend to decrease as a result of the vertical lapse rates of air temperature and atmospheric vapour pressure (recall from Block 2, Part 1, that air temperature generally decreases by about 0.7 °C per 100 m vertical ascent). The importance of Q_{NR} in the energy balance indicates the importance of snow cover in melting, hydrology and mass balance, through its albedo (the average albedo of the Earth is 34%, but fresh snow can have an albedo of 90%). In the study of Aletschgletscher described above, albedo varied from 0.66–0.88 with a mean of 0.74 in the accumulation area, and from 0.21–0.42 with a mean of 0.27 in the ablation area. The energy available from solar radiation at the surface is larger by a factor of about three where snow is absent (Figure 2.1). This also indicates that summer snowfall is particularly important in reducing melting rates and runoff; one strong snowfall at Hintereisferner close to Aletschgletscher in July 1958 reduced July runoff by 27% and annual runoff by 7%.

Figure 2.1 A view of the 4.5 km-long Svalbard glacier *Scott Turnerbreen*, demonstrating the contrasting albedos of snow-covered (above the dashed line) and snow-free (below the dashed line) surfaces.

2.2 The cryosphere and hydrology

2.2.1 Drainage of water through snow

Although we often think of snow as a uniform blanket, snow covers are actually very heterogeneous in character, being influenced by the history of snow accumulation (did the snow fall in gentle showers or during blustery storms?), the snow flake (grain) size and water content of the deposited snow, the number of cycles of diurnal (daily) melting and refreezing at the surface between snowfalls, the number of freezing rain events forming ice layers of variable extent and thickness, and the extent of dense, hardened crusts formed by wind (Figure 2.2). Snow covers therefore tend to develop vertical and lateral variations in grain size and shape, density, porosity and permeability. As the snow ages and as melting occurs, larger grains grow at the expense of smaller ones. Smaller grains are progressively eliminated, and the edges of larger grains are smoothed to give a spherical shape. Hence old snow tends to be wetter and grainier than freshly deposited snow. These processes strongly affect the amount of water that can be stored in a given snow cover, and the ways in which water flows through a snow cover.

Snow is a porous medium, and in general water flows through it by percolation. The rate of percolation is affected by a range of factors, but as in soil, rates of water flow generally increase as snow (soil) becomes saturated. However, even saturated percolation is a slow process compared to surface runoff from ice (see below). Because of the slowness of the percolation process, inhomogeneities in snow covers tend to be very important in the development of drainage pathways.

(a) (b)

Figure 2.2 The heterogeneity of snow, which is of great importance for water drainage. (a) Wind crusts formed at the surface of Scott Turnerbreen (Figure 2.1) at the end of the 1992–1993 winter. Each of the layers is 1–2 cm thick. (b) Ice layers exposed in a pit dug into the snow cover covering another Svalbard glacier: there is even a pipe of ice joining the two layers, showing up in front of the ice axe head. This snow cover formed over the 1998–1999 winter, which in Svalbard was relatively warm, with numerous episodes of melting. These ice layers will have formed following such episodes.

As the snow ages it becomes more permeable, and the formation of coarse-grained flow pathways within the maturing snow enhances percolation rates, in turn reinforcing the formation of these pathways, even though the snow cover density increases as it ages. Water flow through snow covers is also affected by low-permeability horizons (e.g. a wind-packed crust buried by subsequent snowfall) and ice layers (e.g. from freezing rain or an episode of melting: Figure 2.2) that inhibit vertical percolation, causing local variations in water saturation and rerouting of flow. However, sufficient fluxes of meltwater can ultimately force vertical drainage paths through ice layers, increasing permeability. These drainage paths can continue to grow by melting, due to the heat energy associated with flowing water and heat from the atmosphere conducted through the snow cover.

2.2.2 Drainage of water through glaciers

By virtue of their accumulation areas (Section 1), glaciers can provide water from the melting of snow throughout a summer — when a seasonal snow cover might otherwise have disappeared — and add water from the melting of glacier ice into the bargain. Glaciers have their own distinctive drainage systems. Snowmelt can either enter a glacier drainage system directly at the base of the snow cover, or discharge at the snowline over the glacier surface, mixing with icemelt. (Above the snowline snow covers the glacier surface; below it bare ice is exposed.) Meltwater can generally penetrate ice in two ways:

- slow seepage through tiny (micrometre-scale) veins between ice grains, which subsequently feeds tubes and conduits at large (millimetre) scales; and
- rapid flow into metre-scale drainage sinks such as **crevasses** (glacier surface fractures resulting from glacier movement, Figure 2.3a) and **moulins** (vertical shafts directing meltwater into the glacier, often a former crevasse kept open by the heat of flowing meltwater, Figure 2.3b and c).

Crevasses and moulins are significantly more important for water drainage. An idealized diagram of water movement through a glacier is given in Figure 2.4.

Figure 2.3 (a) The crevassed surface of Feegletscher, Saas Fee, Switzerland; (b) and (c) moulins in the surface of Hessbreen, Svalbard, each is 1–2 m wide, arrows indicate water drainage.

Meltwater is directed via **englacial** (within the ice) conduits to the glacier bed. The way in which meltwater flows in the **subglacial** environment (at the glacier bed) strongly affects the relationship between input of energy (for melting) and output of energy (as discharge from the glacier), the ability of the meltwater to erode and transport sediment at the glacier base (deposition by meltwater at the glacier base is probably not very important because sediment transport by meltwater requires very little energy), and the relationship between the flux of water and its pressure: an extremely important control on the motion of glaciers (see below). The material found at the glacier base is potentially important. Glaciologists have typically thought in terms of 'hard' (solid rock) and 'soft' (unconsolidated sediment) beds. However, drainage systems developed over both types of bed seem to share important characteristics, and distinctions are often not as great as theory might suggest. For instance, drainage is likely to form at the surface of saturated sediment as if it were impermeable rock, while the majority of glacier bases are likely to share both 'hard' and 'soft' bed characteristics. We shall therefore not distinguish glacier bed types. The way in which water flows at the base of a glacier is largely influenced by the flux of water, which changes seasonally.

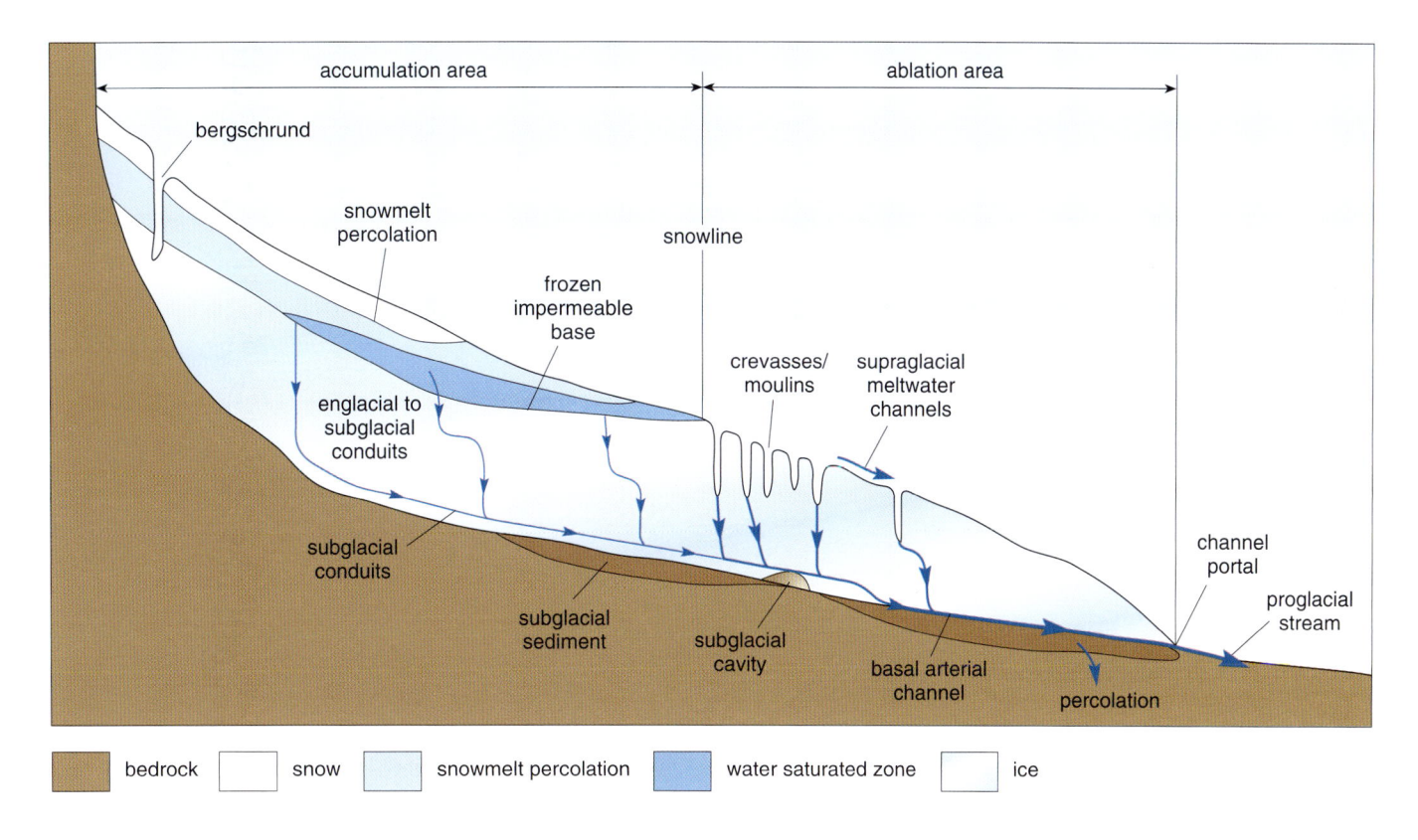

Figure 2.4 Long-section view of an idealized glacier drainage system.

Glacier drainage systems are generally very variable in space and time, and respond to fluctuations in water input and ice configuration such as those described in Box 2.1. Different types of these systems may underlie different parts of a glacier at any given time. For example, a linked-cavity system supplied by small inputs of slowly percolating snowmelt may underlie the upper glacier, feeding into a system of channels supplied by greater inputs of icemelt via moulins downglacier.

Box 2.1 Subglacial drainage systems

There are two general types of subglacial drainage system.

Distributed (inefficient, low water-flux systems)

At the very lowest meltwater fluxes (those resulting from basal melting alone, which may be caused by geothermal heating or pressure of the ice), a water film < 4 mm depth at the ice–rock/sediment interface, or pore flow in sediment, is likely to occur. At greater meltwater fluxes (early in the melt season, as surface melting starts to supply meltwater), a tortuous system of cavities formed in the lee of bedrock bumps, or in the lee of rocks lodged within a sediment layer, linked by narrow orifices (the 'linked-cavity' system, see Figure 2.5a) is likely to form.

Channelized (efficient, high water-flux systems)

At typical melt season water fluxes, when meltwater supply is at a peak, a few large and relatively straight channels incised upwards into ice, or downwards into bedrock, or 'canals' incised downwards into sediment are likely to form (Figure 2.5b). These are all essentially similar, with the exception that the sediment walls of canals tend to deform, yielding higher water pressures than do channels for a given water flux.

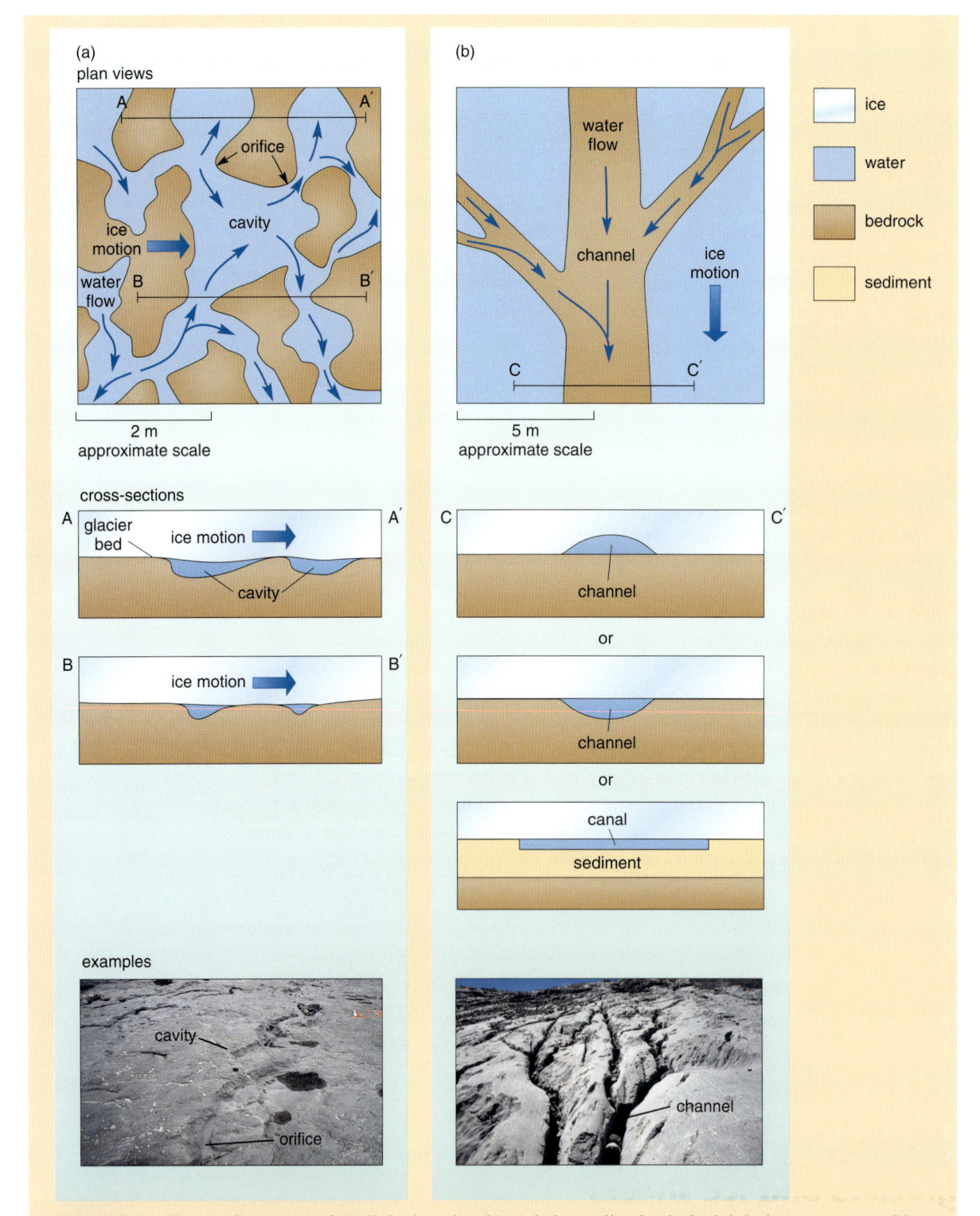

Figure 2.5 The configuration of (a) linked-cavity (b) and channelized subglacial drainage systems. Plan views (blue-shaded areas indicate where ice is in contact with the glacier bed), cross-sections, and examples from the recently deglacierized forefield of Glacier de Tsanfleuron in the Swiss Alps: drainage structures are, unusually, preserved by solution of the limestone bedrock. Note that the ice flow direction and scale are different for each configuration: for the cavities, ice flow is from left to right; for the channels, ice flow is out of the page.

2.2.3 Characteristics of glacial runoff

Even shallow snow and ice covers can modify the level of water discharge, the lag between precipitation and resulting discharge, and the seasonal distribution of discharge, compared with a catchment without snow or ice. In glacierized catchments, runoff consists of snowmelt, icemelt, precipitation and the release of water stored within the glacier and its snow cover. Other sources of glacial water include melt from geothermal heat from within the Earth, and heat generated by processes such as friction during the movement of the glacier ice, although wherever surface melting occurs these make a trivial contribution. The percentage of annual precipitation falling as snow increases by around 3% per 100 m elevation increase; at altitudes of 3500–4000 m in Europe, almost 100% of precipitation is in the form of snow.

○ You have seen in Block 3, Part 1, that the maximum discharge of UK rivers is typically during winter, when most rainfall occurs; at what time of the year does the maximum discharge from a glacier occur?

● Maximum discharge would occur during the summer. If the climate is capable of sustaining a glacier, it is likely that winter precipitation will be predominantly as snowfall, which will not contribute to discharge until it melts the following summer.

An important distinction between glacierized and non-glacierized catchments is that precipitation generally has a short-term negative influence on glacier discharge, because incoming solar radiation is reduced by clouds during storms, and because new snow has a high albedo, which reduces rates of melting. Variations in glacier discharge follow approximately the reverse pattern of a rainfall-dominated discharge regime. Hence a **compensating effect** operates in catchments that have glacierized upper regions experiencing a melt-driven discharge regime and lower regions experiencing a rainfall dominated regime. In a regime that combines inputs from both rainfall and meltwater the variability of streamflow is reduced, which is of great importance in water resources management of mountain catchments (e.g. for agricultural irrigation or hydro-electric power supply: see Section 5).

A number of studies have emphasized the influence of glacierization in the hydrology of catchments. In a study of the North Cascades, Washington, USA, it was found that the presence of glaciers increased runoff volume in comparison with unglacierized basins under similar precipitation regimes. Glacierization at 20% yielded a 10% increase in annual runoff and 50% increase in summer runoff compared with an unglacierized basin, since water is stored as ice and released seasonally through melting, with particularly important contributions in drought years.

2.3 Creeping and sliding: the flow of glaciers and ice sheets

Glaciers and ice sheets flow by two principal mechanisms:
- internal deformation, in which the ice deforms under its own weight as a plastic or a highly viscous liquid (Figure 2.6a); and
- basal motion, which describes the effect of high subglacial water pressures in sliding the glacier along, or in deforming a subglacial sediment layer (Figure 2.6b and c).

These two mechanisms are described in Box 2.2, and in general the internal deformation is much slower than the basal motion.

Box 2.2 The mechanisms of glacier flow

Flow by internal deformation, basal sliding or subglacial sediment deformation occurs in varying combinations in different glacial environments, illustrated here in Figure 2.6.

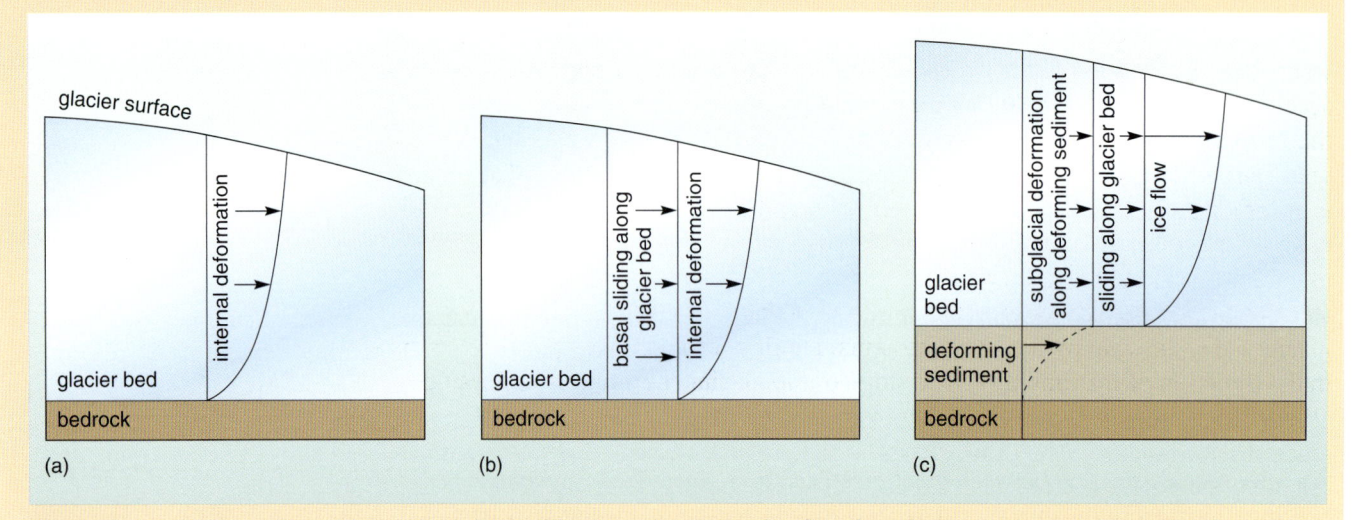

Figure 2.6 In the coldest, high-latitude/high-altitude environments ice flow largely occurs through internal deformation, as meltwater is unlikely to be generated at the bed or to reach it from the surface (a). In warmer glacial environments, basal motion can occur where meltwater is present at the glacier bed. This could take the form of sliding (b), sediment deformation, or a combination of the two (c), depending on the nature of the substrate. Flow by internal deformation will also continue. Flow observed at the glacier surface could therefore result from the combination of several processes.

As you have seen with subglacial drainage systems, researchers have tended to distinguish mechanisms of basal motion between 'hard bed' and 'soft bed' cases. However, the two have much in common. For example, conventional sliding has been observed over an unconsolidated sediment bed. Furthermore, basal water pressure is also important in the deformation of sediment layers beneath glaciers, since high water pressures in pore spaces weaken sediments significantly. Subglacial sediments can deform more readily than ice. It has been suggested that Ice Stream B in Antarctica, which flows at over 800 m yr^{-1} despite a driving stress only 20% of that typical of most glaciers, does so because sediment deformation takes place at its base (see below).

2.3.1 Slow but steady: internal deformation of ice

If mass is added to a glacier in its accumulation area and removed in its ablation area (Section 1), the surface of the ice mass will steepen. This cannot continue indefinitely, some process will act to restore the equilibrium of the glacier. Ice flow can therefore be seen as a response of the glacier to changes in ice thickness and slope caused by accumulation and ablation. In an equilibrium state, the rate of mass gain above the equilibrium line is equal to the rate of mass loss below it, with the rate of mass transfer across the equilibrium line compensating for these changes. In other words, excess mass is shifted from the accumulation area and removed in the ablation area. Glaciers that experience high winter snowfalls and warm summers (e.g. those in the Cascade Range on the mild, wet northwest coast of the USA) generate a higher rate of mass turnover than glaciers that experience small winter snowfalls and cold summers (e.g. in the cold, dry Canadian High Arctic islands), and will correspondingly be more active, exhibiting faster rates of ice flow.

This response of glacier ice to mass balance change is a major reason why defining the sensitivity of an ice mass to climate change is not simply a case of increasing its accumulation or decreasing its ablation (Section 1).

The rate of flow due to internal deformation is proportional to the **driving stress**, τ_b:

$$\tau_b = \rho g h \sin \alpha$$

where ρ is the density of ice (which is essentially a constant), g is gravitational acceleration, h is ice thickness and α is ice surface slope. Observations suggest that the value of the driving stress tends to fall in a relatively narrow range for a wide variety of glacier configurations, an important result that implies that $h \sin \alpha$ is approximately constant, i.e. where the ice surface is steep, a glacier will tend to be relatively thin, and where the ice surface is shallow, a glacier will be relatively thick. Furthermore, it should be noted that the driving stress is influenced by the slope of the ice surface, rather than the slope of the bed beneath. This allows glacier ice to flow up reverse bed gradients. Ice flow should therefore always be in the direction of maximum surface slope of the glacier, a useful result because it allows flow directions to be predicted from contour maps of the surfaces of glaciers and, especially, large ice sheets, which are not constrained by topography (i.e. they are not pinned between valley walls). The constancy of and the influence of ice surface slope on flow direction define simple rules for models or simulations of movement of glaciers and ice sheets, which are the basis of predictions about their behaviour under changing climates.

2.3.2 Fast but jerky: motion at the base of glaciers

Basal motion is intimately connected with the flow of water at the base of the glacier. Certain subglacial drainage configurations yield high basal water pressures, which either offset the pressure from the overlying ice and help the glacier slide along, or weaken subglacial sediments, allowing them to deform beneath the glacier. Essentially, distributed drainage systems generate higher basal water pressures than channelized systems. When water is distributed over the glacier bed (as in Figure 2.5a), the heating effect of flowing water, which would tend to enlarge channels or cavities by melting their ice roofs, is dissipated over a wide area, meaning that features such as linked cavities tend to be quite stable for a certain range of (low) water fluxes. When the water flux increases the cavity grows slowly, so the pressure will tend to increase relatively quickly. Water is therefore driven from large, high-pressure cavities to small, low-pressure cavities, and water remains distributed over the glacier bed, because there is no tendency for large cavities to grow at the expense of small ones. However, the reverse is true of channels: because channels have proportionally small roof areas relative to cavities, the heating effect of flowing water is relatively concentrated, and therefore effective. This means that, above a certain water flux threshold, channels can grow quickly. Large water fluxes destabilize cavities and rapidly enlarge channels; the water pressure then drops in the enlarged channel. The largest subglacial channels tend to have the lowest basal water pressures, and capture drainage from smaller channels with higher water pressures. There is therefore a strong tendency for channelized subglacial drainage systems to evolve towards an arborescent (tree-like) pattern with a series of smaller channels successively feeding into fewer and fewer larger ones (hence the inherent efficiency of this system). Overall, this means that water occupies a larger proportion of the glacier bed in a distributed subglacial drainage system than in a channelized system. The greater the proportion of the bed occupied by water, the greater the extent to which bed roughness is submerged, reducing friction between the glacier and its bed, while ice deformation is enhanced in places where the bed remains in contact with ice, since these have to support proportionally more of the glacier's overburden pressure. The rate of basal sliding, u_b, takes the general form:

$$u_b \approx \tau_b^n N^{-1}$$

where n is usually in the range 3–4 and N is the **effective pressure** given by:

$$N = P_I - P_W$$

where P_I is the ice overburden pressure (the weight of overlying ice, tending to press the glacier down on its bed) and P_W is the basal water pressure (tending to lift the glacier up off its bed, hence the sometimes-used term, **hydraulic jack**).

2.3.3 Glacier advance and retreat in general

The advance and retreat of glacier termini is probably the most obvious manifestation of glacier response to climate change (see Section 6). Glacier retreat is a familiar situation these days, but historical records such as registers of taxable farmland, provide evidence for glacier advances and their effects on human populations. Between 1710–1735 Nigardsbreen, an outlet glacier of the Jostedalsbreen ice cap in Norway, advanced at an average rate of over 100 m yr^{-1}, bulldozing buildings and rendering pasture unusable; such advances were widespread in Norway during the Little Ice Age (1350–1870 AD). The position of the terminus at any time is a function of two competing processes, ablation, which by removing ice leads to retreat, and the forward motion of the ice, which in the absence of ablation would cause the glacier to advance at the rate of ice flow. Whether a glacier retreats or advances depends on the relative strength of each process. Many glaciers make small seasonal advances in winter, when ablation is minimal or absent altogether. Changes in the rate of ablation tend to be felt quite quickly by glaciers. Sustained high rates of ablation near the Brenner Pass on the Italian–Austrian border uncovered the now famous mummified corpse of 'Oetzi', a Copper-Age hunter. However, changes in the rate of accumulation must be transferred through the length of the glacier as changed ice flux, responding to driving stress via changes in the thickness and surface gradient of the glacier (as described above). A time lag therefore exists between climate change and the corresponding effect on a glacier, known as the **response time** (strictly speaking, *the time required for the glacier to reach equilibrium with changed climate*). Any glacier theoretically has a characteristic response time for a given magnitude of climate change, and smaller, rapidly flowing glaciers should have shorter response times than do larger, slowly flowing glaciers (Table 1.2; Section 6). Glaciers in a temperate maritime climate with a thickness of 150–300 m and annual ablation at the terminus of 5–10 m have estimated response times of 15–60 years (e.g. North Cascade Range of the USA, or Iceland), while high-latitude ice caps with a thickness of 500–1000 m and an annual ablation of 1–2 m have estimated response times of 250–1000 years (e.g. the Canadian High Arctic islands, eastern Svalbard or the Russian High Arctic).

Ice on land

2.3.4 Ice streams

Patterns of ice flow within large ice sheets are not uniform: certain regions typically develop very rapid flow compared with the bulk of the surrounding ice, and are known as **ice streams** (Figure 2.7). An ice stream can be defined as 'a region in a grounded ice sheet in which the ice moves much faster than in the regions on either side. Most ice streams lie in deep channels, with beds below sea-level, and end either as a floating glacier tongue or by becoming part of an ice shelf'. Despite comprising only 13% of the Antarctic coastline, ice streams and outlet glaciers drain as much as 90% of the interior accumulation, so that the state of balance of the ice sheets is to a large extent controlled by the flow behaviour of ice streams.

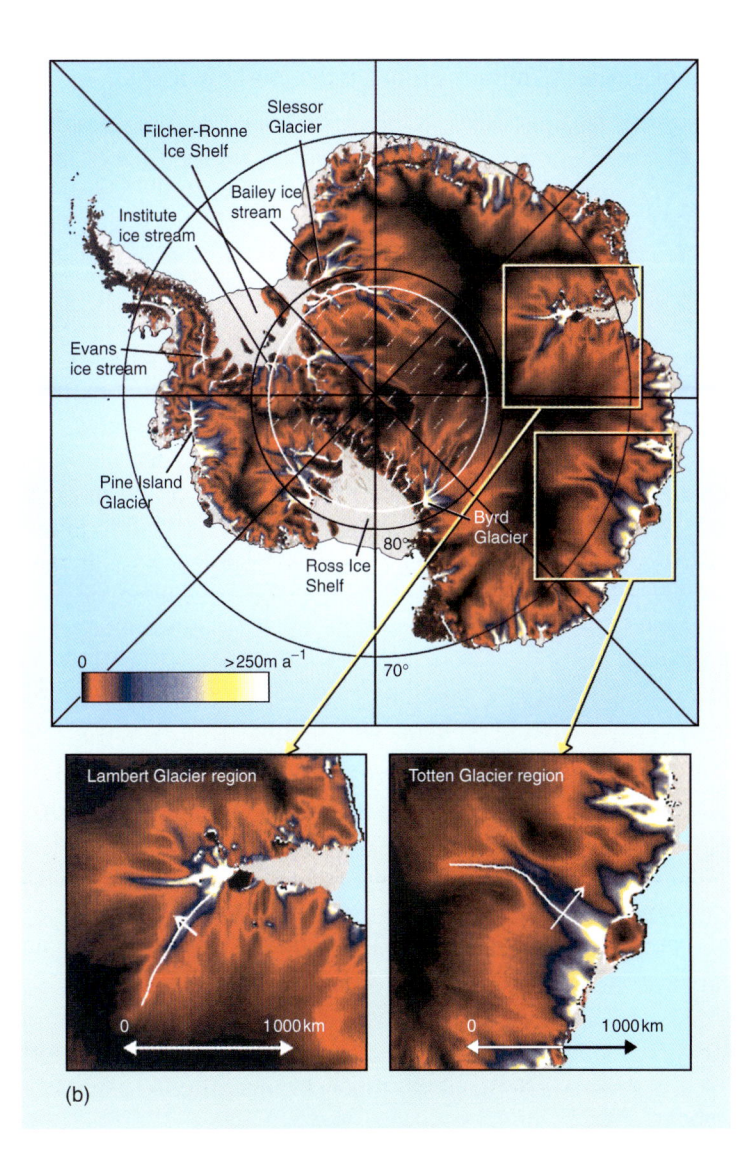

Figure 2.7 (a) Daugaard–Jenssen Gletscher, a fast-flowing ice stream at the head of the Scoresby Sund fjord system in east Greenland. (b) Estimated ice flow velocities in Antarctica. Note how much of the interior ice is very slow moving (dark colours), but how flow gathers into fast streams fringing the continent (bright colours).

Two notable ice streams that have attracted much attention from glaciologists are *Jakobshavns Isbrae* in west Greenland and *Ice Stream B* on the Siple Coast of West Antarctica. The drainage basin of Jakobshavns Isbrae extends from the highest part of the Greenland Ice Sheet, a distance of about 550 km from the glacier terminus. The final 14 km takes the form of a floating glacier tongue confined by fjord walls. Inland from the **grounding line**, the point where the glacier base is no longer in contact with seawater, it extends for about 75 km as a 6 km-wide ice stream with heavily crevassed margins. Ice thicknesses are 1900–2600 m and the whole bed of the ice stream is below sea-level. The surface is relatively steep and driving stresses are unusually high at 200–300 kPa (2–3 times typical glacier values). The velocity is 4 km yr^{-1} around the grounding line, rising to 7 km yr^{-1} at the floating terminus: the highest constant recorded glacier velocity. Very unusually, internal deformation — enhanced by the steep gradient and huge ice thicknesses — probably dominates the motion of this fast-moving glacier, as seasonal flow rate variations are not observed.

Ice Stream B is about 300 km long and 30–80 km wide, with 5 km-wide shear zones separating it from slow-moving ice on either side. Ice thicknesses are 1000–1500 m and the bed is below sea-level. Measured velocities are as high as 830 m yr^{-1}, compared with less than 10 m yr^{-1} in the ice ridges on either side. In contrast to Jakobshavns Isbræ, the high velocities on Ice Stream B occur at unusually low driving stresses (mainly due to a low surface gradient) of no more than 20 kPa, indicating that basal motion must be responsible for the flow of the ice stream, as this level of driving stress would produce no significant deformation in ice. Samples of sediment recovered from beneath the ice stream appear to deform significantly at only 2 kPa, so it has been suggested that subglacial sediment deformation accounts for most, if not all, of the motion. As this sediment is so weak, it has been suggested that the motion of the ice must be resisted by so-called subglacial **sticky spots**, where bedrock protrudes through the sediment layer.

2.3.5 Ice shelves

Bays at the marine margins of large polar ice sheets are often filled by large, flat expanses of ice, known as **ice shelves,** which are supplied by flow from inland accumulation centres (Figure 2.8). An ice shelf can be defined as a large, thick sheet of ice floating on the sea but attached to land or to a grounded ice sheet at the grounding line. Ice shelves are composed of glacier ice, not sea ice. Ice is supplied to ice shelves primarily by flow from ice sheets, in addition to surface accumulation from snowfall or basal accumulation from freezing of seawater onto the underside of the ice shelf. Calving of icebergs and melting at the base are the normal means of ablation.

There are some small ice shelves in the Arctic, but the most important examples are found in Antarctica. The largest is the Ross Ice Shelf, at about 530 000 km^2. Ice thickness varies from over 1000 m at the grounding line to about 250 m at the seaward margin — a 30 m ice cliff. Accumulation is 0.07–0.25 m w.e. yr^{-1}, and basal melting is believed to be about 1 m w.e. yr^{-1} at the seaward margin. Typical velocities are a few hundred metres per year, with a maximum of about 1000 m yr^{-1} at the ice front. It has been suggested that ice shelves are important in controlling the stability of ice sheet margins, protecting them from disintegration by rapid iceberg calving, although it is fair to say this is a contentious issue; it is considered further in Section 6.

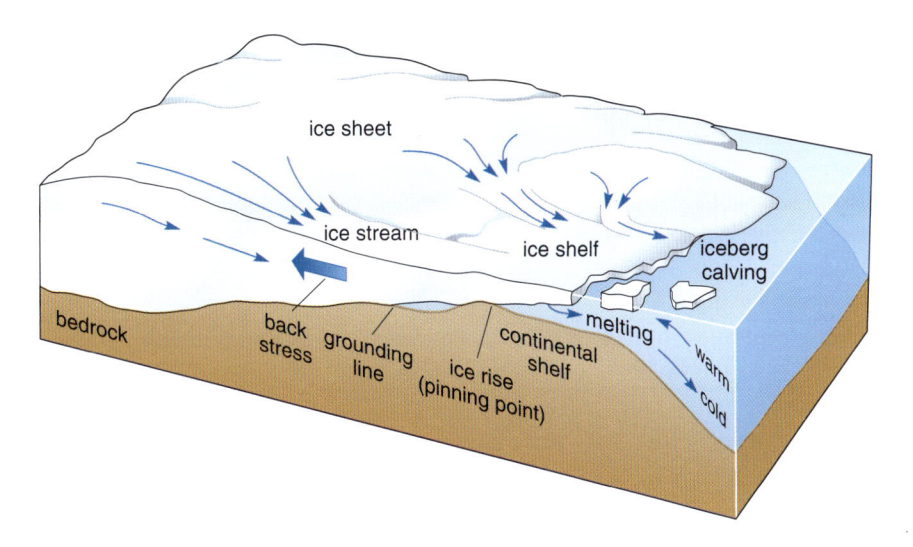

Figure 2.8 Schematic view of ice sheet–stream–shelf relationship. The mass balance of the ice sheet depends on snow accumulation and ablation, which are often wind-driven (surface melting is negligible in much of Antarctica, but more important in Greenland), and subglacial melting and freezing. Under floating ice shelves, circulating ocean waters can drive melt rates in excess of 10 m yr^{-1}. The form of the ice sheet also depends on ice flow, which may increase more than one hundredfold in the distance between the slow interior and the rapidly flowing ice streams.

2.4 Glaciers and the landscape

If you have visited an area which contains glaciers, you cannot have failed to notice the jumble of rock debris that accumulates around the glaciers' margins (Figure 2.9). If you haven't visited a glacier then you will still be familiar with the idea of glaciers carrying rock material which grinds away at mountains and wears them down over time. Much of the UK (among other places) is covered by the sedimentary deposits of former glaciers (Figure 2.9), which have often been important sources of sand and gravel for the aggregates industry. The purpose of this section is to look at the influence of glaciers on the landscape, and particularly to ask how effective glaciers are in eroding the land surface. As ever, the task of answering this question is complicated by variations in the response of glaciers to climate change.

(a)

(b)

Figure 2.9 Glaciers are major agents of change in the landscape. (a) Modern (at Hessbreen, Svalbard) and (b) ancient (at Tyndrum, Scotland) glacial moraines. The 'hummocky' moraine at Tyndrum was deposited about 10 000 years ago by retreating ice of the Loch Lomond Stadial, a short, cold period of about 12 000 years ago, in a similar environment to that exhibited at Hessbreen today.

2.4.1 Rates of denudation of the landscape by glaciers

Although techniques for investigating processes beneath glaciers have progressed enormously in recent years, direct observation remains difficult. Rates of erosion are therefore frequently inferred from the **sediment yield** of a glacier, that is, the mass of sediment per unit area transported from the glacier by meltwater. This is relatively easy to measure and should reflect the rate at which sediment is produced by erosion beneath the glacier. Rates of erosion estimated from glacial sediments yield vary enormously in different glacial environments. (Figure 2.10). The slowest rates are found in polar glaciers and thin, temperate plateau glaciers on crystalline bedrock (0.01 mm yr^{-1}), while the highest rates are found for large and fast-moving temperate valley glaciers in the tectonically active (therefore erodable) ranges of southeast Alaska (10–100 mm yr^{-1}). It is still difficult to interpret the erosion from sediment yield and three factors have to be taken into account (Box 2.3).

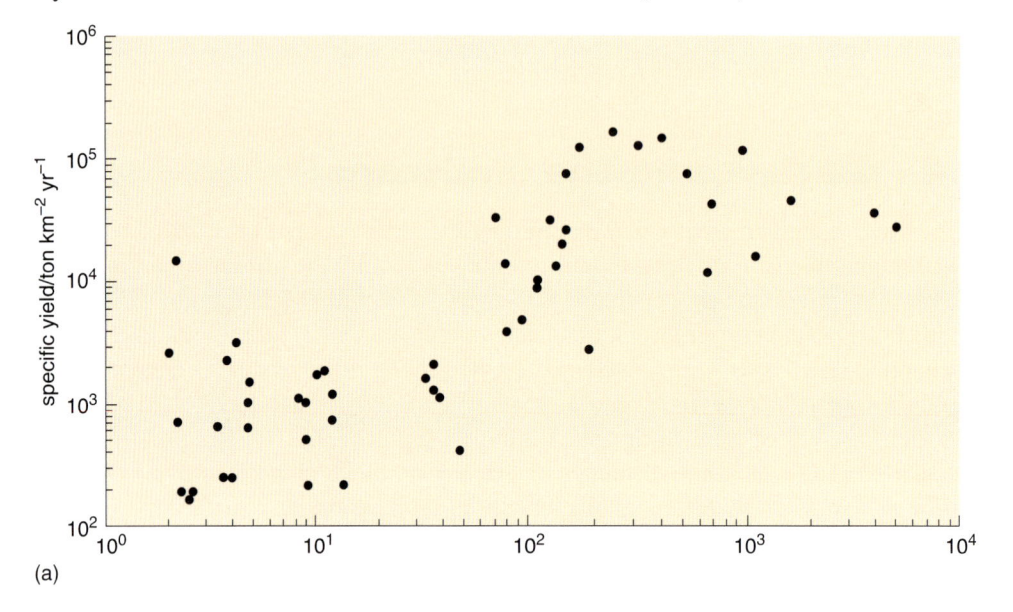

Figure 2.10 Sediment yields (a), and equivalent erosion rates (b) for a range of glaciers worldwide; note that the horizontal and vertical scales on both figures are expressed in powers of 10. Low sediment yields and rates of erosion (bottom left-hand corners of the figures) tend to be characteristic of smaller, less active glaciers on stable, resistant bedrock; high yields and rates tend to be characteristic of larger, more active glaciers in tectonically active areas.

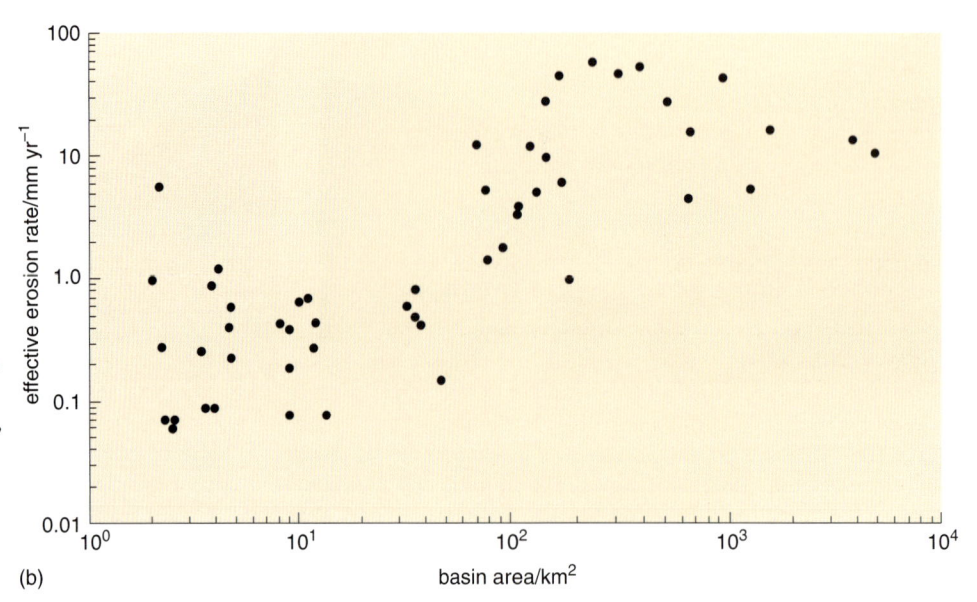

Box 2.3 Relative rates of contemporary glacial and non-glacial erosion: problems of interpretation

(1) *Defining a measure of glacial erosion intensity that is more useful than the proportion of the catchment that is glacier-covered.* The rates of glacial erosion and sediment output from a glacier are influenced by a wide range of variables, including sliding speed, ice thickness and flux, and meltwater production and routing. From an analysis of the characteristics of glaciers and their catchments in the Valais region of Switzerland, Gurnell found that glaciers based on unconsolidated sediment gave significantly higher sediment yields than those based on coherent bedrock. It seems that the nature and evolution of the glacial drainage system have a significant impact on the amount and timing of sediment transport: stable subglacial channel systems are associated with lower rates of sediment transport than the more distributed and rapidly changing subglacial systems.

(2) *Recognizing the importance of sediment storage and paraglacial sedimentation.* The relationship between present sediment yields and present glacial erosion is likely to be masked by changes in subglacial and subaerial (i.e. at the surface) sediment storage. Church and Ryder coined the useful concept of paraglaciation, the idea that modern landscapes could still be influenced by past glacial episodes (Figure 2.11). Contemporary sediment yields are likely to reflect the redistribution of sediments eroded during former glacial advances. Conceptually, sediment yields are elevated following the onset of glacier retreat because of the uncovering of abundant, readily transported sediments, and then decrease at a diminishing rate, eventually converging with background (non-glacial) denudation rates. Estimates of the relaxation time required for postglacial erosion rates to merge with non-glacial rates range from a century to many tens of thousands of years, depending on the size of the catchment, since larger catchments have greater capacities for storing sediment (Figure 2.11). Church and Slaymaker drew attention to this effect from a study of British Columbia, where they suggested that there is a 'disequilibrium' of Holocene (post ice age) sediment yield.

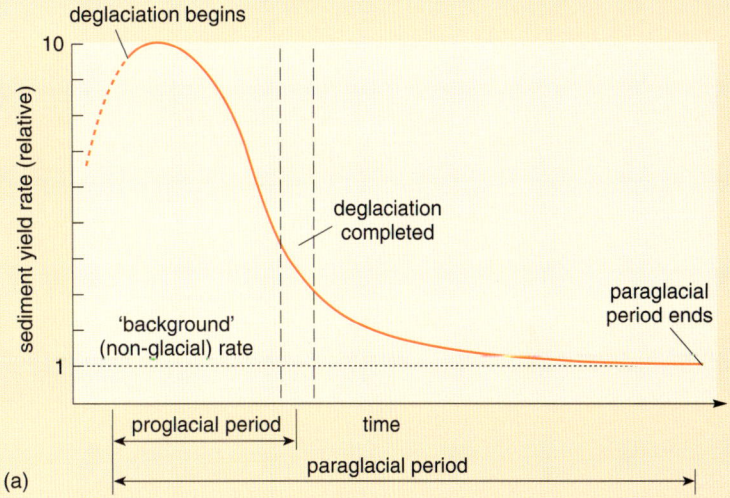

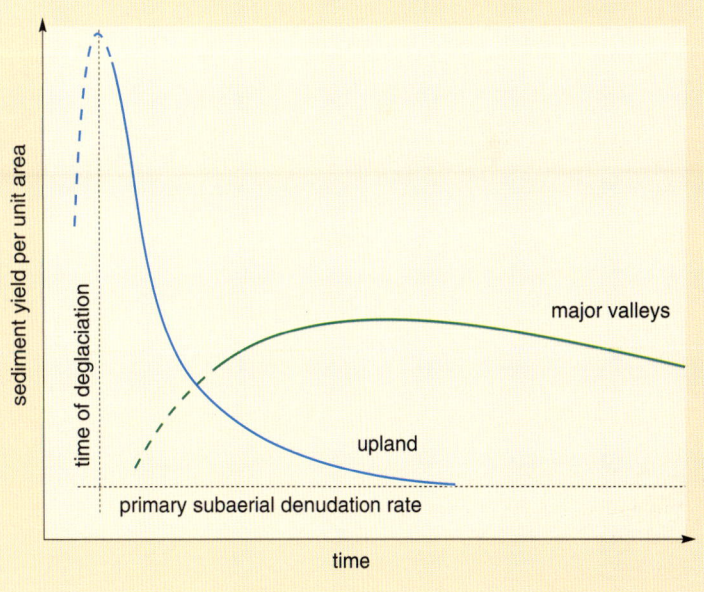

Figure 2.11 The paraglacial concept. (a) The paraglacial cycle of sediment yield, proposed by Church and Ryder. (b) A modification of the cycle proposed by Church and Slaymaker, accounting for the effect of spatial scale (catchment area). Larger catchments store more sediment than smaller ones, and take longer to complete the paraglacial cycle, many unglacierized, mid-latitude catchments can still therefore be described as paraglacial.

(3) *Accounting for the inherent variability of sediment output from glacierized basins.* Calculations of sediment yield typically rely on measurements of the mass of sediment transported from the catchment by suspension in turbulent meltwater. It is simple to acquire and analyse such samples, but much more difficult to ensure that these samples are representative of the sediment transport regime. In particular, meltwater discharge and sediment transport can change enormously over a few hours, or often less, while there can often be significant differences in rates of sediment transport from one melt season to another. Because most glaciers are relatively remote, the sort of intensive, long-term monitoring required to quantify this variability is rarely implemented. However, comprehensive sediment 'budgets', like that shown in Figure 2.12, are valuable for understanding sources, transport and storage of sediment in glacierized catchments.

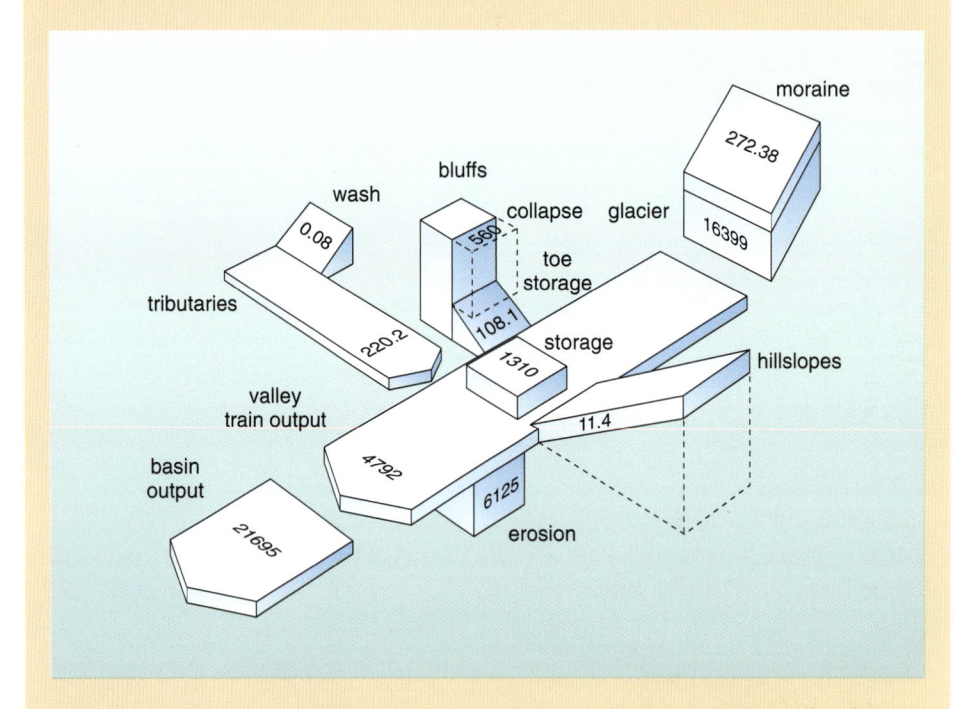

Figure 2.12 Sediment budget for the Bas Glacier d'Arolla catchment, constructed by Warburton. All inputs, stores and outputs of sediment have been quantified for a period of just over 2 months (25 May–30 July, 1987). Glacial inputs to the system are shown at the top right, fluvial outputs at bottom left. There was a net loss of sediment from the catchment, largely from fluvial erosion of the valley train. All values are in tonnes.

From this consideration of the impact of ice at the surface of the Earth, we now turn our attention to ice below the surface.

2.5 Summary of Section 2

1 Snow and ice assume great importance in the hydrology of catchments where they occur, altering the response of a catchment to meteorological inputs such as precipitation, and strongly influencing the seasonal pattern of runoff. Snowfall, for instance, accumulates during winter and melts during spring–summer, so that runoff is low or even absent in winter, increases greatly with the onset of spring–summer, then dwindles again as the snow cover diminishes.

2 The precipitation and runoff pattern described above is the opposite of that experienced in many catchments where wet winters and dry summers give higher and lower rates of runoff during the year, respectively. Catchments with partial glacier cover typically experience a compensating effect, whereby runoff is supplied by winter rainfall in the non-glacierized parts of the catchment and by summer melting in the glacierized parts, yielding a relatively constant streamflow regime valuable for water resource exploitation.

3 The way in which water drains at the base of a glacier has a significant impact on the flow of ice, with high subglacial water pressures promoting rapid motion by sliding (in which high basal water pressure offsets the overburden pressure of the ice) or by the deformation of sediments (in which high basal water pressure weakens permeable sediments). Low subglacial water fluxes generally flow in a spatially extensive, distributed pattern via films, linked cavities or porous flow in permeable sediments. High subglacial water fluxes generally flow in a spatially restricted, channelized pattern via conduits incised into the base of the glacier, or eroded down into sediment/bedrock.

4 Rapid ice flow is a key feature of ice streams — fast-flowing regions that develop within otherwise slow-flowing ice sheets. These dominate the transfer of interior accumulation to the margins of the large ice sheets, and are therefore important in determining the mass balance of these ice sheets.

5 Glaciers erode, transport and deposit rock debris and can have a dramatic impact on the landscape.

Question 2.1

In what ways does the hydrology of a glacierized catchment differ from that of a non-glacierized catchment?

Question 2.2

Given what you have learned about the relationship between water flux and glacier drainage, how would you expect a subglacial drainage system to develop in the course of a melt season?

Learning outcomes for Section 2

Now that you have completed your study of Section 2 you should be able to:

2.1 Identify those variables that determine melt rates and account for variations in melt rates.

2.2 Describe and account for the differences between distributed and channelized drainage systems.

2.3 Outline the mechanisms of glacial flow by (i) internal deformation, (ii) basal sliding, and (iii) subglacial sediment deformation, and identify situations in which each is likely to occur.

2.4 Identify the difficulties in the calculation of the sediment yield from a glacier.

Ice underground

3.1 The periglacial environment

Snow and glaciers are very obvious manifestations of the cryosphere at the Earth's surface, but the cryosphere also extends below the surface. In the UK, weather forecasters will often warn of 'ground frost' on sufficiently cold nights, when soils will freeze to some depth, potentially causing problems for gardeners, for sporting events that rely on turf pitches or courses, or in extreme cases for utility pipelines. Such conditions are usually short-lived in the UK, but in colder parts of the world sub-surface freezing can be much more persistent, and often permanent, and is important in the weathering of rocks or buildings, and in the formation of deeply frozen ground. Environments strongly affected by sub-surface freezing are usually described as **periglacial**. Generally, there are two important features of a periglacial environment: intense frost action (e.g. weathering of soil or bedrock), and perennially frozen ground, or **permafrost** (see below). Approximately 25% of the Earth's land surface currently experiences periglacial conditions. During times of maximum ice extent (which last occurred about 12 000 years ago), an additional 20% can be affected to varying degrees. This section will therefore focus on periglacial environments, where ice is often out of sight but nevertheless highly influential in terms of hydrology and landscape-forming processes.

The periglacial environment can be defined as that where the mean annual air temperature (MAAT) is less than +3 °C, but may be further subdivided into areas in which frost action dominates (MAAT less than –2 °C), and in which frost action occurs but is not necessarily dominant (MAAT between –2 °C and +3 °C). Processes in periglacial environments can be thought of in terms of:

- those unique to the environment (relating mainly to the formation of permafrost);
- those that are effectively unique by virtue of their high intensity and efficacy under periglacial conditions (segregation ice — described below, seasonal frost action, frost weathering, various forms of mass movement); and
- otherwise ubiquitous processes which exhibit distinctive aspects in periglacial environments (such as river and wind processes).

Regions of intense frost action and areas underlain by permafrost do not coincide perfectly, although they generally tend to be closely related. Periglacial environments extend over two major vegetation zones: the subarctic and northern forests, and the Arctic tundra and ice-free polar desert zones. However, there are also various high-altitude periglacial areas in many major mountain ranges, the largest being in western China on the Qinghai-Xizang or Tibetan Plateau. High-latitude periglacial environments tend to be a Northern Hemisphere phenomenon, simply because there is very little high-latitude land in the Southern Hemisphere beyond the Antarctic continent, which is of course almost entirely ice-covered.

3.2 Permafrost

The formation of permafrost is the most important of the periglacial processes and particularly important in terms of economic development. Construction in areas of permafrost can cause disturbance and melting of frozen ground, often leading to serious subsidence, which may require expensive preventative or remedial work (Figure 3.1). Permafrost can be defined as ground that remains at or below 0 °C for at least two years. However, it is not necessarily synonymous with 'frozen ground', since ground may be below 0 °C in temperature and still unfrozen, as a result of freezing point depression caused by factors such as solute-rich groundwater, pressure, or the presence of clay minerals. Furthermore, significant quantities of unfrozen water can persist in otherwise frozen ground (this is explored further below). Therefore, to differentiate between the temperature and state (i.e. frozen or unfrozen) of permafrost, the terms **cryotic** and **non-cryotic** have been proposed, referring solely to the temperature of the material, irrespective of its water or ice content. Permafrost is perennially cryotic ground, and may be unfrozen, partially frozen or completely frozen. It underlies more than 20% of Earth's land surface, including 80% of Alaska, 50% of Canada (almost 6×10^6 km^2), 50% of the Former Soviet Union (11×10^6 km^2) and 20% of China (more than 2×10^6 km^2) (Table 3.1, Figure 3.2). Non-cryotic material is above 0 °C, irrespective of ground ice/water content.

Figure 3.1 The trans-Alaska oil pipeline. For much of its course the pipeline is routed above ground because of the presence of permafrost. The units on top of the vertical pillars are automatic refrigeration systems which help maintain the permafrost. The pipe is able to move laterally on the supporting, horizontal beam in the event of ground movements.

Table 3.1 Extent of permafrost.

Major permafrost regions	Area of region/10^6 km^2	Permafrost area/%
Former Soviet Union	17.1	50
Canada	10.0	50
China	9.6	22
Greenland	2.2	100
Alaska, USA	1.5	82
World land area	148.4	20–25
Northern Hemisphere:		
Continuous permafrost	7.6	
Discontinuous permafrost	17.3	
Mountain permafrost	2.3	

In the very coldest regions permafrost is continuous, but as warmer, lower latitudes or altitudes are approached, its occurrence becomes discontinuous and subsequently sporadic. Its equatorward limit corresponds broadly with MAAT = −1 °C. Climates associated with permafrost generally have daily air temperatures below 0 °C for at

Ice underground

Northern Hemisphere permafrost zones

- sporadic and mountain
- discontinuous
- continuous
- subsea
- ice cap

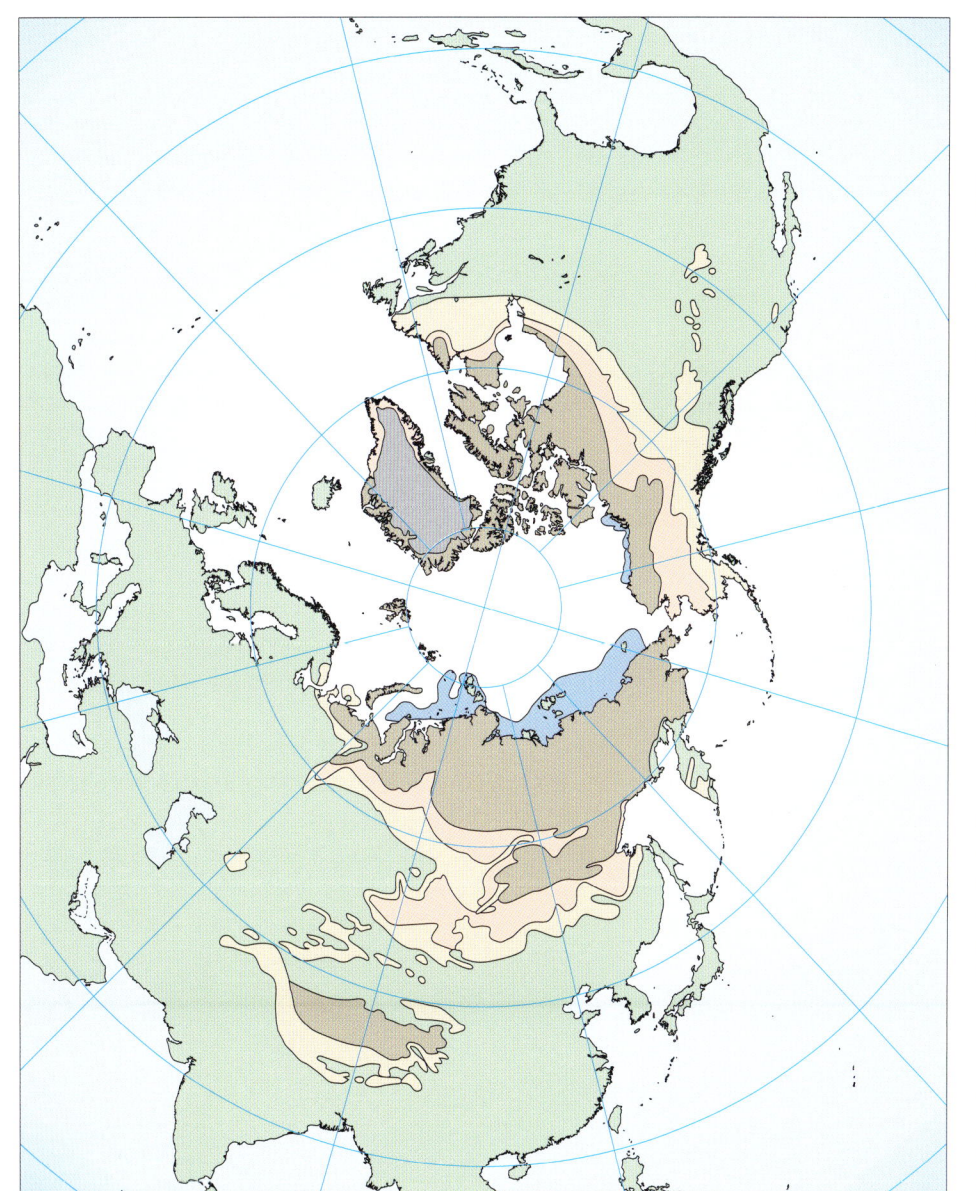

Figure 3.2 Location of permafrost in the Northern Hemisphere. Note the large area of permafrost on the Tibetan plateau in western China, and how permafrost extends much further south in the continental interior of Asia, compared with the relatively mild coast of northwestern Europe. There is little land mass at high southerly latitudes, other than that covered by ice in Antarctica.

least nine months of the year, and below −10 °C for at least six months of the year. Air temperatures rarely exceed 20 °C at any time, and annual precipitation is generally less than 500 mm. Permafrost is found much further south in the Asian continental interior than in milder, maritime-influenced northwest Europe (Figure 3.2). The minimum elevation at which permafrost occurs generally increases equatorwards, but small areas can be found on the highest peaks even at very low latitudes (Figure 3.3), for example above 4170 m on Mauna Kea, Hawaii, at 19 °N. Permafrost can reach a depth of more than 1500 m; it accretes very slowly (a few centimetres a year), so thick, present-day layers must have taken several thousands of years or more to develop. This suggests that permafrost depth and temperature is likely to be influenced by past as well as present climate conditions. Factors that influence temperature conditions in the ground and therefore the depth of permafrost include the presence and type of soil and vegetation, the existence of water bodies, surface insulation by snow or ice, and exposure to the sun and wind (Table 3.2).

Topic 4 Cryosphere

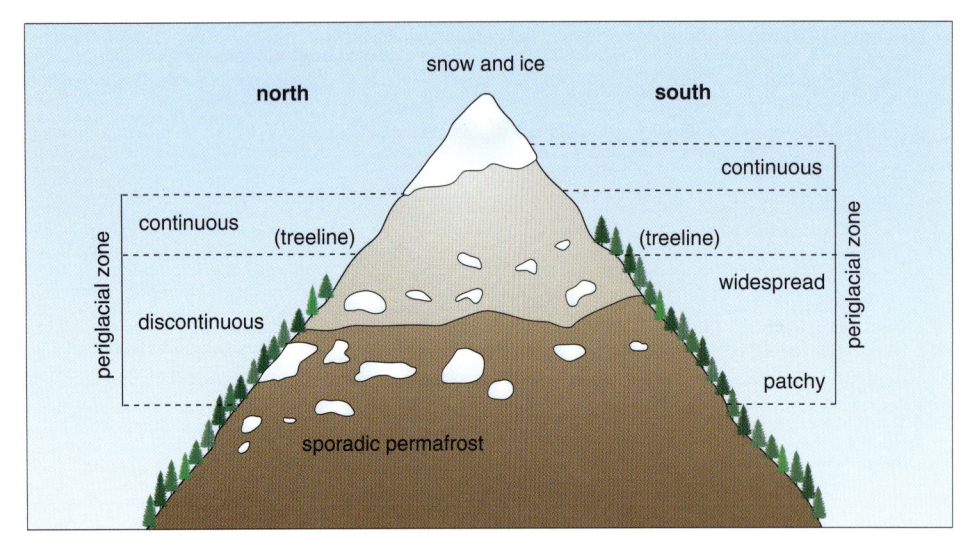

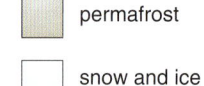

Figure 3.3 Schematic illustration of the distribution of permafrost in mountain environments. The effect of altitude broadly corresponds to the effect of latitude, giving a progression from sporadic to continuous permafrost with increasing height. North facing slopes receive less solar radiation than south facing ones in the Northern Hemisphere and therefore tend to be colder at equivalent altitudes, so permafrost can penetrate to lower elevations.

Table 3.2 Permafrost depth and mean annual air temperature at selected localities in the Northern Hemisphere.

Location	Latitude	Permafrost zone	Mean annual air temperature/°C	Permafrost thickness/m
Canada:				
Resolute, NWT	74° N	continuous	−12	390–400
Inuvik, NWT	69° N	continuous	−9	100
Yellowknife, NWT	62° N	discontinuous	−6	60–100
Schefferville, PQ	54° N	discontinuous	−4	80
Thompson, Man.	55° N	discontinuous	−4	15
Alaska:				
Barrow	71° N	continuous	−12	304–405
Umiat	69° N	continuous	−10	322
Fairbanks	64° N	discontinuous	−3	30–120
Bethel	60° N	discontinuous	−1	13–184
Russia:				
Nord'vik	72° N	continuous	−12	610
Ust'Port	69° N	continuous	−10	455
Yakutsk	62° N	continuous	−10	195–250
Qinghai-Xizang (Tibet) Plateau:				
Fenghuo Shan	34° N	widespread	−6	110
Wudaoliang	35° N	widespread	−5	40

In most areas of permafrost, the surface air temperature exceeds 0 °C for at least a brief period each year. Consequently there is usually a thin, so-called **active layer** overlying the permafrost, which is up to 3 m in depth, and freezes and thaws seasonally. As a result of unequal downward freezing rates in winter, unfrozen water may be trapped above the permafrost and below the frozen active layer. Such water pockets, called **taliks** (Figure 3.4), develop partly because of the release of latent heat of fusion as water freezes, and partly because of the build-up of cryostatic pressure, which is exerted as a result of volume expansion by freezing water. The significance of taliks is explored further below, in the context of permafrost hydrology.

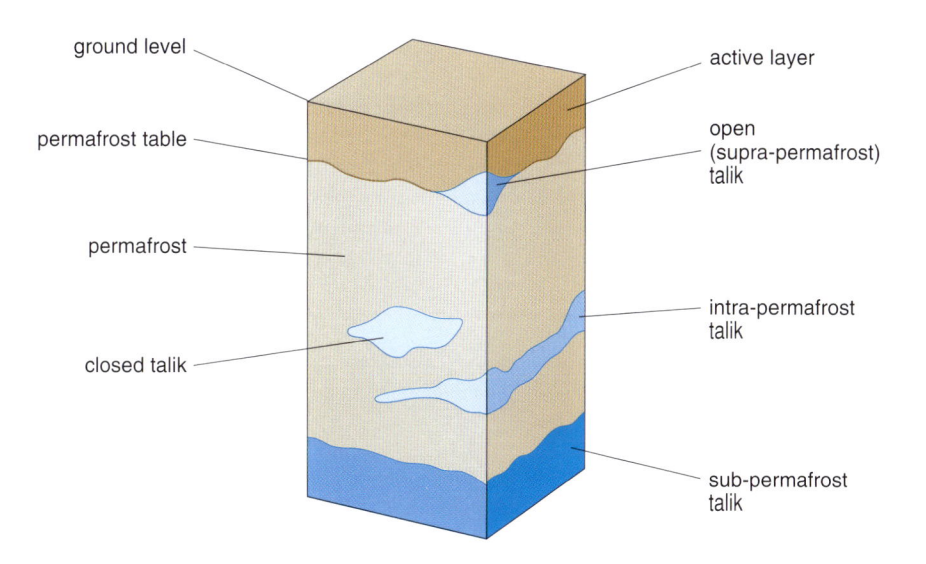

Figure 3.4 The relationship between permafrost, the permafrost table, the active layer and taliks in various locations.

So permafrost is often (but not automatically) associated with ground ice, which takes a wide variety of forms, ranging from a cement in soil pores to segregated masses of almost pure ice in veins, lenses and wedges. Ground ice is generally confined to the upper part of a permafrost layer, the maximum depth recorded being only about 50 m. Pore ice or segregated ice is formed depending on the rate of freezing, soil or sediment grain or pore size, and the potential for capillary suction (the ability of water to move towards the point in the soil where it is freezing — the freezing front). Coarse ground materials (e.g. gravel) are highly permeable but have low suction potential; fine materials (e.g. clay) with much smaller pores have low permeability but high suction potential. Intermediate materials (e.g. silt) provide the best balance for segregated ice formation. Grain or pore size also influences the temperature at which ice will form. The proportion of water remaining unfrozen in soil or sediment at a given temperature varies inversely with grain size, since the surface area on to which absorbed water is bound is greater for small particles. In clays only half the total water content may be frozen at a temperature of −2 °C, whereas in sands virtually all the water freezes at 0 °C. In any material, ice forms initially in the larger pores and only as temperature falls is it able to form in the smaller ones. Ice crystals grow within the material in the direction in which heat is being most rapidly conducted away, this is usually towards the surface. This is an important factor in the movement of soils or sediments on slopes, and is explored further below.

By now you will realize that there are many terms with the word 'glacial' in them. Box 3.1 describes their use.

Topic 4 Cryosphere

> **Box 3.1 Periglacial, paraglacial, proglacial: distinguishing ice-influenced environments**
>
> We have now accumulated a few terms with the suffix 'glacial'. It is important not to confuse them! The term periglacial refers to processes and features of cold, non-glacierized landscapes, *irrespective of their relationship to actual glaciers*. The term **paraglacial** (Section 2) refers to the persistent influence in deglacierized landscapes of glacier-related features (sediments and landforms). The term **proglacial** refers specifically to processes and features at glacier margins. These three similar-sounding terms therefore refer to distinctly different environments, although it is possible for them to overlap in space.
>
> As always, equilibrium with climate is a critical issue in these environments. Equilibrium periglacial environments, insofar as these can be found at all, are most likely to be found in cold regions which remained unglacierized throughout the last 15 000 years or so. The most extensive such areas occur in central and eastern Siberia, the northwestern Canadian Arctic and central Alaska. The Tibetan Plateau, also thought to have remained unglacierized (although very cold, this is also a very dry region) cannot be regarded as an equilibrium periglacial landscape, since it is relatively young, the result of exceptionally rapid crustal uplift of about 10 mm yr^{-1} for the last 130 000 years. Most of the boreal forest, tundra and high Arctic regions have only recently been deglacierized and are therefore unlikely to be in equilibrium.

3.3 Processes and features of periglacial environments

3.3.1 Frost weathering

○ Apart from becoming solid, what happens to water when it freezes?

● It also increases in volume (actually by about 9%).

The *in situ* mechanical breakdown of rock by repeated freezing and expansion of water within confined pore spaces, joints and bedding planes ('frost shattering') is considered to be an important feature of periglacial environments (Figure 3.5), yielding extensive surfaces of angular rock fragments (**felsenmeer**, from the German 'sea of rock', Figure 3.6). This simple 'freeze–thaw' model of frost weathering may be inadequate. Laboratory experiments have shown that freezing-induced fracturing occurs at temperatures between –3 °C and –6 °C, in the absence of freeze–thaw. A more realistic model, especially for soils, is the **segregation ice model** (which was referred to above in the context of ground ice), in which expansion is primarily the result of water migration to growing ice masses, and only secondarily to volume expansion. Weathering then results from the progressive growth of void spaces wedged open by ice growth. There is some experimental evidence to support this model, but a satisfactory degree of control over thermal and moisture conditions, and a sufficiently accurate measure of frost-induced weathering, are currently too difficult to achieve for entirely conclusive results.

Figure 3.5 A pervasively frost-weathered rock in Svalbard (lens cap for scale). This rock is a relatively young shale, which is soft and layered and therefore very susceptible to weathering.

Weathering is most effective when physical and chemical processes combine: rocks and soils are rarely subjected to a single mechanism. Solutes in water appear to be particularly important in enhancing the effectiveness of frost weathering. The damage caused by salt and frost together to roads, buildings and vehicles is well known. There are a number of reasons for this: salts accumulated in the outer layers of rocks by evaporation may block pores and seal the surface, preventing freezing pressures from being relieved; salt crystals may grow in combination with ice crystals; and the depressed freezing point of solute-rich water allows more time for susceptible materials to hydrate, expand, and disintegrate, while slower rates of freezing may allow larger ice crystals to grow.

Figure 3.6 *Felsenmeer*, an extensive surface of angular rock fragments, in Svalbard: an Arctic fox in its summer coat in the centre of the image gives a scale.

3.3.2 Impact on the landscape

Frost heaving is the vertical movement of ground materials due to the formation of ice. The pressure generated by ice crystallization (which, as discussed above, is probably more important than volume expansion), acts predominantly parallel to the direction of maximum temperature gradient, which is more or less towards the ground surface. Factors that control the susceptibility of materials to segregated ice formation also therefore control their susceptibility to frost heaving and thrusting, with silts generally being most vulnerable. Heaving is most evident where moisture is abundant and there is little vegetation to insulate the ground. A slow, downslope movement of soil and sediment takes place by frost creep and gelifluction (Figure 3.7), which are often difficult to distinguish. Frost creep occurs when particles are heaved up perpendicular to a slope by frost, but then fall back under the influence of gravity, the net result being a movement parallel to and down the slope. Gelifluction is a slow slippage of water-saturated soil or sediment.

Figure 3.7 An example of periglacial mass movement: gelifluction. In this case, thawed active layer material has slid *en masse* across the permafrost table. This type of movement is common in late summer in permafrost areas such as Svalbard, where this photograph was taken.

○ Why does the ground in permafrost areas often become saturated?

● Because of restricted drainage promoted by permafrost or a seasonally frozen water table, and water supplied by thawing snow and ice during summer.

Less-steep slopes, which tend to retain moisture, seem to experience faster rates of movement than steeper, freer-draining slopes. Rates of movement associated with gelifluction and frost creep are both generally below 50–100 mm yr^{-1}. Rates decline rapidly with depth, and movement seems to be confined to within 1 m of the surface.

Figure 3.8 Frost-crack polygons in peatland, Hudson Bay Lowlands, Manitoba. Thermal contraction of the ground initiates the growth of ice wedges. Brown edged polygons mark the location of massive ice wedges that extend from the surface down to 2–3 m.

Frost cracking is the fracturing of the ground by thermal contraction at sub-freezing temperatures, and is probably the main cause of the polygonal fracture patterns that are common in periglacial environments (Figure 3.8), although similar features can form by other processes. Frost crack polygons, best developed in areas of frozen ground rich in ice, are typically 5–30 m across. Crack frequency appears to decrease markedly with an increase in the depth of insulating snow. **Nivation** is localized denudation of the ground surface by a combination of frost weathering, gelifluction, frost creep and meltwater flow in association with snow patches, leading to the formation of nivation hollows (Figure 3.9). As the snow patch melts the zone of maximum nivation will follow its contracting margin. Data on denudation by nivation are scarce, but exceptionally large nivation hollows in Quebec, Canada, may have been excavated at a rate exceeding 1500 m^3 yr^{-1}.

Present-day periglacial areas characterized by high aridity are favourable environments for aeolian (wind) activity, and there is much evidence of intense wind action around the margins of the Northern Hemisphere ice sheets when they were at their maximum extent. Steep atmospheric pressure gradients from the centres to the margins of the ice sheets produced consistently strong winds, while low levels of precipitation and low temperatures supported minimal vegetation cover, and finely ground glacial debris was abundant. The result of this activity was the thick blankets of fine **loess** sediment, up to 100 m thick and tens of thousands of square kilometres in area, that cover many areas in North America and Eurasia located on the southern margins of the former ice sheets.

3.3.3 Permafrost hydrology

Although surface water in periglacial environments is frozen for much of the year, and precipitation totals are generally low, river processes still dominate periglacial landscapes. River regimes in such environments are highly seasonal, with very large discharges being sustained for short periods during spring snowmelt. The impermeability of permafrost typically causes a very high percentage of rainfall and snowmelt to contribute to surface runoff. Major lowland rivers may continue to flow throughout winter under an ice cover, but at greatly reduced discharges.

Figure 3.9 A nivation hollow in Okstindan, northern Norway. Denudation occurs around the margin of a contracting snowpatch by a combination of frost-related processes, leading to the formation of a hollow in the slope, which retains more snow in subsequent years, promoting further nivation in a self-perpetuating process.

○ Widespread flooding is common on the *northward*-flowing rivers of Arctic Canada and Siberia. How might seasonal freezing and thawing contribute to this?

● It is likely that these rivers largely freeze over during winter. The thaw occurs first in the southern section of the rivers, and meltwater becomes temporarily dammed behind longer-surviving ice accumulations further north.

A brief interval of high runoff yields higher river energy than a more extended period of lower runoff. Despite low annual precipitation totals, seasonal rivers in periglacial regions can therefore erode and transport sediment very effectively. The 32 km² Bayelva catchment near Ny-Ålesund on the northwest coast of Svalbard (Figure 3.10) is 54% glacierized, with permafrost some 150 m deep beyond glacial limits. With a MAAT of –6 °C and precipitation varying from 400 mm at the coast to 1000 mm at the watershed (with the very great majority falling as snow in winter) runoff occurs on average for 109 days of the year at a rate of 3.4 m³ s⁻¹, with a daily peak of 32 m³ s⁻¹. Annual transport of sediment suspended in turbulent river water varies from 6600–17 000 tonnes. (Look back at Section 2 and the comments on the variability of sediment yields!) Unglacierized high-latitude catchments do not experience runoff sustained by icemelt in late summer, and therefore exhibit even more seasonal regimes. For example, mean annual precipitation in the River Mechan catchment in Arctic Canada is only about 140 mm, half of which falls as snow. Some 80–90% of total annual river discharge is concentrated in a 10-day period in the summer months.

Figure 3.10 The Bayelva catchment near Ny-Ålesund in Svalbard, June 2000. The catchment extends from a glacierized mountain watershed at over 700 m in altitude to a permafrost-influenced fjord coast at sea-level. In the middle of the image is the shallow terminus of the glacier Austre Brøggerbreen.

Permafrost forms an impermeable layer in the ground, restricting the movement of groundwater to various taliks (see Section 3.2), which may be either cryotic or non-cryotic, and open or closed, depending on whether they are completely surrounded by permafrost or reach the seasonally thawed zone. Open taliks are common even within the continuous zone; most result from local heat sources such as lakes and rivers. Closed taliks usually result from a change in the ground thermal regime, such as might occur following the drainage of a lake and the downward accretion of permafrost.

The winter discharge of rivers generally decreases with increasing latitude, since an increasing percentage of runoff is stored as **icings**, sheet-like masses of ice which form at the surface in winter where water issues from the ground (Figure 3.11). Groundwater icings are commonly generated by taliks, which can sustain perennial water flow. River icings may develop where water is driven from the river channel as the water surface freezes and becomes loaded with snow, or where shallow, braided rivers freeze to the bed. Icings start to form as soon as streams are frozen, but the major period of growth (in the Northern Hemisphere) is from mid-winter to early spring. Small monthly accretions during early winter result from continuing discharge in river channels or through sub-river taliks, which freeze only in late winter. Maximum ice accretions involve sub- or intra-permafrost waters and occur in February and March when the ground is fully frozen and before the spring melt begins. During the early summer, icings usually melt away completely. Icings are also often associated with high-latitude glaciers which sustain winter discharge from stored summer meltwater or from basal melting.

Where active-layer freezing restricts perennial discharge from taliks, **frost mounds** can develop. Seasonal frost mounds are usually 1.0–4.0 m high, formed through the doming of seasonally frozen ground by the subsurface accumulation of water under high pressure during freezing of the active layer (a distinct process from the formation of segregated ice). Most frost mounds are seasonal, but some may survive one or more summers, if insulated by a soil cover. Similar forms may also occur at the surface of icings (Figure 3.11).

(a)

(b)

(c)

(d)

Figure 3.11 Images from an icing associated with Scott Turnerbreen in Svalbard (Figure 2.1). (a) The appearance of the icing at the end of winter (May 1993; snowmobile for scale). (b) A frost mound on the surface of the icing, created by the subsurface accumulation of water under high pressure. (c) 2–3 Months later than the first two images, summer melting has given rise to streams which are incised into the icing, in which accumulation layers can be seen. (d) Characteristic needle-like ice crystals (about 30 cm in length) exposed in the deteriorating icing during summer.

Having looked in these early sections at ice on and under the land surface, we shall now turn to ice in the oceans.

3.4 Summary of Section 3

1. As much as 25% of the Earth's surface is affected by periglacial conditions. These are characterized chiefly by intense frost weathering and by permafrost, or perennially cryotic ground (that is, ground at or below 0 °C, irrespective of water/ice content).

2. Permafrost may be continuous, discontinuous or sporadic, and takes thousands of years to reach equilibrium depths. Much current permafrost at high northern latitudes may be relict from the last time of maximum ice extent about 12 000 years ago.

3. Periglacial environments exhibit many distinctive landscape processes, including frost weathering, frost heaving, frost creep, frost cracking and gelifluction. The growth of bodies of segregated ice now seems to be a more important process in the weathering of materials than the volumetric expansion of water on freezing.

4. The hydrology of permafrost-affected regions is typically highly seasonal with runoff restricted to a few weeks in spring/summer, and is influenced by unfrozen groundwaters circulating in taliks, river freezing/thawing and consequent flooding, and by the formation of icings, which provide an important redistribution of water between winter and summer.

Question 3.1

What are the main differences between the hydrology of a temperate environment and a permafrost region?

Learning outcomes for Section 3

Now that you have completed your study of Section 3 you should be able to:

3.1 Provide a definition of periglacial environments at the Earth's surface in terms of MAAT and state.

3.2 Outline the differences between the 'freeze–thaw' model of the frost weathering and the ice segregation model.

3.3 Identify features of the landscape that result from periglacial and permafrost conditions.

3.4 Describe features of the hydrology of permafrost regions, including the seasonality of runoff, and the role of taliks and icings.

4 Ice in the oceans

4.1 Why the Arctic Ocean can be crossed on foot

You have probably come across the term 'the polar ice caps' in the media, most likely in relation to global warming and sea-level rise. This term implicitly suggests that similar, homogeneous bodies of ice lie at both poles: this is far from the case. A fundamental but surprisingly often overlooked fact is that the South Pole is located on an ice sheet overlying a continent, while the North Pole is located over an ocean, the surface of which is frozen.

There are two main types of ice found in the oceans, and they are very different. Sea ice is formed by the freezing of ocean surface water, it covers enormous areas at high latitudes, is highly seasonally variable in extent, and is very important in climate and ocean processes. The other type of ice encountered in ocean waters is the iceberg, famously the nemesis of the *Titanic*. Icebergs break, or 'calve', from the margins of glaciers that terminate in tidal water, usually in fjords, and make their way out into open water on tidal and ocean currents. The relationship between ice dynamics and iceberg calving is an enormously important factor in the stability of glaciers and ice sheets that terminate in tidal water. Though often encountered together, sea ice and icebergs are formed by entirely different processes, and each is important to the environment and to human activity in very different ways. The covering of the Artic Ocean by these components of the cryosphere means that you can actually walk across the frozen sea. The purpose of this section is to look further at the form and significance of the oceanic cryosphere.

4.2 Icebergs

4.2.1 The significance of iceberg calving

Tidewater glaciers (those which terminate in tidal water) tend to occupy sheltered locations in narrow bays with steep sides, called fjords, rather than flowing out into open waters, since iceberg calving limits their extents. The same is true of the large ice shelves in Antarctica (Sections 2 and 6), which are confined to bays.

- ○ With reference to Table 1.2, how important is iceberg calving as a mechanism of ablation of the major ice sheets?

- ● Iceberg calving is very significant, it accounts for almost all mass loss from the Antarctic ice sheets, which are the largest in the world.

Although it is easy to envisage blocks of ice calving from a glacier terminus, the specific mechanism is still not well understood, probably because of the difficulty (not to say danger) of making direct measurements, and the apparent randomness of major calving events. When a glacier reaches open water, wave and current action tends to remove ice from the terminus and induce calving by undercutting. Tidewater glaciers and ice shelves therefore terminate in characteristically near-vertical ice cliffs (Figure 4.1). If a glacier does for some reason advance into deeper water, the terminus may begin to float. This appears not to be a feature of temperate glacier environments, such as Alaska and Patagonia, where tidewater glacier termini remain grounded on the floors of fjords, but is relatively common in high latitude environments, such as Greenland and Antarctica. There are many implications for glacier dynamics stemming from whether the glacier terminus is grounded or floating, but it is probably fair to say that these are not yet fully understood.

4.2.2 The iceberg calving mechanism

Crevasses near glacier termini are important areas of weakness for calving, and tend to dictate the size of icebergs (Figure 4.2). Many icebergs calved from grounded tidewater glaciers are typically quite small (a few metres to a few tens of metres) and irregular in shape, although much larger icebergs may be calved during certain conditions of rapid retreat (see below). Large icebergs may be calved from the floating termini of high-latitude ice streams (Figure 4.2).

Figure 4.1 (a) An aerial view of the tidewater terminus of Daugaard–Jenssen Gletscher, Nordvestfjord, Scoresby Sund, east Greenland (see also Figure 2.7); (b) Panoramic fjord-level view of the tidewater terminus of Daugaard–Jenssen Gletscher, about 1000 m wide and 30 m high.

(a)

(b)

Figure 4.2 (a) Small (about 10–20 m) icebergs in Gaase Fjord, Scoresby Sund, east Greenland. (b) Large (about 500 m) iceberg in NordVest Fjord, Scoresby Sund, east Greenland. This iceberg was calved from the outlet glacier.

Calving from the large Antarctic ice shelves tends to take the form of very large, flat-topped, tabular icebergs as much as 10^4 km^2 in area (Figure 4.3). Iceberg calving appears to be a seasonal process in locations where surface melting occurs (generally outside Antarctica). Outlet glaciers in Greenland calve in summer, but advance during winter, when calving doesn't restrict the forward flow of ice. Likewise, Columbia Glacier, Alaska, calves only in summer, when subglacially emerging water undercuts and forms embayments within the ice front.

Calving usually occurs when stresses at the glacier terminus exceed the strength of the ice. Several theories of iceberg calving have been proposed, but in general terms the rate of calving can be related to the water depth:

$$u = c(H - h)$$

where u is the calving speed (the volume of icebergs calved per unit time per unit vertical area of glacier terminus) and $(H - h)$ is the water depth minus the thickness of the glacier at the terminus, both averaged over the width of the glacier and a year. The parameter c is a constant which varies from glacier to glacier. Figure 4.4 shows the relation between calving speed and water depth for a series of glaciers.

Deep tidal water promotes rapid and unstable calving, which is why glaciers and ice shelves are not able to advance out into the open ocean. The relationship between water depth and iceberg calving rate appears to give rise to a distinctive pattern of advance and retreat of grounded tidewater glaciers in Chile and Alaska, discussed in Box 4.1.

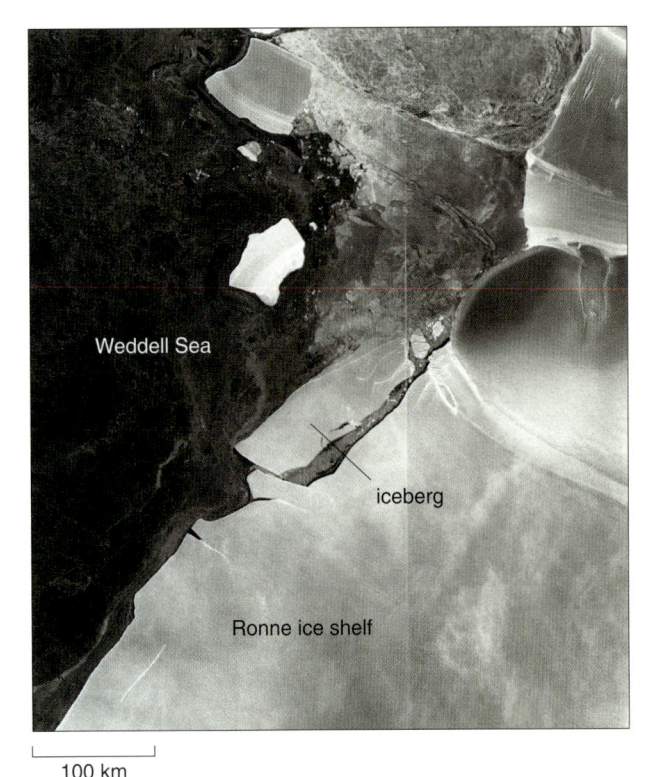

Figure 4.3 Satellite image of a large tabular iceberg calved from the Ronne Ice Shelf in Antarctica. The length of the iceberg is about the equivalent of the distance between London and Birmingham.

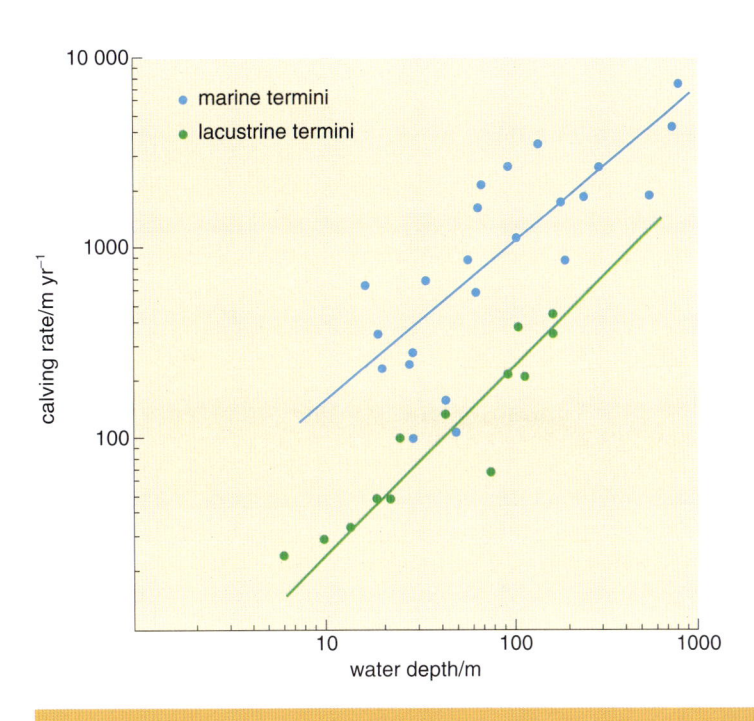

Figure 4.4 Water depth and calving rate for a selection of marine-terminating calving glaciers (blue circles) and lake-terminating calving glaciers (green circles). The rate of iceberg calving is clearly strongly related to water depth in both cases.

Box 4.1 The advance and retreat of tidewater glaciers: a consequence of iceberg calving

Many tidewater glaciers in fjords in Alaska have retreated dramatically since the early explorations of the region about two centuries ago. Ice extended to the mouth of Glacier Bay in 1750, but retreat had begun when Vancouver mapped the area in 1794 and has continued since, one glacier retreating by as much as 100 km. On the other hand, some Alaskan tidewater glaciers have been advancing steadily over the same period. The rates of advance, typically 20–40 m yr^{-1}, are much less than the rates of retreat, 200–2000 m yr^{-1}, although advances appear to have much longer durations than retreats. The advance–retreat cycle can be explained by variations in water depth at the terminus. Glaciers with stable termini are located in shallow water, usually at the head of a fjord, but sometimes on an underwater bank of sediment in otherwise deeper water. Because this stabilizes the terminus, calving is reduced and the glacier starts to advance. The advance can continue into deeper water only if the sediment bank is also moved forward, to restrict the calving rate. This can be achieved by erosion of the upstream side of the bank, with deposition of the eroded material on its downstream side. This is a slow process, so the advance is slow. Ultimately, the glacier's accumulation area is not able to supply ice at a sufficient rate to maintain the advance, while the increase in size of the ablation area resulting from the advance increases the rate of surface melting. A small retreat from an extended position moves the terminus back into deep water (particularly since a bank cannot migrate back up the fjord), the calving rate increases and the retreat accelerates unstably. Termini grounded in deep water are therefore usually retreating rapidly.

Figure 4.5 is an image of the terminus of the Perito-Moreno glacier; when the glacier front reaches the lake, it calves and retreats. It is estimated that advances may continue for perhaps 1000 years whereas retreats may be completed in a century or less. Because this advance–retreat cycle is not climatically controlled, adjacent glaciers may be in different stages of the cycle.

Figure 4.5 An image of the terminus of the Perito-Moreno glacier, Patagonia.

4.3 Sea ice

4.3.1 The formation of sea ice

Any open body of water will freeze at sufficiently low temperatures, and the sea is no exception. However, while fresh water generally freezes at 0 °C, the presence of salt reduces the freezing point of seawater to approximately −1.9 °C. Salt also increases the density of seawater, and cold, salty water tends to sink relative to warm, fresher water. This introduces a complication: because relatively cold water sinks, displacing relatively warm water to the surface, the entire depth of water has to be cooled to freezing point before any freezing can actually occur. At least, this would be the case without **stratification** (layering) of the water. In the Arctic Ocean, there is a low-salinity surface water layer, arising from the inflow of fresh water from major rivers, and this is relatively isolated from denser water at greater depths. This means that just the uppermost 50 m of water have to be cooled for ice to form.

The initial form of sea ice is a slurry-like suspension of ice crystals called **frazil ice** (Figure 4.6), which subsequently coalesce into other forms. As the ice crystals form, the great majority of the sea salt is rejected into the surface layers of the ocean below the ice, because ice of any form can't hold much solute in its crystal structure.

Figure 4.6 Newly forming sea ice (frazil ice) in the Greenland Sea. The viscous appearance is typical of freezing water.

This so called 'salt rejection' causes a local density increase in the surface water, which can lead to deep water formation under certain conditions (see below). Frazil crystals subsequently freeze together into a skin called **ice rind** or **nilas**, while the constant movement of water and wind break this layer into **pancakes** a few centimetres in diameter, with upturned edges resulting from multiple collisions. Further movement is dampened as pancakes coalesce into forms about 3–5 m in diameter and 0.5 m in thick. Ultimately **ice floes** (Figure 4.7), in excess of 10 m in diameter are formed. There is a standard but complex set of descriptive terms for sea ice types, which has been developed by the World Meteorological Organization, but by far the most common distinction is between **first year** and **multiyear** ice. First year ice is generally considered to be the product of one year's growth, up to approximately 0.75 m thick. Multiyear ice is that which has survived one or more summers, increasing in thickness each winter up to approximately 5 m thick. The quantity of salt trapped in the crystal structure of multiyear ice is even less than that of first year ice, as the surface of the ice may melt in summer, flushing out incorporated salt.

Figure 4.7 Ice floes in Scoresby Sund, east Greenland; these floes are of the order of 10 m across.

4.3.2 The distribution of sea ice

The contrasting characteristics of the Arctic and Antarctic have a considerable effect on the type of sea ice occurring around each pole. The central Arctic Ocean is an enclosed basin and retains a large, permanent, multiyear ice cover. At its minimum in September the ice covers an area of approximately 8×10^6 km^2, rising to a maximum of approximately 15×10^6 km^2 in March as the ice expands from the Arctic Basin to cover Hudson Bay, Baffin Bay, the Canadian Archipelago and the Kara Sea (Figure 4.8). Large portions of other peripheral seas are also covered, such as the Greenland Sea, Barents Sea and smaller areas of the Labrador Sea and the Sea of Japan.

On the other hand, Antarctica is surrounded by open ocean and the permanent sea ice cover is restricted to small regions of the Weddell Sea, the Bellingshausen Sea and the Amundsen Sea. Because of the smaller regions of permanent ice cover, Antarctica has a much smaller percentage of multiyear ice. At its minimum in February the ice covers approximately 4×10^6 km^2, but the comparative increase in area in winter is much greater, with the ice expanding from the continent to cover an area of almost 19×10^6 km^2 (Figure 4.9). The global sea ice coverage can best be observed remotely by satellites (Figures 4.8 and 4.9).

To calculate the volume of water stored in sea ice we must know its thickness and, compared to its extent, this is a very difficult measurement to make. It is especially difficult to obtain a representative mean. Most thickness measurements from the Arctic Ocean have been obtained from submarines using upward-looking sonar, and so far the measurements of thickness of the Antarctic sea ice have all been made via holes drilled from the surface. Near the beginning of Section 1, it was suggested that the cryosphere is enormously variable, and that its various components are important in different ways. Compared with the water stored in the world's glaciers and ice sheets, the volume of water stored in sea ice is negligible. Sea ice covers a vast area of the Earth's surface. In the Arctic winter, sea ice can occupy almost 5% of the entire Northern Hemisphere, and in the Antarctic winter, sea ice can occupy as much as 8% of the Southern Hemisphere. On a global scale, approximately two-thirds of the ice-covered areas of Earth are covered by sea ice, which exerts a significant effect on climate, as discussed below.

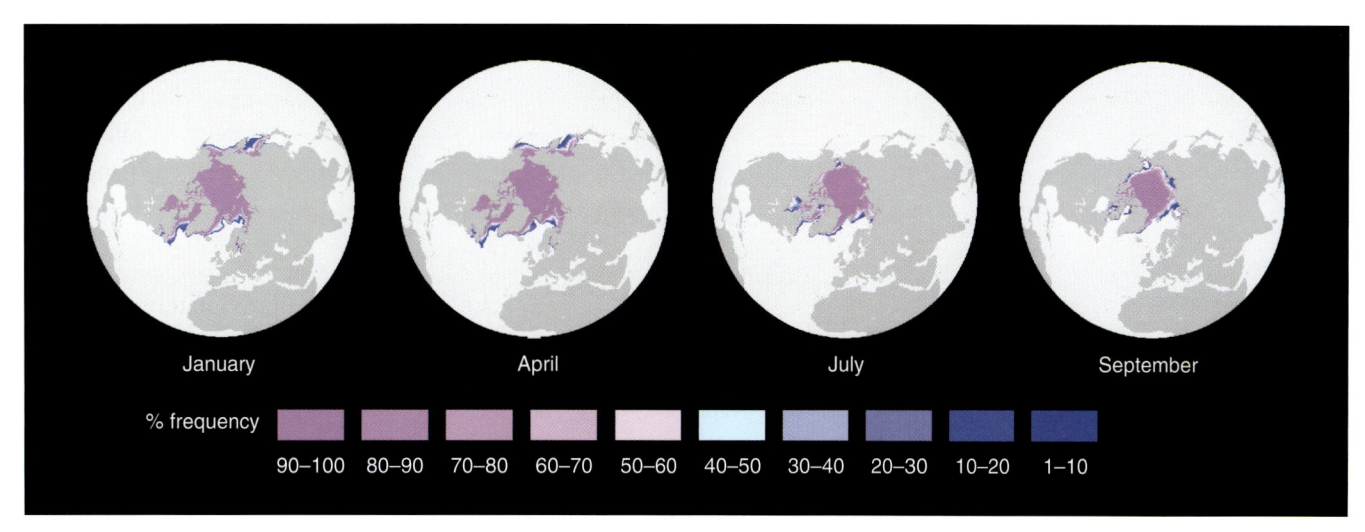

Figure 4.8 Mean seasonal variations in sea ice extent in the Arctic and surrounding oceans, 1978–1995. The key shows the frequency of ice coverage in given months. The colours show the extent of ice cover in any given month.

Topic 4 Cryosphere

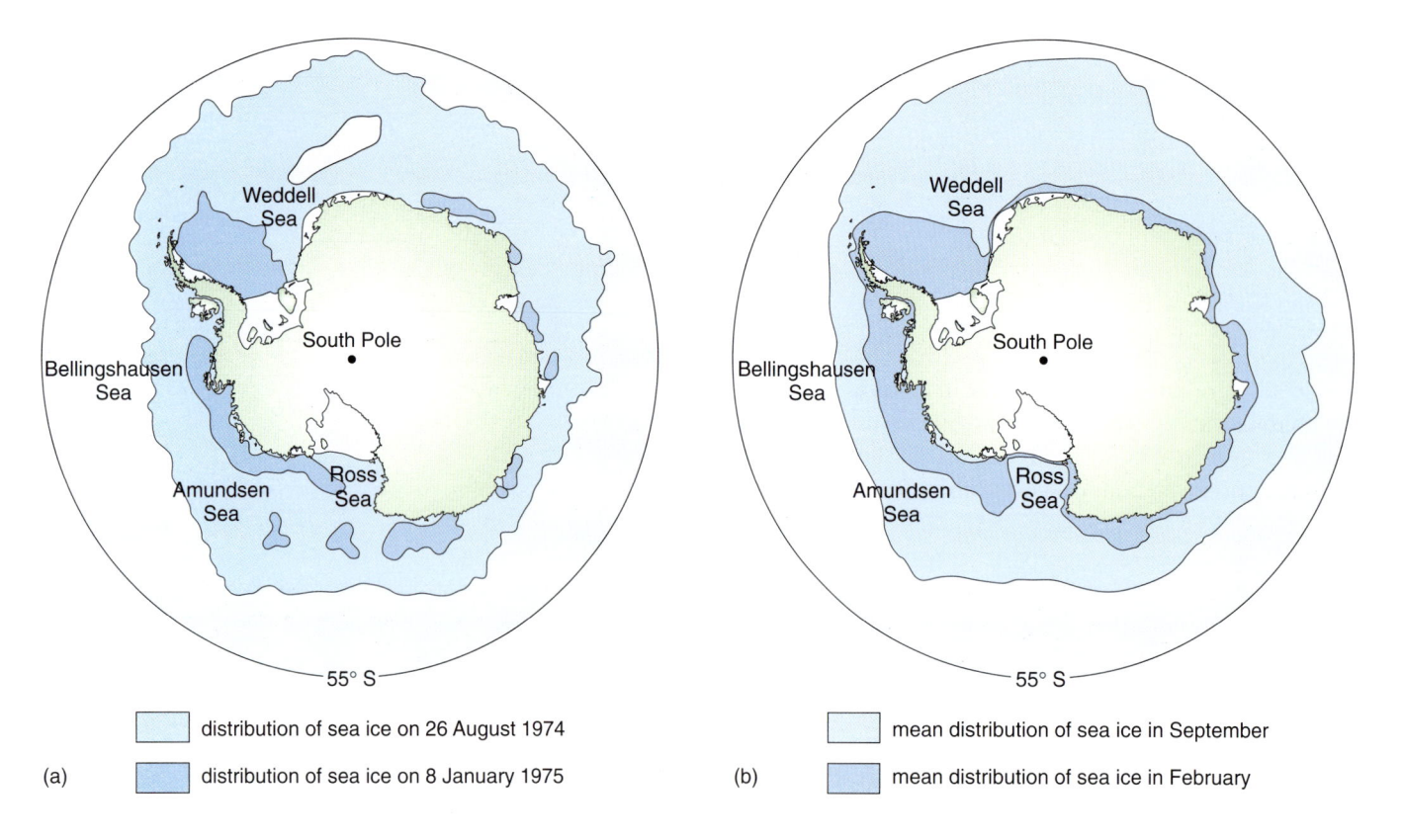

(a) distribution of sea ice on 26 August 1974
    distribution of sea ice on 8 January 1975

(b) mean distribution of sea ice in September
    mean distribution of sea ice in February

Figure 4.9 Seasonal variations in sea ice extent in the Antarctic: (a) August and January 1974–1975, (b) September and February 1978–1987 average.

4.3.3 Sea ice and climate

At the most simple level, a cap of sea ice seals off the ocean surface, limiting exchanges of energy and mass between the atmosphere and ocean, and stabilizing the climate system, reducing heat loss from the ocean in winter.

○ Sea ice covers large areas of the Earth's surface; in what way does it affect receipts of energy at the surface?

● The ice cover reflects a large proportion of incoming solar radiation — up to 55% for bare sea ice and as much as 90% for snow-covered sea ice (see Section 2), which has a significant impact on the global atmospheric energy budget.

The reduced light penetration and increased albedo of the frozen ocean surface have the effect of further depressing temperatures, imparting a stronger seasonality to ocean life than would be observed otherwise. Reduced sunlight curtails photosynthesis and therefore biological productivity. An important point to make is that the ice cover of the polar oceans is not continuous and complete. In the summer of 2000 reports of open water at the North Pole led to concern about the effects of global warming. However, this is actually not an unusual occurrence. Sea ice is a complex and flexible medium and is moved, broken and piled up by a combination of winds, ocean currents and (at the edge of the ice cover) wave energy. Pressure ridges up to 10 m high form from the collision of ice floes in the Arctic Ocean; Antarctic ridges tend to be a little smaller due to the more open nature of the southern ocean and the lower concentration of multiyear ice. These forces move the ice to create open water, even in winter, in the form of linear cracks called **leads** (Figure 4.10) and large areas of persistent open water called **polynyas**. The winter open water, although small in area (up to 5% of the winter pack), is of great importance for the heat balance of the polar oceans. Rates of heat and moisture loss to the atmosphere, ice growth and salt rejection to the underlying ocean are as much as two orders of magnitude greater over a refreezing lead than over multiyear sea ice. Vigorous heat and moisture fluxes between the relatively warm water in leads and the much colder atmosphere cause the characteristic 'steaming' effect known as **frost smoke** (Figure 4.11).

Figure 4.10 A lead system in the Arctic ice photographed from the air.

Figure 4.11 Frost smoke forming in a region of open water in the Arctic Ocean.

4.3.4 Sea ice and ocean circulation

Oceanographers now believe that the circulation of water through the world's oceans is linked through a series of strong currents displacing water globally, driven by **deep water formation** as a consequence of salt rejection during sea ice formation in the polar seas and heating of water in the tropical seas of the central Pacific Ocean. This circulation is called the **thermohaline circulation** (*thermo*, temperature and *haline*, salt), or the **ocean conveyor belt**, because of the way in which water slowly flows back and forth at shallow and deep levels (Figure 4.12). The formation of sea ice and salt rejection (see above) are probably the principal force driving the deep water formation at the 'cold end' of the oceanic conveyor belt, and are also thought to be important in the exchange and absorption of carbon dioxide from the atmosphere to the ocean. In the central Greenland Sea, the ocean loses heat to the atmosphere, rapidly forming a distinctive sea ice feature known as the *Odden Ice Tongue*. Associated salt rejection from the Odden is believed to drive deep convection, and over thousands of years this leads to the circulation of cold, dense water from the Arctic Ocean to the Pacific Ocean via deep currents through the Atlantic and Indian Oceans, returning warm waters to the Arctic.

Topic 4 Cryosphere

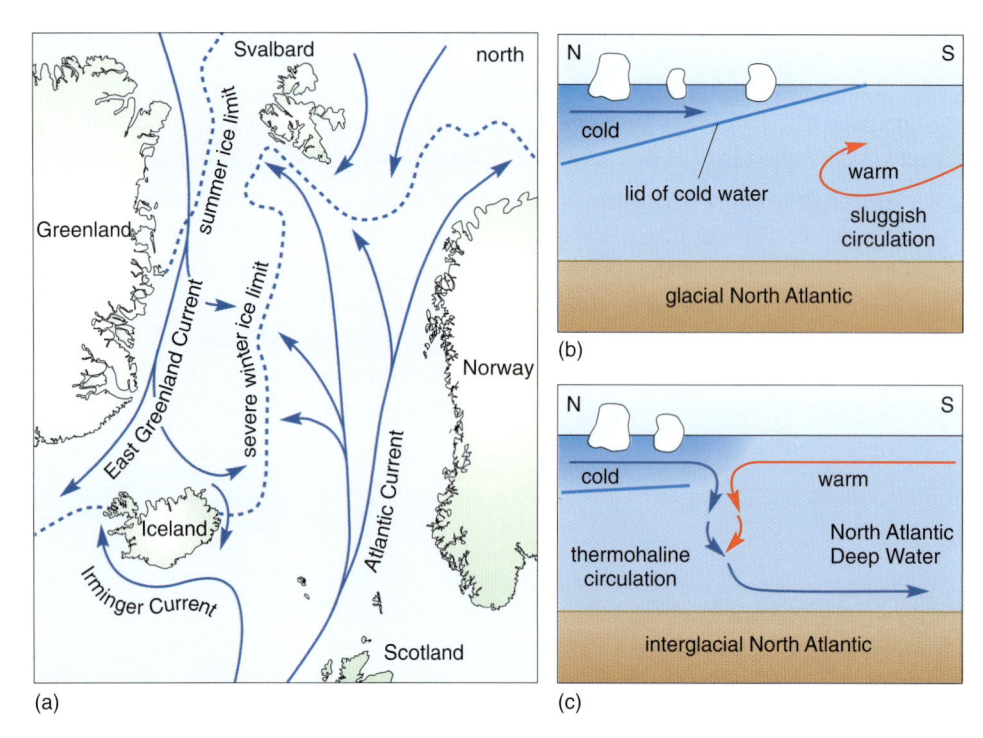

Figure 4.12 (a) The thermohaline circulation in the North Atlantic, and its relation to (b) cold (glacial) and (c) warm (interglacial) climate episodes.

There is much evidence from records of past climates (such as ice cores recovered from the centre of the Greenland Ice Sheet) to suggest that there have been repeated, rapid variations in the Earth's climate of several degrees Celsius in the space of a few decades over the past 100 000 years or more, which have been linked to changes in the pathways or even the modes of operation of the thermohaline circulation (Figure 4.12a). The release of heat energy to the atmosphere associated with deep water formation in the North Atlantic ocean is equivalent to up to one-third of the solar energy input to this region: an enormously important contribution to the relatively mild climate of western Europe. This effect is believed to be less strong during ice-age episodes of major glacial advance (Figure 4.12b) than during warm intervals such as the present (Figure 4.12c). A broad community of researchers has therefore turned its attention to features that may perturb the thermohaline circulation. Given its importance in deep water formation, it is quite plausible that changes are associated with modifications of the sea ice cover.

Finally, in addition to the scientific issues discussed above, routine observations of both sea ice, and iceberg locations and behaviour, are required for operational purposes in the shipping, fishing and offshore hydrocarbon extraction industries, and for meteorological forecasting.

We now go on to look at the interactions between the cryosphere and human populations.

4.4 Summary of Section 4

1. Icebergs, calved from glaciers in fjords or from ice shelves, and sea ice, formed by the freezing of seawater, are the major components of the cryosphere in the oceans.

2. The stability of tidewater glaciers and ice sheets is related to the calving of icebergs. Deep water promotes rapid and unstable calving, placing a limit on the extent to which glaciers and ice shelves can expand into the ocean, and controlling the advance–retreat cycle of tidewater glaciers.

3. Sea ice is a complex, highly seasonally variable medium, which constitutes about two-thirds of the global cryosphere in terms of area, although it is very thin compared to glaciers and ice sheets. Sea ice has a major impact on the global albedo and on energy exchange between the atmosphere and oceans.

4. Salt rejection during sea ice formation gives rise to deep water formation, which is a critical component of global oceanic circulation.

Learning outcomes for Section 4

Now that you have completed your study of Section 4 you should be able to:

4.1 Describe the relationship between water depth and iceberg calving rates.

4.2 Outline how sea ice forms, and comment on its seasonal variation in both the Northern and Southern Hemispheres.

4.3 Describe the effects of sea ice on albedo, and heat and moisture exchange, and the relevance of these to climate.

4.4 Outline the role of sea ice in deep water formation.

5 Ice and people

5.1 For better or for worse

'Dangerous glaciers impinge directly on the lives of people in mountain regions, and some have been responsible for huge loss of life … despite this, glaciers provide considerable benefits to human society, notably in terms of providing a reliable water supply and energy resources, and not least because they are of considerable value to the tourist industry.'

Hambrey and Alean, 1992

Unless you venture into the mountains in winter, the cryosphere in the UK tends to be associated with inconvenience rather than genuine hazard (most obviously with snow and ice disrupting transport). Likewise, little use is made of the cryosphere as a resource in the UK, except perhaps by mountaineers and skiers, and for the occasional playground snowball fight. Nevertheless, there are many other parts of the world, particularly mountainous regions, where the cryosphere is of great, practical importance in the day-to-day lives of people, not all of whom are glaciologists or ski instructors. Interactions between the cryosphere and people will be examined in this section by dividing them into the negative interactions, which constitute hazards, and the positive interactions in which some aspect of the cryosphere constitutes a valuable, sometimes indispensable, resource.

5.2 Cryospheric hazards

5.2.1 Glacier advances and unstable glaciers

The destruction of farmland during the Little Ice Age by advancing glaciers was mentioned briefly in Section 2. Though the majority of glaciers worldwide are currently retreating, glacier advances in historical times have caused great problems for human populations, and could conceivably do so in future. During the Little Ice Age destruction of settlements and farmland was widespread as glaciers advanced in the Alps in the first half of the 17th century and in Scandinavia in the late 17th to early 18th centuries. This appears to have occurred suddenly in some cases; for example, records show houses being exchanged in Le Châtelard in 1600, though the village was almost completely destroyed by the advancing Mer de Glace by 1610. In Iceland, settlements around Vatnajökull that had been in existence since AD 900, such as Fjall, had to be abandoned around 1695, and were buried under ice by 1709. Probably more ruinous than direct glacier advances were the effects of expanded melt streams. Cultivated land and productive fishing rivers were often ruined due to erosion or to the deposition of infertile sand and gravel. One Norwegian farm was recorded to have 'suffered very great damage from the terrifying river which comes from the glacier … which has washed away most of their best arable and pasture, leaving nothing but rocks, scree and grit, which can never be cleared or be of any use'.

As a rule, even the most destructive glacier advance at least has the virtue of occurring at a relatively steady rate. However, there are important and very dangerous exceptions. Surges are cyclic, rapid motion events which occur at about 4% of glaciers worldwide, largely concentrated in clusters in Alaska–Yukon, the Karakoram, Pamir, Tien Shan, Caucasus, Andes, Iceland and Svalbard. There are no surge type glaciers in, for example, the Rockies or the European Alps. Surges are caused by internal instabilities of the glacier system and entail the relatively sudden, rapid redistribution of mass in a glacier which has generally been flowing steadily and unremarkably. Enhanced ice flow rates of up to 100 m d^{-1} have been observed. This redistribution often (though not necessarily) leads to a brief, rapid advance of the glacier. References to events interpreted as glacier surges can be found in traditional stories. One such story suggests that a surge of the Lowell Glacier in the Yukon Territory was caused by a shaman seeking revenge on the people of a village who had insulted one of his elders. The Alsek River was dammed by the advance, and the subsequent flood killed the entire population of that village. During the winter of 1936–1937, the 3 km-wide front of Black Rapids Glacier in Alaska, which was formerly smooth and gently sloping, advanced 66 m d^{-1} as a heavily broken ice cliff 100 m high. The advance stopped just short of damming a major river and severing the Richardson Highway (which links Fairbanks to the rest of North America). As surges are known to be cyclic phenomena, a future surge of Black Rapids Glacier is anticipated. Any future advance would threaten the trans-Alaska oil pipeline, so a programme of monitoring is underway. However, most concentrations of surge type glaciers are fortunately relatively remote from human populations.

5.2.2 Ice avalanches

Ice avalanches, in which blocks of ice break from a glacier and fall down a mountainside, are rare compared with snow avalanches, but less predictable, and therefore potentially highly dangerous. The principal threats are to mountain settlements, ski resorts and hydro-electric power installations. Aerial photographs have been used to identify and map the effects of former ice avalanches, particularly their **run-out distance** (the distance a detached ice block is likely to travel).

- ○ Would you expect run-out distances to be greater in winter or in summer?

- ● Run-out distances tend to be greater in winter, when a snow cover smoothes the path of the ice avalanche.

Certain glaciers generate frequent, small ice avalanches which serve as a guide to likely hazards. However, many ice avalanches occur without warning and are consequently catastrophic. One of the most notorious ice avalanche disasters occurred in 1965, during the construction of a dam for a hydro-electric power plant at Mattmark, Valais, when about a million cubic metres of ice broke from Allalingletscher (Figure 5.1), sweeping through a construction workers' camp and killing 88 people. A positive outcome of such events is that at least they serve as warnings for future hazards and suggest courses of remedial action. When a hydro-electric power dam was proposed at Mauvoisin, Valais, estimates were made of the waves that might be generated by ice avalanching into the lake from Glacier de Giétro, and the dam constructed so that they could be contained.

Topic 4 Cryosphere

Figure 5.1 Allalingletscher, August 2000. In 1965, during the construction of a dam for an HEP plant at Mattmark, Valais, 88 construction workers were killed by an ice avalanche from this glacier, which has since significantly retreated. However, the glacier is clearly very steep, and the light grey mix of ice and rock debris at the foot of the slope (indicated by an arrow) testifies to a continuing hazard.

The Rio Santa valley in the province of Ancash in the Peruvian Andes has experienced many terrible, glacier-related disasters, including massive ice avalanches. The highest mountain of the Cordillera Blanca, Nevado Huascaran (6768 m), is particularly unstable. In 1962 about 4000 people were killed in an ice avalanche, then in 1970 a large earthquake triggered an even greater avalanche. Approximately 50×10^6 m^3 of rock and ice broke off and buried an area up to 16 km from its source in under three minutes. The town of Yungay was buried, with about 18 000 people killed there alone.

5.2.3 Glacier floods

Flooding from ice-dammed lakes has been enormously important in landscape development. Flows of the order of 10 000 000 m^3 s^{-1} have been estimated from the glacial Lake Missoula during the last ice age, forming the 'channelled scablands', a 5000 km^2 area of intense scouring in eastern Washington State, USA.

○ How may contemporary glaciers impound water to form lakes?

● Glaciers might store water within their mass or at their margins, between the ice and valley sides, or between the ice and moraine ridges downvalley.

Lakes tend to fill steadily during the melt season, and perhaps over many melt seasons, until sufficient head is generated to force a route out beneath, through or over an ice dam. Sudden draining of glacier-dammed lakes or of water stored within glaciers has been reported from Iceland, Norway, Switzerland, Pakistan, Greenland, Alaska, Canada, South America and New Zealand. Probably the most common situation of lake damming is where a glacier blocks a stream draining a side valley, although a glacier advance out of a side valley can also block drainage from a major valley and unstably dam a large lake.

Surges of the Plomo Glacier in the Argentinian Andes have led to the formation of ice-dammed lakes, which have subsequently broken through the ice dam to cause flooding downstream. In 1934, seven bridges and 13 km of the Mendoza–Santiago railway were destroyed, and several people killed, when peak flood discharges reached fifteen times normal levels. The worst recorded glacial lake outburst disaster occurred in the Rio Santa valley in the Peruvian Andes (also mentioned above); 6000 people in the market town of Huaraz drowned in a flood in 1941.

The Icelandic word **jökulhlaup** ('glacier runs') is often used to described outburst floods from glaciers, reflecting the frequency and importance of such events from the ice-clad volcanoes of that country. Jökulhlaups drain regularly from six subglacial, geothermal areas in Iceland. From the Grímsvötn volcanic caldera in Vatnajökull, the largest ice cap in Iceland (length 150 km, average thickness 420 m, maximum thickness 1000 m), jökulhlaups have occurred at a 4–6 yr interval since the 1940s, with peak discharges of 1000–10 000 m^3 s^{-1} and 2–3 weeks duration. Prior to that, about one jökulhlaup occurred per decade with an estimated total discharge of 5 km^3 of water and a peak discharge of approximately 30 000 m^3 s^{-1}. Jökulhlaups do not always coincide with eruptions; the high heat flow from the volcano can melt large volumes of ice on its own. The peak flow in 1934 from Grímsvötn was 50 000 m^3 s^{-1} or about one-quarter the flow of the Amazon, flooding an area of 100 km^2. Even larger jökulhlaups may have occurred in the past. At Grímsvötn, a geothermal area melts ice at the base of the ice cap to form a subglacial lake, creating a depression in the ice surface which also collects surface meltwater. The lake, almost entirely covered by a floating 'ice shelf' about 220 m in thickness, drains catastrophically to the coastal plain through a subglacial tunnel 50 km in length. Drainage stops when the lake level has fallen about 100 m, which is much less than the full depth of the lake. The water regains the draining level every five or six years. Water drains at an increasing rate for about ten days, then drops away rapidly in about two days. It has been suggested that the lake starts to drain when the water becomes deep enough to float ice separating the lake and tunnel, although this has been questioned. The speed of these processes can be increased through the interaction of glacial ice with volcanoes (Box 5.1). A particularly dramatic jökulhlaup occurred in November 1996 on Skeidarasandur following a very violent eruption from beneath Vatnajökull. Unusually, discharge rose in only a few hours to a peak of 45 000 m^3 s^{-1}, then receded over a few days. This extreme flood caused up to US$15 million damage to roads, bridges and power lines.

Box 5.1 Glaciers and volcanoes

The juxtaposition of great heat from volcanoes and great cold from snow and ice gives rise to some spectacular, but dangerous, interactions. Volcanoes supporting glaciers are common in the Andes of South America, in North America's Cascade Range, on the highest peaks in Mexico (e.g. Citlatepetl, 5700 m), throughout the Siberian Kamchatka Peninsula, the Aleutian Islands and, as noted above, in Iceland. The world's southernmost active volcano is Mount Erebus in Antarctica (3794 m), which is entirely glacier-covered. Many other volcanic peaks support a snow cover at altitude (e.g. Mount Fuji, Japan, 3776 m; Mauna Kea, Hawaii, 4206 m).

A 1986 eruption of Nevado del Ruiz in the Colombian Andes would have caused little damage but for its glacier cover. Ice and snow that melted during the eruption mobilized huge quantities of volcanic ash in a mud flow. This flowed at great speed into the heavily populated valley below the volcano, completely burying the town of Armero with the loss of 30 000 lives. This was all the more tragic since the hazards associated with an eruption had been accurately predicted, but no remedial action taken. Probably the most famous volcanic eruption of recent times was that of Mount St Helens in the Cascade Range of Washington State in 1980. In the violent eruption the height of the mountain was reduced from 2949 to 2549 m, and 70% of the mountain's glacier volume was lost. In this case, damage from mud flows was limited, as precautionary measures had been taken, including lowering the water level of a reservoir used for hydro-electric power generation so that it was able to contain flood waters. Mount St Helens is neighboured by Mount Rainier, 4391 m in altitude with more than 90 km^2 of glacier cover, including the largest American glacier outside Alaska, Emmons Glacier (11 km^2). Although dormant since the mid-19th century, there is evidence of Holocene mud flows up to 110 km in length, filling valleys to depths of up to tens of metres.

5.2.4 Snow avalanches

Almost certainly, the most notorious hazard associated with the cryosphere is the snow avalanche. There are three basic types of snow avalanche (Figure 5.2).

- Powder avalanches consist of a powdery mass of partially airborne snow, usually involving only recently deposited snow.
- Slab avalanches consist of slabs of settled snow that become detached and slide downhill. They usually contain a greater volume of snow and travel longer distances than loose powder avalanches. Slab avalanches may be dry or wet, depending on the water content of the snow. Dry slab release occurs when the stress exerted by an overlying snow cover exceeds the strength in a weak, buried snow layer, leading to failure of that layer and detachment of the slab. Most dry slabs are released by new snow loading (i.e. following heavy snowfalls), or may be deliberately or accidentally triggered by people. Wet slab avalanches tend to occur mainly towards the end of winter when air temperatures are high enough to give melting and rainfall, and are due to increased loading with increasing snow cover wetness, and a reduction in the strength of snow layers following water penetration.

Figure 5.2 (a) Powder snow avalanche, Colorado, USA; (b) slab avalanche, also in Colorado; (c) small avalanches at the head of Scott Turnerbeen, Svalbard (see also Figure 2.1) contributing to accumulation (surveying pole and target in foreground); (d) cornice collapse, Bolterdalen, Svalbard. A cornice is an overhanging wind-blown snow mass at the crest of a slope.

- Slush avalanches are relatively rare, flow-like movements of water-saturated snow. These almost always occur in high latitudes and tend to initiate upon impermeable surfaces following rapid melting. They may be important in opening Arctic glacier drainage systems early in the melt season.

Any individual avalanche results from the complex interaction of a range of factors, but two broad types of avalanche environment can be recognized overall: maritime and continental. Maritime climates have high precipitation totals and relatively mild temperatures (the type of conditions that give rise to a high mass balance gradient: Section 1). Intense orographic precipitation results in heavy snow cover loading during storms, while rain may occur at any time during the year. This may produce dry or wet snow avalanches depending on the nature of the precipitation. Avalanche release in maritime regions therefore tends to occur during or following storms in the fresh, less stable recent snow layers. Mild temperatures promote intergranular binding, which enhances the strength of the snow cover, so that deep-seated instability is relatively rare in maritime snow covers. When it does occur, it usually results from the freezing of rain into ice layers which do not bind firmly to the surrounding snow and act as potential sliding surfaces (Figure 2.2). In continental climates, lower precipitation quantities and intensities give a less close relationship between storms and avalanching. Instead, instability may persist long after storms have passed because of buried weak layers in older snow, meaning that an understanding of snow cover structure is as important as weather conditions for avalanche forecasting.

The formation of weak layers at depth within a snow cover is crucial for avalanching. Ice layers may form in maritime climates, but the conversion of snow crystals into new forms with low cohesion is common in continental climates. In dry snow, there are two main processes by which crystal form can be altered. Overburden pressure can pack snow layers, increase binding and promote snow strength, but this is only important in very dense snow. The other process depends on the existence of vapour pressure gradients, which arise because of environmental temperature differences between snow layers. This occurs in dry, alpine snow covers, as these are warmer at their bases (due to geothermal heat) than at their surfaces (which are chilled by the atmosphere). Water vapour therefore diffuses towards the surface, evaporating from the top of one crystal and condensing at the base of one above, adding facets to the base of that crystal. This process is favoured in continental climates, where suitable temperature gradients are common because of colder air temperatures and thinner snow covers. **Faceted crystals** therefore tend to form weak layers in the snow cover. Such crystals may form at the very top of the snow cover where local temperature gradients are high, and subsequently become buried by new snowfall. **Surface hoar** develops at night when a temperature inversion above the snow cover produces a flux of water vapour into the snow cover. Buried ice and faceted crystal layers are therefore the major sources of weak layers in snow. When these layers are loaded by new snow following storms, avalanche release, including accidental release by people (Figure 5.3), becomes likely.

	Avalanche Danger Level Descriptors **United States**		
Descriptor	**Probability of Avalanches** **Natural and Man-Made**	**Terrain Considerations**	**Decision Making Considerations**
LOW (green)	Natural - Very Unlikely Human Triggered - Unlikely	Generally stable, but isolated areas of instability are possible	Generally safe. But remember to always use caution and to use safe travel practices.
MODERATE (yellow)	Natural - Unlikely Human Triggered - Possible	Areas of instability are possible on steep terrain	Use extra caution on certain aspects and/or slope angles as described in the current bulletin.
CONSIDERABLE (orange)	Natural - Possible Human Triggered - Probable	Areas of instability are probable on steep terrain	Be extra cautious in steep terrain, consider avoiding areas of highest hazard as described in the current bulletin.
HIGH (red)	Natural - Likely Human Triggered - Likely	Areas of instability are likely on a variety of aspects and slope angles	Travel in avalanche terrain is unwise. Stay on lower slope angles or safer aspects as defined in the current bulletin.
EXTREME (black)	Natural - Certain Human Triggered - Certain Both - Widespread	Instability certain and widespread. Large avalanches are likely	Avoid avalanche terrain, including runout zones. Seek out low angle terrain unthreatened from above.

Avalanche Safety is up to You!

Avalanches don't happen by chance, and most people who get caught trigger the avalanche themselves. Even small slides can be dangerous or fatal, so watch for "terrain traps". Select your route wisely and carry avalanche rescue gear. Base your decisions on the "big picture" which should incorporate all information and observations. Be willing to modify your big picture as new observations are made and/or conditions change.

Take an avalanche course from a professional. A list is available on line at http://www.csac.org/Education/

Cyberspace Snow and Avalanche Center - www.csac.org

Figure 5.3. Avalanche danger level description system used in the USA. Note that natural and human-triggered avalanches are distinguished, and that the latter are always more likely than the former.

An important point is that avalanching should not be thought of as rare and exceptional, but as a normal and regular mountain process. It is therefore possible to install avalanche protection in areas that are known to be vulnerable (Figure 5.4). Furthermore, avalanching leads to the redistribution of snow, and can have important hydrological implications. In particular, avalanches are a potentially important source of accumulation in glacier mass balance, as observed in Section 1, but also cause important changes in the properties of the deposited, post-avalanche snow, compared to the pre-avalanche snow, which are liable to influence melt rates and the hydrology of the snow cover. There are four possible changes to consider.

1 Avalanched snow is deposited at a lower altitude than that at which it accumulated, and will therefore experience higher air temperatures with greater energy for melting.

2 The snow deposit is reshaped, affecting the surface area exposed to the atmosphere. The depth of the deposited snow is generally greater than that of the pre-avalanche snow.

3 There is an increase in the density, and therefore water-equivalence, of the deposited snow (typically about a third — this is why it's extremely difficult to dig yourself out of an avalanche).

4 The avalanched snow typically has a lower albedo than the pre-avalanche snow, because of the inclusion of entrained debris (e.g. underlying soil and vegetation), and increased water content, resulting from intergranular collisions and heat production during the avalanche itself.

Figure 5.4 Summer view of avalanche defences above the Swiss village of Saas-Fee. Fences are placed along known avalanche tracks in order to break any avalanches up before they reach buildings and ski pistes below.

○ Which of the above four factors will enhance, and which will reduce, the rate of ablation?

● Factors (1) and (4) accelerate melting (see Section 2.1), while factors (2) and (3) retard melting, as they generally result in a more compact and denser snow deposit with less surface exposed to the atmosphere.

Overall, factors (1) and (2) are considered to be most important, because density changes (3) are closely linked to the surface area of snow in the runout zone, and the albedo (4) of undisturbed snow declines anyway as it ages.

5.3 Cryospheric resources

It's probably easy to envisage the downside of the cryosphere: avalanches crashing down mountainsides, glaciers bulldozing their way across the land, icebergs bearing down on ships. However, the cryosphere also brings considerable, practical advantages for people. Snow and ice are critical water sources in many parts of the world, not only those actually covered by snow and ice themselves. Mountain environments cover about 20% of the Earth's land area and are often dominated hydrologically by glaciers. The characteristics of glacial runoff were discussed in Section 2. Mountain environments are sensitive to climate change because small changes in precipitation, solar radiation or air temperatures can have significant impacts on glacier mass balance and hydrology. A number of the world's desert regions, including northwestern China, the Thar Desert of northwestern India and Pakistan, the coastal desert of Peru and the wine-growing area of Mendoza in Argentina, all receive meltwater from adjacent mountain ranges. Himalayan snow and ice meltwater is supplied to the Rajasthan canal system, irrigating the Thar Desert via hundreds of kilometres of artificial channels. Rain shadows in central European valleys such as the Rhône, Valais, Switzerland, limit the productivity of agricultural land, but this can be raised with irrigation supplied by alpine meltwater. A happy fact is that the glacial component of runoff is greatest when it is needed most, in hot, dry periods. The other, major use to which meltwater is put around the world is hydro-electric power (HEP) generation.

Switzerland's HEP stations were mostly built in the 1950s and 1960s, and generate about half of Switzerland's winter energy production. The Massa HEP station, near Brig, runs almost entirely on meltwater from the 23 km-long Grosser Aletschgletscher, the largest glacier in mainland Europe. Attention will now be focussed on one example of an irrigation scheme and one example of an HEP scheme, each of which depends on snow and ice meltwater.

5.3.1 Irrigation: Hopar, central Karakoram, Pakistan

Hopar is a community of five agricultural villages with about 4000 inhabitants, located in the central Karakoram region of northern Pakistan. Most essential food is either grown in the villages or traded with Hopar produce, which consists mainly of apricots, walnuts, wheat, barley, beans, potatoes and alfalfa. The terrain is severe in this area, and agriculture is largely limited to shallow-gradient land below 3000 m altitude. The ultimate constraint on agriculture is water supply. Maximum precipitation (about 1.1–1.6 m yr^{-1}) occurs above 3500 m. The Hopar valley is naturally arid, but has been transformed by over 300 km of channels which supply meltwater from above 3600 m to 280 ha of cultivated terraces and 160 ha of sloping alfalfa fields at altitudes between 2500 and 3600 m. Most of the irrigation water in Hopar is supplied by melt from perennial and seasonal snow and ice between 3600–4900 m a.s.l. The 12 km^2 local catchment is about 2 km^2-covered by ice and 6 km^2-covered by perennial snow; seasonal snow covers the entire catchment in winter. From March–May, melting is irregular and unpredictable, but by late July melting occurs throughout the source catchment (up to 4900 m a.s.l.). The main melt stream carries abundant flow for the period of regular, intensive irrigation from mid-June–early August, and for progressively lighter irrigation until the potato harvest in September. Total seasonal discharge is more reliable than in comparable rain fed basins, because when seasonal snowfall fails, perennial snowmelt and icemelt is enhanced. In contrast to precipitation, which occurs primarily during the cold season, melting coincides well with both maximum air temperature and with crop water requirements.

5.3.2 Hydro-electric power: Grande Dixence, Switzerland

Developed throughout the 1950s and early 1960s, this scheme diverts meltwater from 35 glacierized catchments in the Mattertal and Hérens valleys in the Valais region of Switzerland to a reservoir, Lac des Dix (Figure 5.5), in the Hérémence valley. Intake structures are situated in front of glaciers to capture meltwater (Figure 5.6), which is diverted into a 6 m-diameter underground tunnel, sloping from 2496 m a.s.l. in the east to 2364 m a.s.l. where it enters Lac des Dix more than 40 km away to the west. Almost all of the water intake structures are at greater elevations than the tunnel, so water largely flows without the need for pumping. From the reservoir, water is transferred by tunnel through the Fionnay power station in the Bagnes valley to Nendaz power station in the Rhône valley. These power stations generate about 1800 GWh of power annually, which is more than 5% of Switzerland's HEP and more than 3% of Switzerland's total energy production.

Figure 5.5 An aerial view of Lac des Dix, the major reservoir of the Grande Dixence HEP scheme in the Valais region of Switzerland. Lac des Dix can store over 40×10^6 m^3 of water. At 285 m, the dam wall is one of the world's highest.

Figure 5.6 HEP water intake structure downvalley of Bas Glacier d'Arolla, Valais, Switzerland. This is part of the Grande Dixence scheme, and contributes meltwater to Lac des Dix (Figure 5.5) during summer.

Transfer of water to the reservoir requires careful management. The aim is to fill it to its maximum capacity during the summer at minimum pumping cost. At the height of summer, water from the lower catchments is usually stored during the day in 'compensation basins' and pumped up to the tunnel at night when it is cheaper to do so. Care also has to be taken not to damage the tunnel by over pressurizing it. Coarse sediment must be removed from the water before it enters the tunnel, as this would erode the tunnel walls and lead to sedimentation in the reservoir. Bedload (relatively coarse debris that moves in traction at the stream bed) is mostly trapped in settling basins immediately upstream of the water intakes (Figure 5.6), while suspended sediment (finer material maintained in the water column by the turbulent nature of the water flow) is usually trapped in underground settling tanks just inside the structures. This trapped sediment must periodically be purged by opening the sluice gates and temporarily diverting water away from the tunnel and into the natural stream channels below the intake structure.

Global and regional changes in climate can of course affect the melting of ice masses, and consequently the human use of water resources. These are explored in Box 5.2.

Topic 4   Cryosphere

> ### Box 5.2   Implications of climate change for water resources
>
> Clearly, where communities rely on water supply from snow and ice for any reason, climate change leading to a reduction in the quantity of this snow and ice is a major cause for concern. There have already been major changes over the past century. The glacierized area of the Rhône/Porte du Scex basin (5220 km², altitude range 373–4634 m a.s.l., mean 2130 m a.s.l.) decreased from 16.8% in 1916 to 13.6% in 1968, leading to an estimated 16% reduction in mean April–September runoff (other things being equal); over the period 1876–1973, the glacier cover of the entire Swiss Alps was reduced from 1818 km² to 1342 km², corresponding to a 26% reduction in the 1876 glacierized area, or a 3.4% reduction in annual specific runoff from all Swiss river basins.
>
> The implications of climate change for the water yield from Valais glaciers and for the operation of the Grande Dixence HEP scheme are summarized below, but you should note that the trends here exemplify issues in many glacierized mountain ranges across the world. Meteorological data collected at Bricola in Valais since 1968 display an overall trend of rising air temperatures throughout the summer period, May–September. There is a good correlation between summer air temperature and water yield for the Hérens valley over this period. Before 1982, the total summer water yield from Haut Glacier d'Arolla was always less than 24×10^6 m³, and was often less than 18×10^6 m³. Since 1982, the total summer water yield has *always* been at least 24×10^6 m³. Furthermore, this increase in runoff volume has been reflected throughout the Mattertal and Hérens valleys: the total volume of water supplied by all 35 basins in the Grande Dixence HEP scheme has increased from about 300×10^6 m³ in the late 1960s to about 450×10^6 m³ in the early 1990s.

In the final section, we look in detail at the prospects for change in the cryosphere with future climate change.

5.4   Summary of Section 5

1. The impact of the cryosphere on the lives of people can generally be divided into two categories: hazards and resources.
2. Hazards include glacier advances and unstable glaciers, ice avalanches, glacier floods, glacier–volcano interactions and snow avalanches. Some of these have caused great loss of life in mountain regions around the world.
3. Climate change that leads to a reduction in snow and ice extent is therefore a serious concern in some regions. In the short term, runoff rates will increase, but in the longer term there is the prospect of greatly diminished runoff and water supply.

Question 5.1

Based on what you learned in Section 4, what particular hazards might retreating tidewater glaciers pose?

Question 5.2

What practical problems and solutions might be associated with the use of meltwater for agricultural irrigation in a mountain region, such as at Hopar?

Learning outcomes for Section 5

Now that you have completed your study of Section 5 you should be able to:

5.1   Document the hazards and resources that are associated with the cryosphere.

5.2   Describe and account for the main types of snow avalanche, and outline the hydrological implications of avalanching.

Ice and environmental change

6.1 Global warming

Since pre-industrial times, levels of atmospheric carbon dioxide in the atmosphere have increased by 30%. This increase is of concern because the majority of recent climate simulation experiments indicate a close relationship between levels of atmospheric CO_2 and global temperature: indeed they predict warming of 2–4 °C for a doubling of CO_2 (Box 6.1). This issue has important implications for all aspects of the environment, but there are particular reasons why the cryosphere is worthy of interest and concern. Principally, a greater mean annual warming is forecasted at high latitudes than at low latitudes, which is referred to as a **polar amplification** of climate warming. This effect is the result of two features of high latitudes, first, the feedback between air temperature and the extent of snow and ice (which are of course concentrated in high latitudes), and second, differences in the feedback involving atmospheric lapse rate. The extent of snow and ice decreases with increasing air temperatures, resulting in a reduction of the surface albedo, which, other things being equal, tends to amplify the local temperature response. Furthermore, at high latitudes the lapse rate tends to increase as a result of the stability of cold air masses, which restricts the warming to a shallower layer of the atmosphere.

Most models predict a polar amplification of 2–3 times the tropical response, although some data suggest an amplification factor of 6 or more. In this section we review evidence of change in the cryosphere over about the past century, and look at prospects for further change in the future.

6.2 The cryosphere and climate change

6.2.1 Snow cover

For an apparently simple medium, terrestrial snow cover is deceptively difficult to measure, and this hampers attempts to quantify any natural variability in snow extent and water equivalence, and detection of trends in these quantities. Conventional manual methods of snow surveying are labour intensive and spatially restricted, while drifting snow is notoriously difficult to measure representatively. Remote sensing can provide large-scale coverage, but data can be difficult to interpret and the technique can be of limited utility in areas of forest or steep topography.

While there is no significant trend over the whole 21-year period for which satellite data are available, Northern Hemisphere snow cover extent appears to have decreased in the late 1980s and 1990s. This has been particularly apparent in spring, but summer and autumn snow cover has also been low in recent years, while winter snow cover appears relatively stable (Figure 6.1). The annual mean area of Northern Hemisphere snow cover has decreased by about 10% over the past two to three decades, with similar decreases over both North America and Asia. This decrease in snow cover is closely linked to an increase in air temperature. There has been strong spring warming over northern land areas, which is also reflected in, among other things, earlier lake ice melting, earlier snowmelt-related floods on rivers in Canada, and reduced duration of river ice over the Former Soviet Union (FSU).

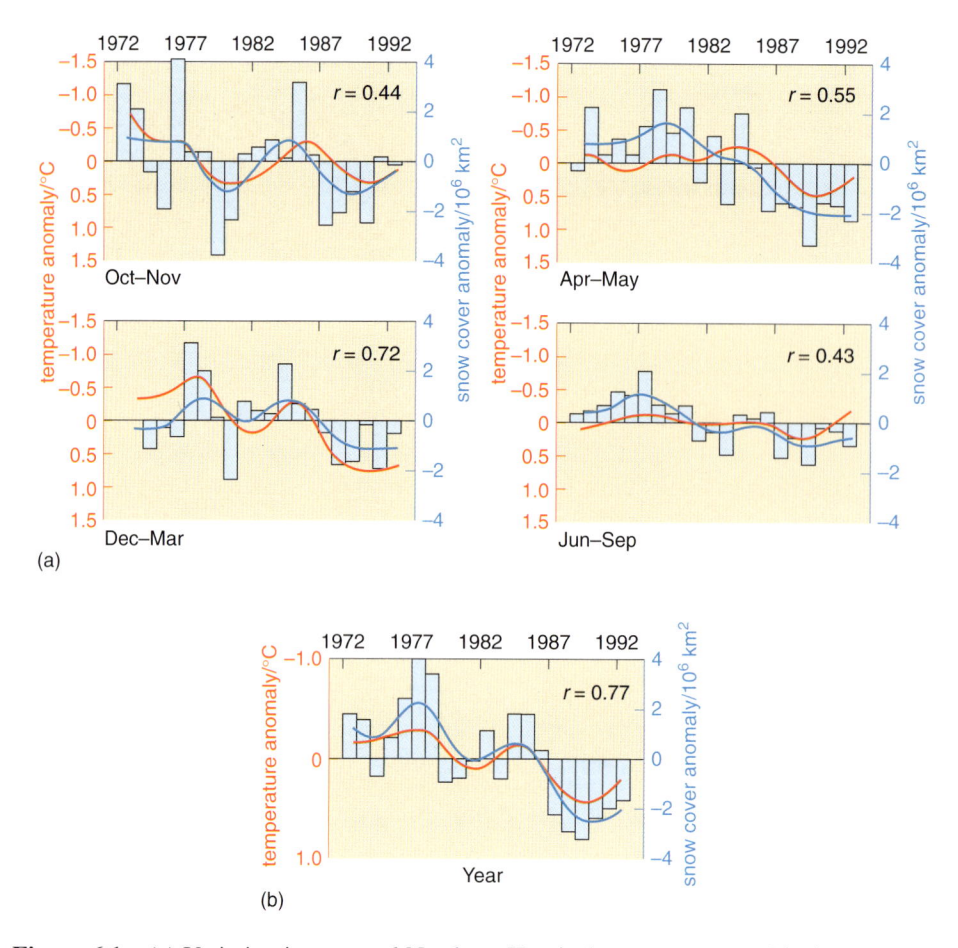

Figure 6.1 (a) Variation in seasonal Northern Hemisphere snow cover (the bars represent yearly values, and the blue line is a smoothed trend) and of air temperature (red line is a smoothed trend). (b) The mean annual variation of Northern Hemisphere snow cover (same symbols).

The detailed picture of snow cover variations from region to region is complex. For periods longer than about the past two or three decades, meteorological station data provide the only information on snow cover variations. These indicate that average snowfall *increased* over the period 1950–1990 by about 20% over northern Canada and by about 11% over Alaska. In southern Canada and the northern USA however, an increase of precipitation has been accompanied by higher ratios of liquid to solid precipitation, as temperatures have also increased. Evidence from the changing thickness of annual layers in ice cores and from other sources indicates that snowfall over Antarctica has increased by 5–20% in recent decades. In contrast, evidence from an ice core in central Greenland suggests that the rate of snow accumulation has decreased by 20% during the past 50 years.

Box 6.1 The warming world

The 1990s were the warmest decade in the series (Figure 6.2). The warmest two years of the entire series were 1997 and 1998, with the latter the warmest at 0.57 °C above the 1961–1990 mean. The seven warmest years globally have now occurred in the 1990s. Analyses of over 400 proxy climate series (from trees, corals, ice cores and historical records) show that the 1990s was the warmest decade of the millennium and the 20th century the warmest century. The warmest year of the millennium was 1998, and the coldest was probably 1601. The Intergovernmental Panel on Climate Change in its most recent report stated: 'the balance of evidence suggests a discernible human influence on the climate system'.

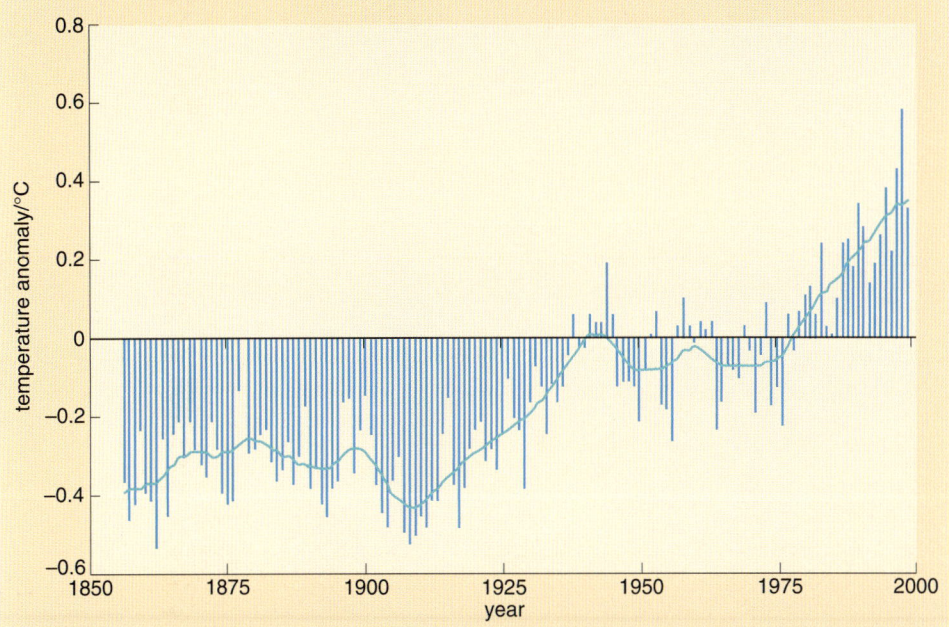

Figure 6.2 A time series of the combined global land and marine surface temperature record from 1856 to 1999, compiled jointly by the Climatic Research Unit, University of East Anglia, and the UK Meteorological Office Hadley Centre.

6.2.2 Mountain glaciers

The overall 20th-century atmospheric warming trend (Figure 6.2) has been accompanied by a general retreat of mountain glaciers (Figures 6.3 and 6.4). However, there is no simple relationship between climate and glacier extent (area, thickness, volume): this varies from glacier to glacier. Overall, flatter glaciers tend to be more sensitive to climate change than steeper ones, because, for a given change in the altitude of the equilibrium line (ELA), a greater proportion of the surface is affected (Figure 6.5). However, there can be any number of variations depending on the configuration of the glacier, and even neighbouring glaciers can therefore have differing responses (Figure 6.6).

Topic 4 Cryosphere

Figure 6.3 Cumulative mass balances, in m w.e. relative to 1890, of well-studied mountain glaciers. The steady 20th-century decline is typical for mountain glaciers worldwide.

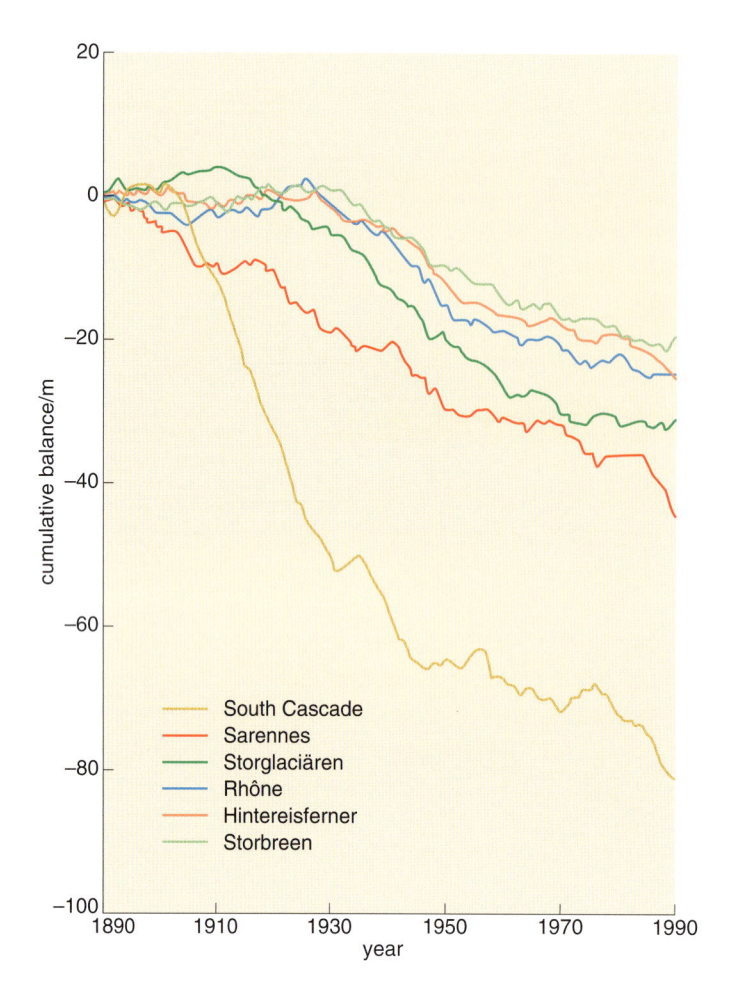

Legend:
- South Cascade
- Sarennes
- Storglaciären
- Rhône
- Hintereisferner
- Storbreen

Figure 6.4 A typical example of 20th-century mountain glacier change, from the French Alps. (a) Glacier d'Argentière, viewed from the village of Argentière, 1896. (b) Glacier d'Argentière, viewed from approximately the same point, in 1997.

Ice and environmental change

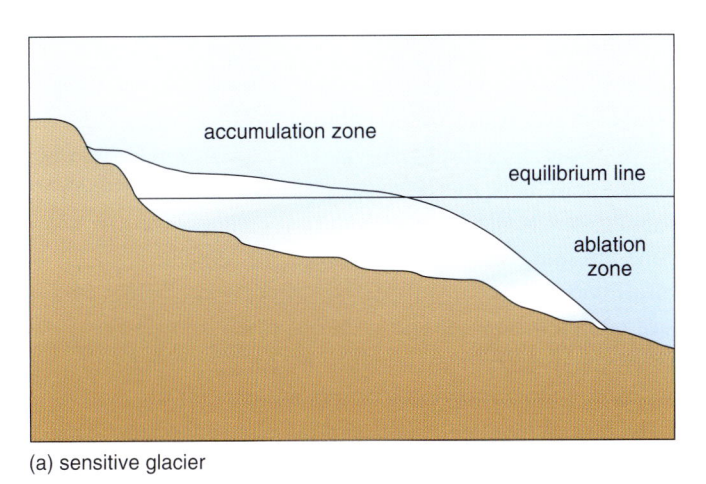

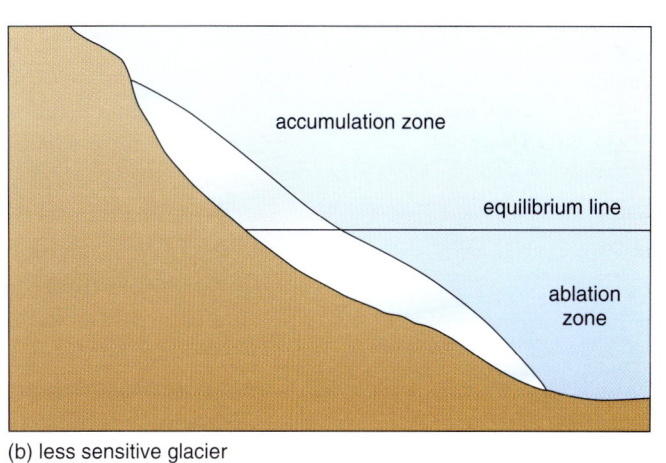

(a) sensitive glacier

(b) less sensitive glacier

Figure 6.5 The dependence of glacier sensitivity to climate change on the distribution of mass around the equilibrium line. Other things being equal, a flatter glacier will tend to be more sensitive, as a given change in the equilibrium line will affect a great proportion of the glacier area.

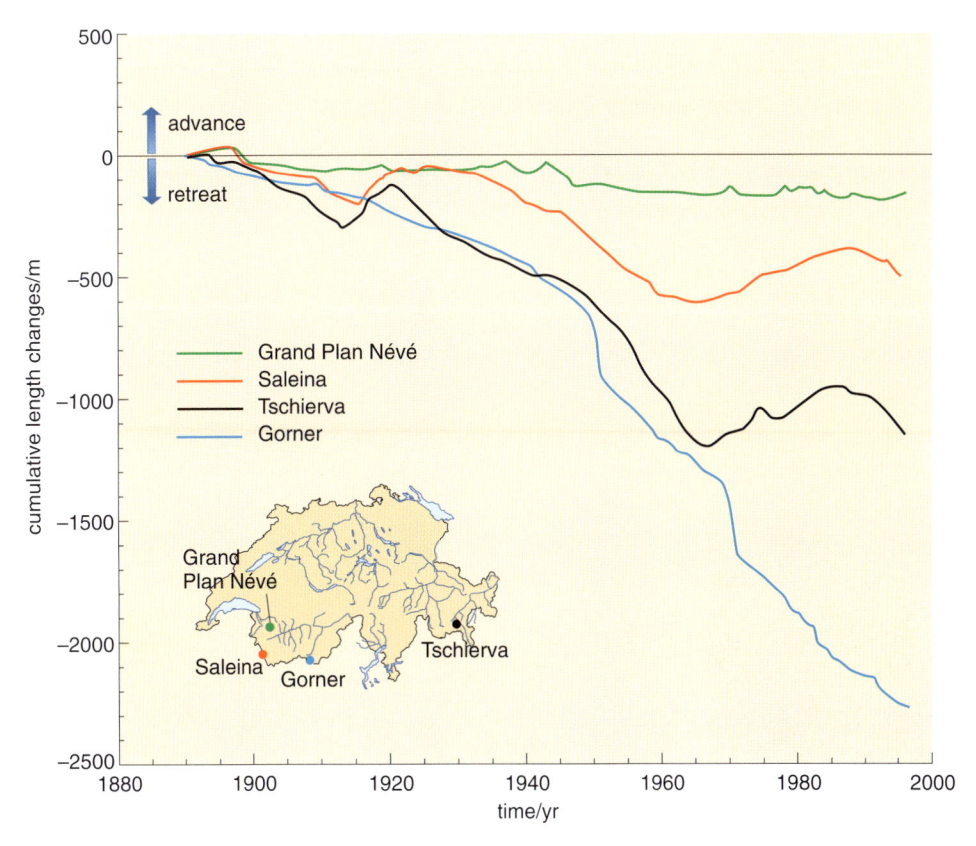

Figure 6.6 Cumulative length changes of four characteristic glacier types in the Swiss Alps since the end of the 19th century. Smaller glaciers such as Grand Plan Névé Glacier have low driving stresses and respond relatively directly to annual mass balance variability. Intermediate-sized glaciers such as the Tschierva or Saleina glaciers flow under higher driving stresses and react dynamically to decadal mass balance variations in a smoothed, lagged manner. Larger glaciers such Gorner Glacier may be too long to react dynamically to decadal mass balance variations, but exhibit strong long-term trends.

The overall climate sensitivity of mountain glaciers is demonstrated by their widespread advance during the Little Ice Age (LIA), starting in the 13th to 14th centuries and culminating between the 16th to 19th centuries. The average LIA air temperature was 0.5–1.2 °C lower, and the ELA was 100–200 m lower than at present. While there are areas where there has been no significant change, such as the Canadian Arctic, glaciers worldwide have generally receded since the end of the LIA, by about the same amount they advanced. There have been fluctuations within this overall pattern. For instance, the proportion of observed glaciers worldwide that were advancing increased from 6% to 55% between 1960 and 1980. This appears to have given way again to a general retreat, but there are notable exceptions. Some Scandinavian glaciers are currently advancing strongly: outlet glaciers of the Jostedalsbreen ice cap in Norway have advanced several hundred metres since the mid-1980s, mainly due to increased winter snowfall (Figure 6.7). Glaciers in Swedish Lapland responded to an increase in summer mean air temperature of about 1 °C in the early 20th century by rapid recession and thinning. Larger, more continental glaciers began to retreat some 10–15 years later than their smaller, more maritime counterparts. Over the 1990s the climate seems to have changed from being generally continental to being more maritime, with increased precipitation, and most Swedish glaciers have actually gained mass over the last ten years. Smaller glaciers (< 2 km^2) in particular have thickened in their accumulation areas and ceased retreating or even started to advance; larger glaciers are still retreating in response to early 20th-century atmospheric warming.

Figure 6.7 Length variations of three outlet glaciers of the Jostedlasbreen ice cap in Norway between 1901 and 1999. Each displays an initial period of stability followed by a decline, then a subsequent period of renewed relative stability, or even an advance. The timing of these changes varies from glacier to glacier, with Briksdalsbreen apparently exhibiting the earliest response.

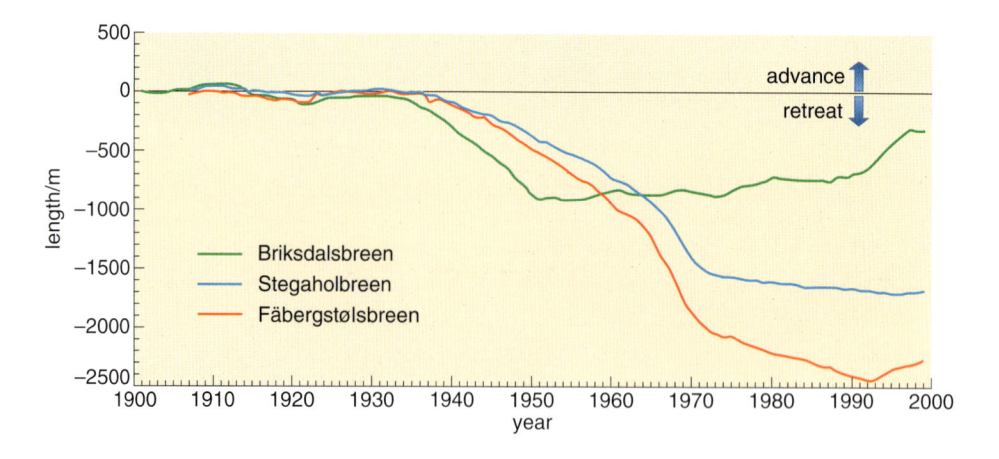

It is estimated that 14–18% of the mean rate of sea-level rise in the last 100 years is the result of melting of 'small' glaciers (i.e. all except the Greenland and Antarctic ice sheets). The contribution of glaciers to sea-level rise has increased greatly since the middle 1980s and even more steeply since the late 1980s, which accords with the rise of global temperature. The melting of glaciers contributes to **eustatic** sea-level change, which results from change in the volume of water in the oceans. This is distinct from **isostatic** change, which results from vertical land movements due to tectonic processes, sedimentation and adjustment to loading. Estimates of sea-level change on the basis of mass loss from glaciers and ice sheets can be obtained by expressing the mass loss as a water-equivalent volume, and dividing by the total area of the oceans.

6.2.3 Ice sheets

There are enormous uncertainties in estimating the net mass balance of the Greenland and Antarctic ice sheets, not least because the response times for ice sheets can be as much as 1000–100 000 years (Table 1.2), so present change may be the result of ancient climate change. Consequently, many alternative interpretations of observed changes are possible.

○ At the end of Section 1 you read of two alternative interpretations of the observed thickening of the interior Greenland Ice Sheet, invoking effects from very different periods of time. What were these?

● The thickening has been attributed both to reduced rates of ice outflow as ice accumulated during the last ice age has become deeply buried, compressed and stiffened (an effect of the last 10 000 years), and to increased accumulation associated with a warmer global warming influenced climate (an effect of the past 100 years).

In addition, the sheer physical size and inaccessibility of the ice sheets, the extreme climates, and the occurrence of long periods of polar darkness, have long rendered the acquisition of representative measurements extremely difficult.

Early measurements from satellite-based instruments, originally designed to measure the elevation of the ocean surface, indicated a thickening of the inland ice of southern Greenland during the 1970s and 1980s. However, these measurements only reached to 72° N and were unreliable at the steep, irregular ice sheet margins. Surface mass balance measurements from central Greenland and also from the Jakobshavns Isbrae basin (Section 2) between the late 1960s to the early 1990s have tended to indicate a slight thickening or no significant change. A general retreat of southwest Greenland outlet glaciers in the early 20th century has recently slowed, although it is not clear how this affects the total volume of the ice sheet. Recent satellite surveys over southern Greenland have shown little net elevation change, but large spatial variations. Elevation changes reconstructed from ice-core records and models of snow densification suggest that the observed changes can be accounted for by variability in snow accumulation, which remains well within long-term natural variability. It therefore appears that any trend has yet to emerge from typical year to year variations.

A number of studies have indicated both thickening and thinning of individual ice drainage basins in different parts of Antarctica. Despite the collapse of certain ice shelves (see below), there appears to be no evidence of significant change in the grounded ice of the Antarctic Peninsula. The influx of ice from East Antarctica to the eastern Ross Ice Shelf (Section 2) appears to exceed outflow at the ice front, while the western part of the ice shelf appears to be close to steady state. However, the west Antarctic ice flowing into the same ice shelf appears to be gaining mass at present. Of the Siple Coast ice streams, B (Section 2) is thinning slowly, although a complex pattern of thinning and thickening exists throughout its drainage basin; C, on the other hand, is thickening at the accumulation rate, mainly because of its very low flow velocity. Summaries of available data for the whole of Antarctica have tended to find positive net balances overall, but with high degrees of uncertainty. Recent results indicate that Pine Island Glacier, which drains about 10% of the West Antarctic Ice Sheet, thinned by up to 1.6 m yr^{-1} between 1992 and 1999, as a result of changes in glacier dynamics. It is not yet clear if there are implications for the interior ice.

It is not clear whether the large ice sheets have contributed to sea-level change over the past century or so. The climate sensitivities of the ice sheets is considered to be 0.30 ± 0.15 mm yr^{-1} °C^{-1} for Greenland (ice mass shrinkage due to increased ablation) and -0.30 ± 0.15 mm yr^{-1} °C^{-1} for Antarctica (ice mass growth due to increased accumulation accompanying a warmer atmosphere with increased moisture availability).

6.2.4 Permafrost

The forecasted enhanced warming over land surfaces in high latitudes will affect periglacial and permafrost environments in the 21st century. Feedback effects are likely to amplify the impacts of climate changes, particularly from increased methane fluxes due to the decomposition of organic matter frozen in permafrost.

- ○ What features of periglacial environments are likely to change in a warmer climate?
- ● The depth of the active layer, the temperature of permafrost, the extent and duration of seasonal snow cover, and the freeze-up and break-up of lakes and rivers.

It is still difficult to generalize about environmental change in periglacial regions, which are, like the ice sheets, relatively remote and inaccessible. Significant trends must be distinguished from short-term fluctuations, and this requires high quality, long-term and continuous data, which are rarely available.

There is evidence of permafrost warming from a number of regions. In the European Arctic and the subarctic of Russia, data indicate a rise in the temperature of permafrost at a depth of 3.0 m of 0.6–0.7 °C between 1970 and 1990. Data from northeast Siberia also indicate warming by 0.03 °C yr^{-1} at a depth of 10 m between 1980 and 1991. Temperatures at 20 m depth on the Tibet Plateau have risen by 0.2–0.3 °C during the last two decades, and since the plateau is sparsely populated and human disturbance minimal, this permafrost degradation is most likely to be caused by climatic amelioration. However, the warming trend is not universal: a cooling trend at 20 m depth of about 0.05 °C yr^{-1} between 1988 and 1993 has been observed in coastal northern Quebec.

Increases in ground temperature can lead to a decrease in overall permafrost thickness, and ultimately to its elimination in some cases, accompanied by erosion, subsidence, and extreme slope instability. On the other hand, frost damage to roads and buildings is reduced significantly. In continuous permafrost, a temperature rise leads in due course to active layer thickening, and thinning of permafrost from both top and bottom. In discontinuous permafrost, a temperature rise probably leads to disappearance of the permafrost. Change in total permafrost thickness is an extremely slow process: for example a 4 °C surface warming over 50 yr in northern Alaska would result in active layer thickening from 0.50 to 0.93 m, but permafrost temperatures at 30 m depth would have increased by only 1 °C. Spatial changes in the extent of permafrost under a warmer climate are likely to be significant: about a 10% reduction over 50 yr in regions such as Siberia and Canada for atmospheric warming of the order 2–4 °C (Figures 6.8 and 6.9). As the active layer deepens, there is likely to be increased groundwater flow. Coupled with other effects of a warmer atmosphere, such as more precipitation in the form of rain and changes in the timing and intensity of snowmelt, changes to streamflow regimes are likely.

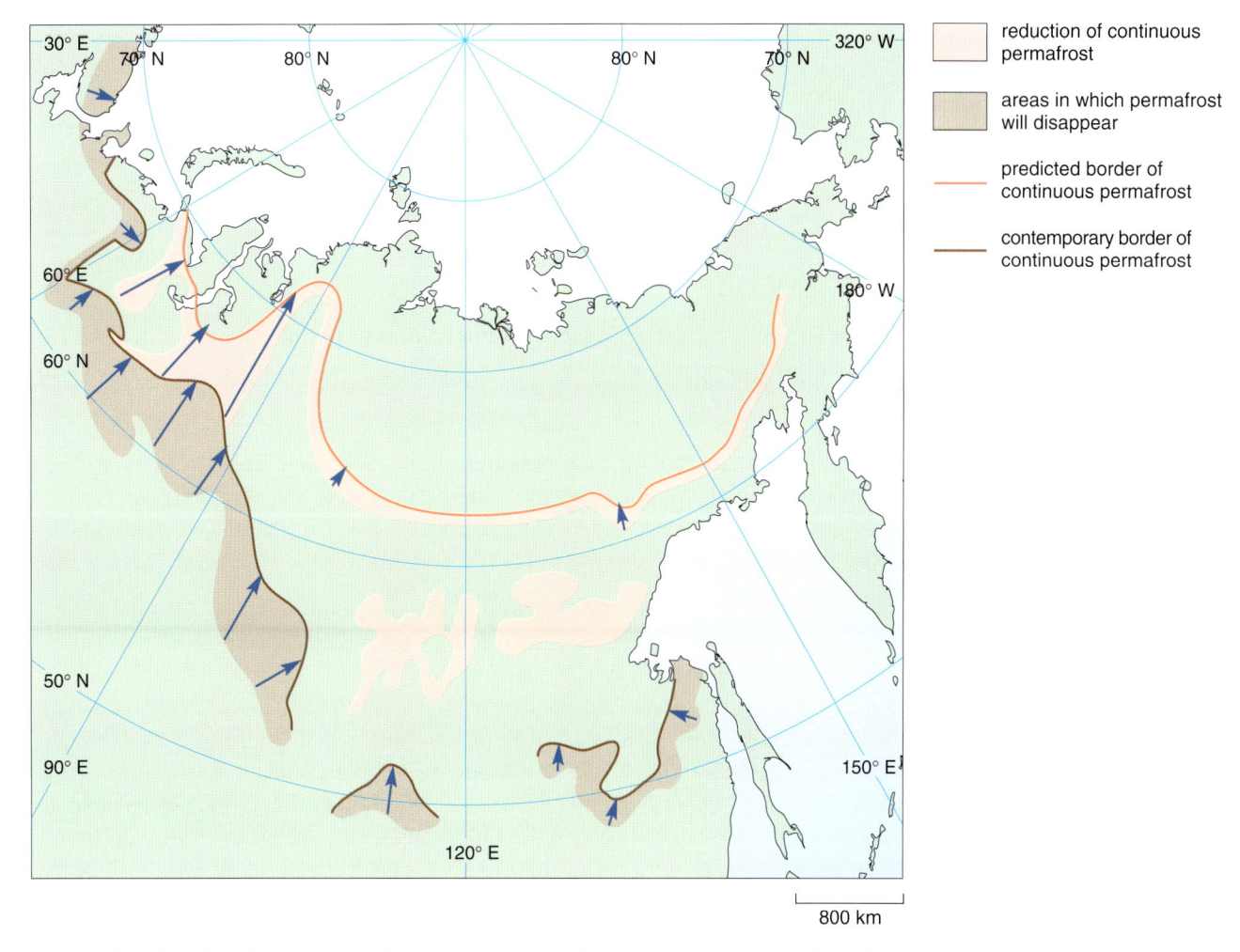

Figure 6.8 Predicted changes in Siberian permafrost distribution following a 2 °C air temperature increase.

Topic 4 Cryosphere

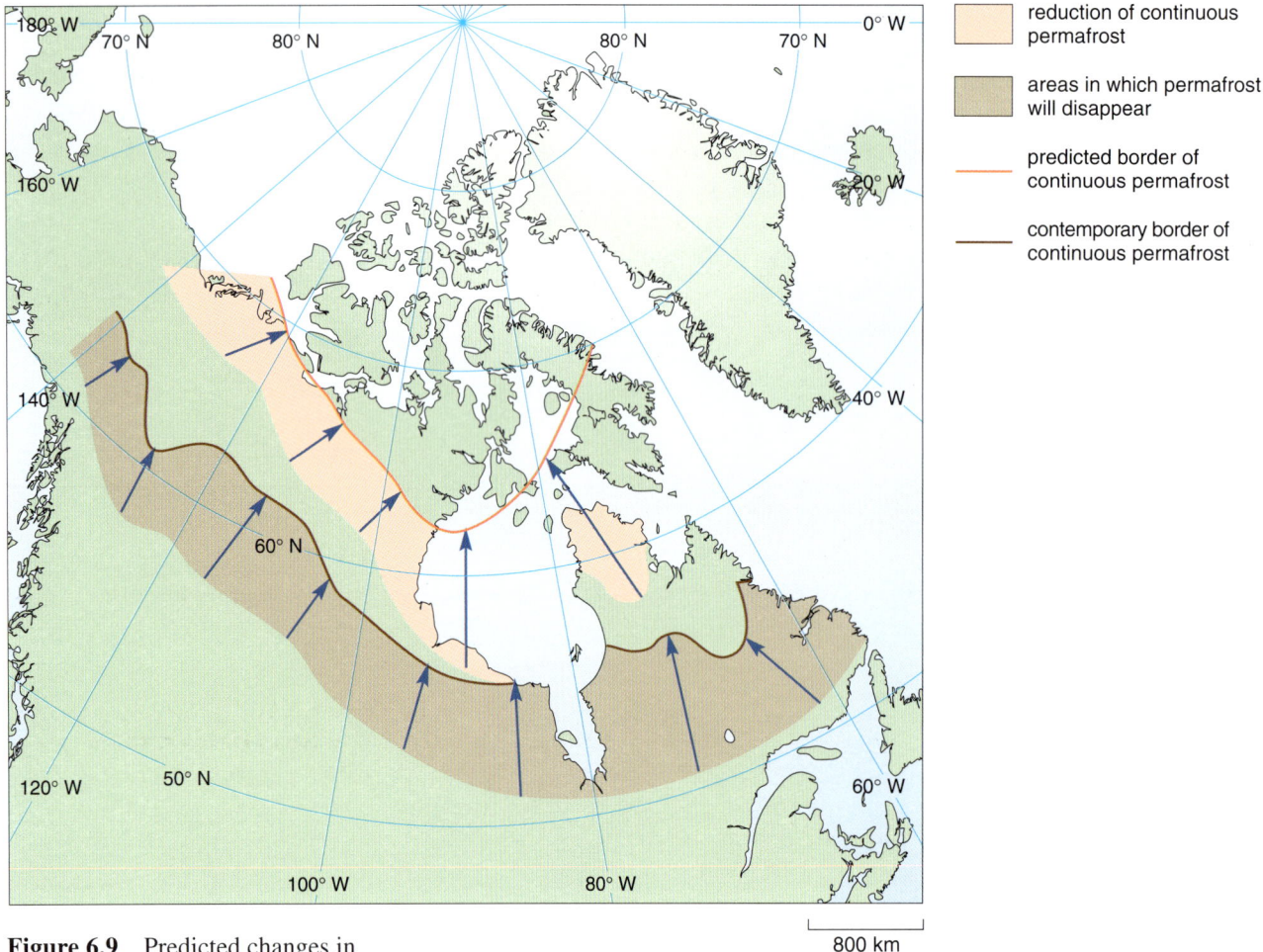

Figure 6.9 Predicted changes in Canadian permafrost distribution following a 4 °C air temperature increase.

Legend:
- reduction of continuous permafrost
- areas in which permafrost will disappear
- predicted border of continuous permafrost
- contemporary border of continuous permafrost

6.2.5 Sea ice

Observations from whaling ships prior to the advent of remotely sensed data indicate that Southern Hemisphere sea ice decreased in area by about 25% between the mid-1950s and early 1970s. However, neither hemisphere has exhibited significant trends in sea ice extent since 1973 when satellite measurements began, although there has been generally below average extent in the Northern Hemisphere in the early 1990s. Coverage in the Southern Hemisphere has remained close to average. The most continuous and consistent records of sea ice extent began in 1978 with the use of satellite microwave sensors. Over the period 1978–1996, these reveal no trend in area during winter and spring, and weak upward trends in summer and autumn (Figure 6.10). Similar techniques indicate a clear downward trend in the area of Northern Hemisphere sea ice over about the same period (Figure 6.11). The total mass of sea ice may be a more sensitive indicator of climate change than its area alone, although representative data, from sources such as upward sonar profiling from submarines and from moored subsurface sonar instruments, have been hard to come by. However, it is estimated that the mean thickness of sea ice over the Arctic Ocean at the end of the melt season has decreased from 3.1 m over the years 1958–1976 to 1.8 m over 1993–1997.

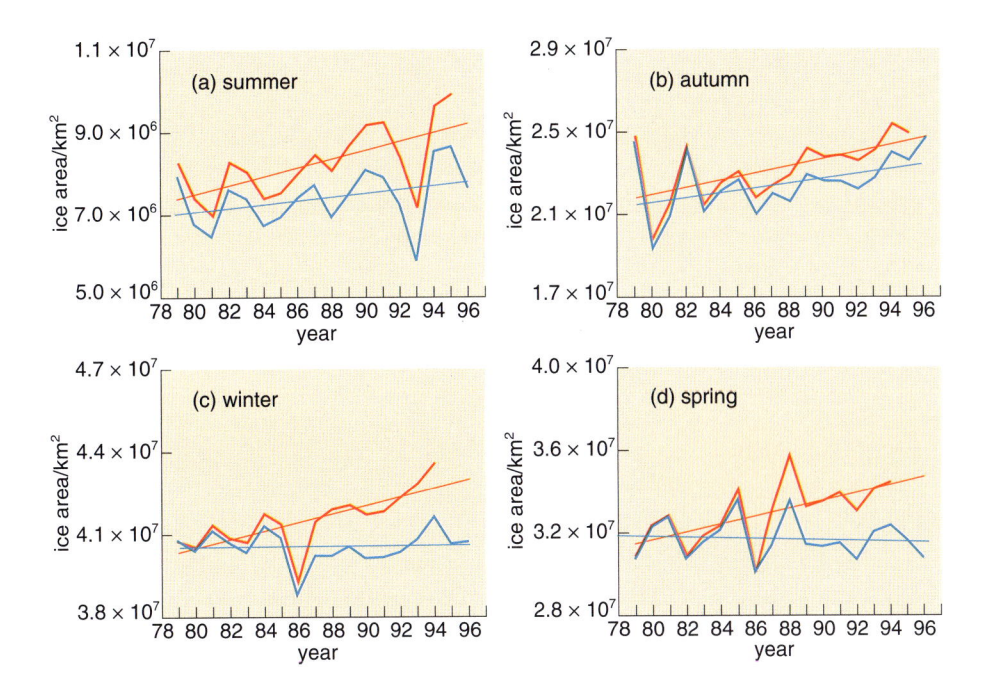

Figure 6.10 Variation in seasonal sea ice area in the Southern Ocean between 1979–1996, from original (red) and re-interpreted (blue) satellite data.

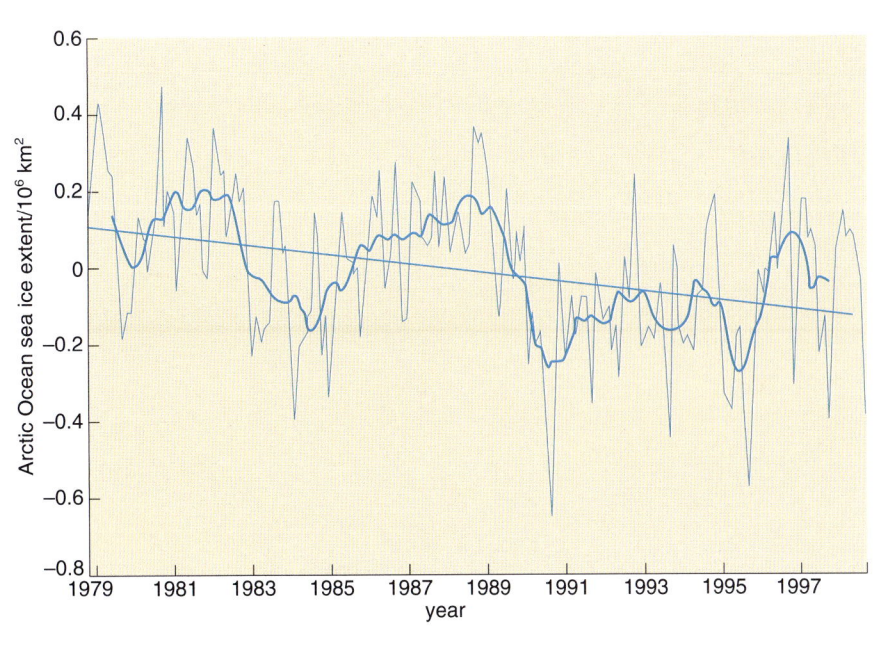

Figure 6.11 Variation in the area of Northern Hemisphere sea ice between 1979–1998.

6.3 Water resources

Changes in the quantity and seasonal distribution of precipitation, in the balance between snow and rain, and in melt season air temperatures, will all influence the extent of snow and ice in mountain regions, where meltwater is often important for water supply (as discussed in Section 5). Future climate scenarios for the world's mountain regions indicate increases in winter and particularly in summer air temperatures (for the Alps, Altai, Himalayas and Rockies), and a slight increase (for Alaska, the Rockies, Andes and Altai) or even decrease (for the Alps and Tien Shan) in winter precipitation. Overall, this is likely to reduce the extent and duration of snow cover, and to lead to continued retreat of mountain glaciers, with reduced winter accumulation, increased summer ablation, and the enhanced retreat of snowlines and equilibrium lines.

In the immediate future, water yields from mountain glaciers are likely to increase. Increased exposure of ice with the further retreat of the transient snowline will lower albedo and probably increase surface roughness, maximizing, in combination with increased air temperatures, radiative and turbulent heat fluxes (Section 2) leading to higher rates of ablation. In addition to the quantity of runoff, its temporal pattern is also likely to change. Elevated summer snowlines will reduce the storage effect of slow meltwater percolation through the snow cover, in favour of rapid runoff from bare ice surfaces (Section 2).

In the longer term of the next few decades, continued retreat and loss of glacier area will ultimately decrease water yields from mountain glaciers. Winter accumulation decreases and summer ablation increases will alter mass balance gradients, and the pattern and rate of ice flow. Thinner glacier ice deforms more slowly and is less likely to close down efficient, high water flux subglacial drainage systems during winter. This may enable subglacial channels to survive all year round, altering the pattern of runoff, particularly in the early summer, when the response of the glacier hydrological system to increased seasonal melting could be much faster than at present.

6.4 The West Antarctic Ice Sheet: 'Threat of disaster'

Antarctica is a continent of superlatives. Not only is it the coldest continent, it is also the highest and the windiest. The Antarctic ice sheet, which would yield a 66 m sea-level rise if completely ablated (see Table 1.2), consists of two contrasting components: the East Antarctic Ice Sheet (EAIS) is a broad dome over 4 km in altitude (for comparison, Mont Blanc is 4.8 km high, while Ben Nevis is 1.3 km high!), based on bedrock mainly above sea-level and drained at the margins by largely radial ice streams; the West Antarctic Ice Sheet (WAIS) consists of three domes rising to about 2 km, based on bedrock that is largely grounded *below* sea-level, making it the world's only existing marine ice sheet (Figures 6.12 and 6.13). The WAIS is also drained by ice streams, many of which flow into ice shelves, the largest of which is the Ross Ice Shelf (Section 2).

There has been ice on the Antarctic continent for tens of millions of years. Scientists disagree as to the exact extent of ice coverage at different times, and therefore how dynamic or stable the ice-sheet system is. There is a lot of evidence that removal of ice sheets can be much more rapid than the accumulation (Box 6.2). It is generally agreed that the EAIS is very stable compared to the WAIS, but the stability of the WAIS itself is a major source of uncertainty in science. This is an important issue because the WAIS contains 3.8×10^6 km^3 of ice, the equivalent of about a 6 m sea-level rise (Table 1.2). If this great volume of ice is vulnerable to climate change, a rapid and catastrophic response may be triggered. An early assessment of the problem noted that rapid sea-level rise of this magnitude would cause 'the oceans to flood all existing port facilities and other low-lying coastal structures, extensive sections of heavily farmed and densely populated river deltas of the world … and large areas of many of the world's major cities'.

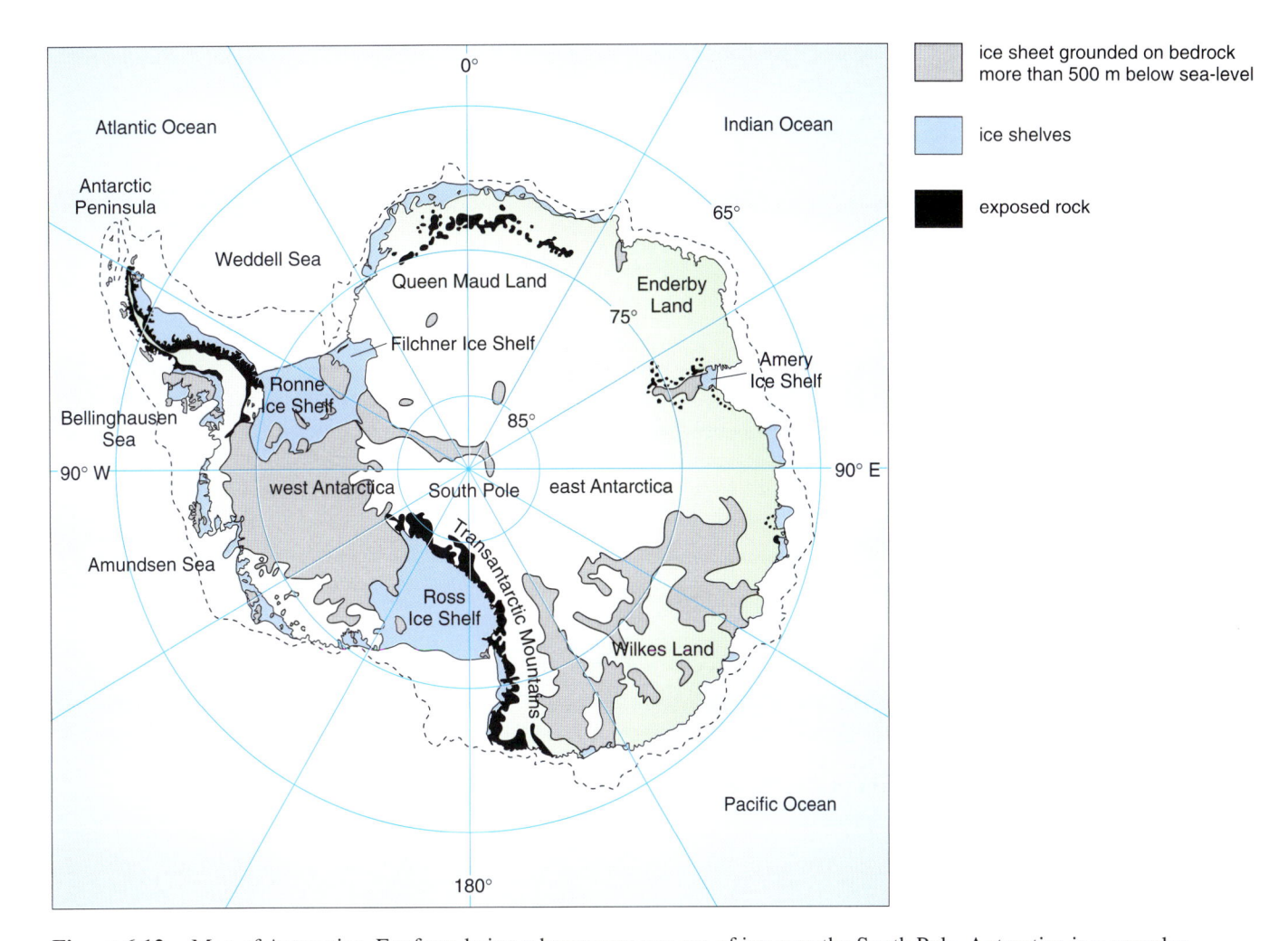

Figure 6.12 Map of Antarctica. Far from being a homogeneous cap of ice over the South Pole, Antarctica is a complex glacial environment with two major ice sheets, largely divided by the Transantarctic Mountains.

Figure 6.13 Satellite image map of Antarctica compiled from Advanced Very High Resolution Radiometer (AVHRR) images. Comparison with Figure 6.12 shows the rough Transantarctic Mountains and smooth, flat ice shelves and east Antarctic interior very well.

> **Box 6.2 Collapsing ice sheets**
>
> In this context, collapse is understood to mean the loss of the vast majority, if not all, of the land based, grounded ice on a timescale that is very much shorter than that required to accumulate it (which is more than 10 000 years at the present rate, other things being equal). Timescales as short as 200–400 years have been suggested. Likewise, instability here means that this collapse of currently grounded ice would occur following only small perturbations to the ice from external sources, in a positive feedback process.
>
> There is a precedent, in the Laurentide Ice Sheet, which covered much of North America until about 10 000 years ago (and at its peak was actually larger than the EAIS itself). This is widely believed to have collapsed rapidly by calving retreat at the end of the last ice age. Combined expansion of the ice sheet to the Atlantic margins of the continent and depression of the crust under the weight of the ice ultimately led to rapid, deep water calving, in a process akin to the rapid retreat of tidewater glaciers discussed in Section 4. There is evidence across the North Atlantic of huge episodes of iceberg-derived sedimentation at this time.

A rapid collapse scenario for the WAIS depends on two factors: the climate sensitivity of the ice shelves, and the presence of suitable deep-water calving bays. Attention on this issue has recently been sharply focused by the rapid disintegration of the Wordie and Larsen ice shelves around the Antarctic Peninsula (Figure 6.14). It has been reported that the five northernmost ice shelves on the Antarctic Peninsula have retreated dramatically in the past 50 years, while those further south show no clear trend. Comparison with air temperature records, which show an atmospheric warming in the region of the Peninsula of about 2.5 °C since 1945, suggests that there is a threshold for ice shelf viability near a mean annual air temperature of −5 °C, above which rapid disintegration occurs.

It is known that the activity of the WAIS is dominated by fast-flowing ice streams that feed into floating ice shelves, and that appear to be able to respond rapidly to changes in the ice sheet–shelf system (see Section 2). However, the effects these dynamic ice streams might have on the stability of the WAIS is debatable. On the one hand, the ability of ice streams to transport ice rapidly from the interior of the ice sheet to the ocean (on a timescale of the order of 100 years), indicates that small changes in ice flux could be amplified rapidly, and therefore that a greatly accelerated discharge of ice could be generated. On the other hand, the short response times of ice streams suggests that any changes can be rapidly removed by normal flow processes, which would tend to promote stability. Recent theoretical analyses have not resolved the question, having lent credence to both views. A theoretical development that has attracted much attention and that suggests dramatic instability of a marine ice sheet is the **binge–purge cycle** (a model of repeated cycles of steady ice sheet growth and sudden collapse) put forward to explain the massive outpourings of icebergs from the Laurentide Ice Sheet at the end of the last ice age. A model study of the WAIS over the last million years that incorporates rapidly flowing ice streams suggests that the ice sheet collapsed in this way in the past.

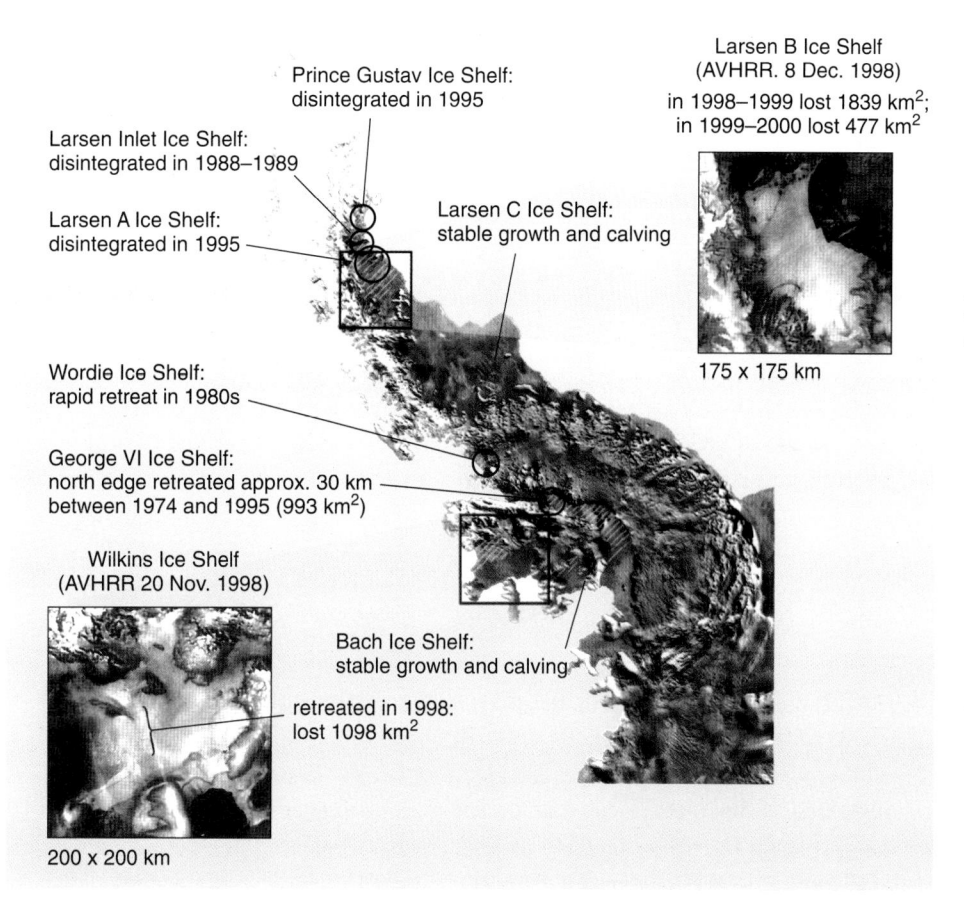

Figure 6.14 Overview of the Antarctic peninsula, summarizing recent ice-shelf changes and highlighting areas where meltwater accumulates seasonally at the surfaces of ice shelves. The base map is a mosaic of AVHRR images acquired between 1980 and 1994. Since this figure was produced, the Larsen B ice shelf has completely collapsed (March 2002).

Recent field evidence is also equivocal. On the one hand, there is evidence suggesting instability. Ice Stream B is currently flowing faster than it should for steady state, while Ice Stream C appears to have stagnated within the past 150 years. Furthermore, there is evidence from subglacial microfossils that this ice sheet has been largely or completely absent at some time after its initial formation, which suggests dynamic behaviour, although the evidence is susceptible to alternative interpretation. On the other hand, there is evidence to support stability, including the current growth, rather than collapse, of the glaciers feeding Pine Island Bay and the lack of evidence for any significant change in the geometry of the WAIS in the Byrd Station area for many thousands of years.

In the most recent IPCC report, it was concluded that, 'Given our present knowledge, it is clear that while the [WAIS] ice sheet has had a very dynamic history, estimating the likelihood of a collapse during the next century is not yet possible… our ignorance of the specific circumstances under which West Antarctica might collapse limits the ability to quantify the risk of such an event occurring, either in total or in part, in the next 100 to 1000 years.' An important point is that, given the very long response time of the WAIS (Table 2, Section 1), if collapse does occur, it will probably be more due to changes of the last 10 000 years rather than to recent CO_2-induced atmospheric warming. Therefore, the fate of the WAIS may already be sealed. However, focus on the WAIS should not detract from other aspects of the cryosphere. Because of its vastness, relatively small changes in the more stable EAIS could have large effects. The direct effect of a 10% increase in precipitation (that might accompany a doubling of atmospheric CO_2 concentration) is a *lowering* of sea-level at a rate of 27 cm per century (static sensitivity: see Section 1). This clearly illustrates the importance of putting research effort into understanding the complex behaviour of the cryosphere and its relationship with climate.

6.5 Summary of Section 6

1. The cryosphere is a key global climate change indicator; the volume of the cryosphere is strongly related to atmospheric temperature.

2. There is evidence of a reduced Northern Hemisphere snow cover over the past decade or so, but there is no strong overall trend during the period of satellite observations.

3. At the start of the 20th century, mountain glaciers were at generally close to their Little Ice Age maximum extents, but they have largely retreated throughout the 20th century. The general retreat of mountain glaciers is estimated to have contributed up to about a fifth of global sea-level rise over this period.

4. Much uncertainty surrounds the mass balance status of the major ice sheets. The errors associated with current estimates are still too great to determine with confidence whether the Greenland and Antarctic ice sheets are growing, shrinking or stable.

5. Atmospheric warming is likely to lead to degradation of permafrost, but rates of change at depth within the ground are very slow.

6. Neither hemisphere has exhibited significant trends in sea ice coverage during the period of satellite measurements, although there appears to have been generally below average coverage in the Northern Hemisphere during the 1990s. More important than area alone may be the volume of sea ice, but it is very difficult to achieve representative measurements of sea ice thickness.

7. Continued reduction in snow and ice cover in mountain regions will increase rates of runoff in the short term, accompanied by changes in the timing, duration and diurnal variation of flow, but ultimately lead to water resource shortages.

8. The world's only remaining marine ice sheet, the West Antarctic Ice Sheet (WAIS), could be unstable. If it were to collapse, sea-level would rise dramatically. However, there is plausible evidence both for and against WAIS instability and it remains an unresolved issue.

9. There remains significant uncertainty about future change in all aspects of the cryosphere. Much of this uncertainty stems from the need to identify significant trends from interannual variability in large and complex systems, particularly the lack of long time-series data from remote areas, and the need to develop comprehensive observational strategies based on satellite remote sensing. This is particularly true for snow and sea ice, the areas of which are key parameters in global climate and hydrology, but exhibit relatively rapid variability.

Learning outcomes for Section 6

Now that you have completed your study of Section 6 you should be able to:

6.1 Describe the main patterns of variation of the different components of the cryosphere over about the past century, and also identify areas of uncertainty.

6.2 Outline the major issues concerning the stability of the West Antarctic Ice Sheet.

Answers to questions

Question 1.1

The mass balance b is given by

$b = c + a$

Substituting the values given into this equation

$b = 1.5 - 3$ m w.e. yr^{-1}

$b = -1.5$ m w.e. yr^{-1}

and so the glacier is losing mass at a rate of 1.5 m w.e. yr^{-1}

Question 2.1

There are important differences in the relationship between water inputs by precipitation, and outputs by runoff. In higher altitude/higher latitude catchments that support glaciers, runoff is at a maximum during summer, due to water supply from melting of snow and ice, and at a minimum in winter, since precipitation will largely be in the form of snow, which is accumulated until the following melt season. The winter maximum and summer minimum of rainfall in typical lower altitude/lower latitude catchments, which do not support glaciers, tend to produce winter runoff maxima and summer minima. Furthermore, summer precipitation tends to reduce runoff rates in glacierized catchments, as it is associated with cloudy conditions which reduce incoming solar radiation and therefore the rate of melting, while fresh falls of snow increase albedo and also therefore reduce the rate of melting.

Question 2.2

As the water flux to the glacier bed increases due to summer melting, there may be a transition from distributed drainage to channelized drainage.

Question 3.1

Compared with temperate regions, the hydrology of regions underlain by permafrost is typically highly seasonal, with runoff restricted to a few weeks in spring and summer, dominated by snowmelt. A very high proportion of precipitation and snowmelt contribute to surface runoff, and flooding is common because of the shallow, impermeable, permafrost table and the formation of ice jams on rivers.

Question 5.1

Retreating tidewater glacier termini can pose particular hazards through rapid iceberg calving.

Question 5.2

Building and maintaining irrigation channels across steep and unstable terrain poses major difficulties. While thin coatings of sediment prevent seepage from channels and their erosion, and supply inorganic nutrients, sediment can infill channels, raise terrace levels and choke plants. A simple rock dam constructed at the junction of melt streams with irrigation channels allows flow to be regulated by altering the density of rocks in the dam, reducing flooding, erosion and misallocation of water. In Hopar, the original melt stream flow of 11 m^3 s^{-1} is reduced to 0.001 m^3 s^{-1} by the time it reaches the crops: flow energy is so reduced that flooding and erosion problems are negligible. Channels can be dug wide and shallow to allow the streamwater to lose energy and therefore deposit sediment, which is occasionally dredged and used to line channels, or mixed with fertilizer and put on crops. Meltwater is, naturally, quite cold and liable to stunt or kill crops. Again, channels can be dug wide, so that water is shallow, meandering and slow-flowing, ensuring maximum radiative and convective heating. In Hopar, water seldom reaches crops at less than 9 °C.

Acknowledgements

Grateful acknowledgement is made to the following sources for permission to reproduce material in this book:

Figures 1.1, 1.7: NASA; *Figure 1.2*: © NSIDC User Services. These data, Northern Hemisphere EASE-Grid Weekly Snow Cover, were obtained from the EOSDIS NSIDC Distributed Active Archive Center (NSIDC DAAC), University of Colorado at Boulder; *Figures 1.3, 1.6, 2.1, 2.2, 2.3, 2.5, 2.7a, 2.9, 3.5, 3.6, 3.7, 3.10, 3.11, 4.1, 4.2, 4.7, 5.1, 5.2c, d, 5.4, 5.6, 6.4b*: © Richard Hodgkins, Royal Holloway, University of London; *Figure 2.4*: Collins, D. N. (1988) 'Suspended sediment and solute delivery to meltwaters beneath an alpine glacier', *Mitteilungen der Versuchsanstalt für Wasserbau, Hydrologie und Glaziologie*, **94**, Swiss Federal Institute of Technology, Zurich; *Figure 2.7b*: Bamber, J. L. *et al.* (2000) 'Widespread complex flow in the interior of the Antarctic ice sheet', *Science*, **287**, AAAS; *Figure 3.1*: © Pekka Parviainen/Science Photo Library; *Figure 3.8*: Copyright © Natural Resources Canada; *Figure 3.9*: Professor James Rose, Royal Holloway, University of London; *Figure 4.3*: © Canadian Space Agency. www.space.gc.ca; *Figure 4.4*: Benn, D. I. and Evans, D. J. A. (1998) *Glaciers and Glaciation*, Edward Arnold, London; *Figure 4.5*: © Bernhard Edmaier/Science Photo Library; *Figures 4.6, 4.10, 4.11*: © Mark Brandon, The Open University; *Figure 4.8*: © NSIDC User Services. These data, Northern Hemisphere EASE-Grid Weekly Sea Ice Extent, were obtained from the EOSDIS NSIDC Distributed Active Archive Center (NSIDC DAAC), University of Colorado at Boulder; *Figure 4.9*: Hansom, J. D. and Gordon, J. E. (1998) *Antarctic Environments and Resources: A Geographical Perspective*, Longman Publishers; *Figures 5.2a, b*: © 1998 Colorado Avalanche Information Center; *Figures 5.2c, d, 5.4, 5.6*: Richard Hodgkins, Royal Holloway, University of London; *Figure 5.3*: © CyberSpace Avalanche Center, / http://www.csac.org/; *Figure 5.5*: © Verband Schweizerischer Elektrizitätsunternehmen; *Figure 6.1*: Harvey, D. D. (2000) *Global Warming: The Hard Science*, Prentice Hall; *Figure 6.2*: © Copyright 2001, Climatic Research Unit; *Figure 6.3*: Warrick, R. A. *et al.* (1996) 'Changes in Sea Level', in *Climate Change 1995: The Science of Climate Change*, Cambridge University Press; *Figure 6.4a*: © Druck und Verlag Engadin Press AG, Samedan; *Figure 6.6*: World Glacier Monitoring Service, Monitoring Strategy, http://www.geo.unizh.ch/wgms; *Figure 6.7*: Nesje, A. and Dahl, S. O. (2000) *Glaciers and Environmental Change*, Edward Arnold, London; *Figures 6.10, 6.11*: Harvey, D. D. (2000) *Global Warming: The Hard Science*, Prentice Hall; *Figure 6.13*: © United States Geological Survey; *Figure 6.14*: Scambos, T. A. *et al.* (2000) 'The link between climate warming and break-up of ice shelves in the Antarctic Peninsula', *Journal of Glaciology*, **46**, p. 154, International Glaciology Society.

Every effort has been made to trace all the copyright owners, but if any has been inadvertently overlooked, the publishers will be pleased to make the necessary arrangements at the first opportunity.

Index

Note: Entries in **bold** are key terms. Page numbers referring to information that is given only in a figure or caption are printed in *italics*.

A

ablation (of ice by glaciers) **232**, 233
ablation areas 232, *241*
accumulated ozone time (AOT) 115
accumulation (of ice by glaciers) **232**, 233
accumulation areas 232, *241*
acetaldehyde 105
acid rain 61
active layer 258–9, *261*
adiabatic lapse rate (ALR) 106
aerenchyma 148, *149*, 196
aerobic respiration 57, 58, *201*
  of POC and DOC 188–91
  in wetland plants 148
  in wetland soils 145–6
aerobic zones in soils 150, 197
aerosols, atmospheric 54, 105
air pollution 52, 54, 56–7, **88**
  see also under ozone
air quality guidelines (ozone exposure) **113**
aircraft, weather observations 25, 26
albedo 237, 238
aldehydes 105
alder (*Alnus* spp.) 178, 182
Aletschgletscher, Switzerland 237, 238
alkanes 102–4
  see also methane
alkenes *102*, **103**, 104
Allalingletscher 277, *278*
Alligator mississippiensis (American alligator) *176*
Alnus spp. 178, 182
ALR (adiabatic lapse rate) 106
American alligator (*Alligator mississippiensis*) *176*
ammonia
  natural sources 55
  reaction with hydroxyl radicals 76
ammonium ions, bacterial oxidation 147
anaerobic bacteria 191–2
anaerobic conditions 58
  in wetlands 140, 144
anaerobic respiration in wetland soils 146, **147**, 190, *201*
Andover tornado 36
animals
  adaptations to wetlands 149–50
  of marshes 168, *169*
  of peatlands 163

  of swamplands 175–7
anoxic environments **58**, 160, 161, 192
Antarctica 271
  effects of climate change in 293–4
  ice flow velocities in *247*
  ice sheets 229, *230*, 298–302
  ice shelves in 248, *249*, 268, 293, *299*, 300, *301*
  ice streams in 244, 248, 293, 298, 300–1
anthropogenic sources (trace pollutants) **89**, 90, 91–3, 103
  see also ozone, air pollution by
AOT (accumulated ozone time), **115**
AOT40 index 115, 116
Arctic Ocean 266, 271, 273
arctic wetlands 142
aromatic hydrocarbons *102*, **103**, 104
  biogenic 105
asthma, effects of ozone exposure 112–13
atmosphere, composition 51, 53–4
atmospheric aerosols 54, 105
atmospheric budgets 83–94
atmospheric chemistry 51–2
atmospheric lifetimes 84–6
  and atmospheric transport 88
avalanches
  ice 277–8
  snow 280–83

B

background atmosphere 55
bacteria
  adaptations to wetland 147–8
  anaerobic 191–2
  denitrifying 192, 195
  methanogenic 192, 195
  methanotrophic 192, 195
  nitrogen-fixing 150
  sulfate-reducing 192, 194, 195
barrier islands 164
Bas Glacier d'Arolla
  sediment budget *252*
  water intake *285*
basin wetlands *141*, 143
Bayelva catchment, Svalbard 263
Beaufort wind scale 12, 13
Belzona, moved by hurricane 26
benzene *102*, 103, 104
Betula spp. 182
bilberry (*Vaccinium myrtillus*) 162

bimolecular reactions 85
binge–purge cycle 300
biomass 57
  burning 122–3
birch (*Betula* spp.) 182
birds, of peatlands 163
Black Rapids Glacier, Alaska 277
black spruce (*Picea mariana*) 162
black-headed gull (*Larus ridibundus*) 169
bogs 158, 159, 161
  vegetation 162
  'bog bodies' 161
bog myrtle (*Myrica gale*) 162
bombs 9
bond energy 69, *70*
boreal wetlands 142
bottomland forest 178, *179*
boundary (mixed) layer 86, 108, 110
box model (ozone formation simulation) **116**–17
brown hawker dragonfly *169*
Butorides virescens (green heron) *176*
buttress roots 175

C

C4 plants 151
Caltha palustris (marsh marigold) *180*
calving (icebergs) **232**, 248, 266–9
carbon cycle (in wetlands)
  carbon gains and losses 185–99
    carbon entering ecosystem 185–8
    carbon release 196–9, 209–11
    carbon transformation in soils 188–95
  carbon storage and accumulation processes 199–201, 204–10
  effects on drainage on 213–14
  global picture 204–11
carbon dioxide 57
  reaction with hydroxyl radicals 76
  released from wetlands 186, 196, 210, 216
  released in respiration 186–7, 190
  uptake by photosynthesis 185–6
carbon monoxide 58, *59*
  control of emissions 119–21
  geographical distribution 93, *94*
  oxidation pathways 75, 77–9
  reaction with hydroxyl radicals 76
  sources and sinks 55, 57, 60, *64*, 123
  varying content in air 83, 91, 92
Carex spp. 162, 167
carnivorous plants 162, *163*

Index

catalysts 120
cation exchange capacity (CEC), wetland soils 144–5
CEC *see* cation exchange capacity
Cercopithecus aethiops (vervet monkey) 176
Chamaedaphne calyculata (leatherleaf) 162
chamaeleon (*Chamaeleo parsoni*) 171
channelized subglacial drainage 241, *242*, 245–6, 251
Chapman, Sydney 72
Cladium spp. 167
climate change 272
    and cryosphere 287–302
    effects
        on carbon storage in wetlands 208
        of glaciers 234, 246
        on peatlands 160
        on wetlands 215–17
    and melting of ice masses 285, 286
    past variations 274
cloud clusters 13–14
Columbia Glacier, Alaska, calving 268
compensating effect in glacierized catchments **243**
cordgrass (*Spartina* spp.) *166*, 168
Coriolis force 14
cotton grass (*Eriophorum* spp.) 162
crabs, saltmarsh 169
crevasses 239, *240*, *241*, 267
critical level concept (ozone) 115
cryosphere 227–30, *231*, 235
    energy balance in catchments 237–43
    hydrology 238–43
    mass balance in 231–4
cryospheric hazards 276–83
cryospheric resources 283–6, 297–8
cryotic ground **256**
cyclones 12
    see also hurricanes
Cyperus spp. 167
Cyperus papyrus (papyrus reed) 137
cypress (*Taxodium* spp.) 170, 175

D

Daugaard–Jenssen Gletscher, Greenland *247*, 267
deep water formation 273, 274
denitrification 189, *190*, **192**–3, *194*
denitrifying bacteria 192, 195
DIC 187
    see also dissolved inorganic carbon
dissolved inorganic carbon (**DIC**) **187**, 188, 199–201
    leaching from wetlands 198
    production *190*

dissolved organic carbon (**DOC**) **187**, 188, 199–201
    aerobic respiration of 188–91
    fermentation of 191–2
    leaching from wetland 198
    oxidation through carbon dioxide reduction 194–5
    oxidation through sulfate reduction 194
    production *190*
distributed subglacial drainage 241, 245, 246
DOC 187
    see also dissolved organic carbon
domed swamps 174
downstream wetlands *141*, 143
drainage of wetlands 213–14
driving stress in glacier **245**
Drosera spp. 162, *163*
dry deposition 60, 97–8, 110
dry intrusions 9, 10, *11*
dryline 31–2

E

East Antarctic Ice Sheet 298, 301
ebullition 196
effective pressure (subglacial) **246**
Egretta alba (great white egret) 169
ELA *see* equilibrium line, changes in altitude
electromagnetic radiation 65–6
electronically excited states 68
ELR (environmental lapse rate) 106
energy balance of catchment 237–8
energy of radiation 65–6
englacial meltwater flow **240**
enteric fermentation 58
environmental lapse rate (ELR) 106
epiphytes 170, *171*
equilibrium line (glaciers) **232**
    changes in altitude 289, 292
ericaceous shrubs **158**, 162
Eriophorum spp. (cotton grass) 162
erosion, by glaciers 250–52
ethanoate (ethanoic acid) 191, 194
ethene *102*, 103
eustatic sea-level change **292**
evapotranspiration 160
Everglades, Florida 167
excited states *see* electronically excited states
explosive frontal cyclones 7–11
exudates from roots **187**
eye of hurricane **17**
eye irritants 100, 105
eyewall cloud 17, 27–8

F

faceted crystals in snow **281**
Feegletscher, Switzerland 240
felsenmeer 260, *261*
fens 158, *159*, 177
    soil *144*
fermentation *190*, **191**, 195
    of POC and DOC 191–2
fiddler crab 176
fire 56–7, 122–3, *201*, 216
first year ice **270**
flooding 5
    glacier 278–9
floodplains 143, 144
floodplain swamps 174
floodplain wetlands 142, 143–4
    denitrification in 193, *194*
    losses 153–4
    restoration 215
forest fires 56–7, 123
formaldehyde 57, *59*, 60, 81–2, 105
frazil ice 270
free radicals *see* radicals
frequency of radiation **66**
freshwater marshes 163, 166–7, 174, *177*
    animals of 168, *169*
freshwater swamps 170, 173–4
    biota of 175–7
frontal systems 7
    see also bombs; explosive frontal cyclones
frost cracking 262
frost creep 261
frost heaving 261
frost mounds 264
frost smoke 273
frost weathering 260–61
Fujita F scale of tornado intensity **34**–5
Fung, Inez 211–12
funnel clouds 29, **30**

G

gelifluction 261
glacial surges 277, 278
glaciated land **229**
Glacier d'Argentière 234, *235*, 290
Glacier de Tsanfleuron, Switzerland *242*
glacier floods 278–9
glacier ice 229, *231*
    distribution *230*
glacierized land **229**
glaciers *235*, *238*
    advance and retreat 246, 269, 276, 289, *290*, 291, 298

and climate change 234, 289–92
   erosion by 250–52
   flow of 243–8, *249*
      by basal motion 243, 244, 245–6, 248
      by internal deformation 243, 244–5, 248
   and landscape 249–52
   mass balance 232–3, *290*
   runoff 243, 284, 298
   tidewater 266–7, 268, 269
   unstable 277
   water drainage through 239–41, *242*
glasswort (*Salicornia* spp.) *164*, 168
gleying 145
global warming 287
   see also climate change
Gorham, Eville 206
Grande Dixence hydro-electric scheme 285, 286
Great Black Swamp, USA 154
great white egret (*Egretta alba*) *169*
green heron (*Butorides virescens*) *176*
green pygmy goose (*Nettapus pulchellus*) *176*
greenhouse gases 54, 138
   see also methane; nitrous oxide
Greenland Ice Sheet 228, 229, *230*, 235, 293
grey heron *169*
gross primary production 186
ground frost 255
ground state 68
grounding line (glacier) **248**
groundwater, in wetlands 141

H

Hadley circulations 87, 108
Hessbreen, Svalbard *240*, 249
Hidrovia project 154
hollows in peatlands **188**
Hopar, Karakoram, Pakistan, irrigation water 284
horizontal vortex roll 32–3
horseshoe crab (*Limulus polyphemus*) *169*
human activities
   carbon releases to atmosphere by 209
   effect on atmospheric composition 89–92
   effect on global climate 289
   effects on wetlands 152–5
   see also cryospheric hazards; cryospheric resources
hummocks in peatlands **188**
hurricanes 5, **12**–29
      in USA 20–29
Hurricane Andrew *6*, 24, 25–9
Hurricane Camille 21

Hurricane Hugo 29
Hurricane Mitch 21, *22*
hydraulic jack 246
hydric soils 144
   see also wetland soils
hydrocarbons 101–3
   biogenic 103, 105, 119
   control of emissions 119–21
   non-methane 92
   reactivity 104
   see also methane
hydrocarbon groups 102
hydrocarbon-limited regime 117–18, 119
hydro-electric power 284, 285, 286
hydrogen-abstraction reactions 77
hydrogen-fuelled vehicles *123*
hydrogen sulfide
   natural sources *55*
   oxidation *59*, 147
   reaction with hydroxyl radicals *76*
   released from wetlands 197
hydrology
   of cryosphere 238–43
   of marshes 166
   of permafrost 262–4
   of wetlands 139–44, 177–8
hydroperoxy radical *77*, 105
hydroxyl radicals 76–8, 79, 85–6, 88, 101–4, 105

I

ice avalanches 277–8
ice floes 270
ice flow *see* glaciers, flow of
ice rind 270
ice sheets 249
   effects of climate change 293–4, 298–302
   flow of 243–4
   mass balance in 233, 234
   see also Antarctica, ice sheets; Greenland Ice Sheet
ice shelves 248, *249*, 266–7, 268
ice streams 244, **247**–8, *249*, 267
icebergs 266–9
   calving 232, 248, 266–9
Iceland
   glacier advances in 276
   jökulhlaups in 279
icings 263–4
insectivorous plants 162, *163*
intertidal zone 164
Intertropical Convergence Zone (ITCZ) 87
irrigation water 284
isoprene *102*, 103, 119
isostatic sea-level change **292**
ITCZ (Intertropical Convergence Zone) 87

J

Jakobshavns Isbrae, Greenland 248, 293
jökulhlaups 278–9
Jostedalsbreen ice cap, Norway 292
Juncus gerardii (rush) 168

K

Keddy, Paul 139
kettle hole bogs 159, *160*

L

Labrador tea (*Ledum groenlandicum*) 162
Lac des Dix, Switzerland 285
lachrymators *see* eye irritants
lakes
   ice-dammed 278
   terrestrialization 158–9, 166
lake-edge swamps 174
landscape
   glaciers and 249–52
   periglacial environments 261–2
   wetland 143–4
Larix laricina 162
Larus ridibundus (black-headed gull) *169*
Laurentide Ice Sheet 300
leaching 198–9, *201*
leads 273
leatherleaf (*Chamaedaphne calyculata*) 162
Ledum groenlandicum (Labrador tea) 162
Limulus polyphemus (horseshoe crab) *169*
linked-cavity subglacial drainage 241, *242*, 245
'Little Ice Age' 234, 246, 276, 291
loess 262
London fogs 98–9, 107
Los Angeles, smog 99–100, 108, 122
low-level inversions 107–8
Lowell Glacier, Yukon Territory 277

M

mangrove swamps *150*, **170**–73
   biota of *176*, 177
marshes 157, 163–9, 177–8, 189, 190
   see also freshwater marshes; saltmarshes
marsh marigold (*Caltha palustris*) 180
mass balance (ice masses) **231**–3, *249*
   and climate change 234, 293
Matthews, E. 211–12
Mattmark, Valais, ice avalanche 277
Mechan River catchment, Canada 263
Melaleuca spp. *143*
Mer de Glace, Switzerland 276
mesocyclones 33
mesotrophic grasslands 181

methane 83
    atmospheric lifetime 84
    bacterial oxidation 147, 197–8
    oxidation 57, *59*, 60, 61–3, 75, 79, 81–2
    reaction with hydroxyl radicals *76*, 85
    released in restored wetlands 215
    released in wetlands 58, 146, 196–8, 210, 214–15, 216–17
    sources and sinks *55*, 58, 64, 89–91
    varying content in air 89–91
methanogenesis *190*, **194**–5, 210
methanogens 192, 195
methanotrophic bacteria **196**–7
methyl radicals 76, 79
microbial respiration 58
mineral soils (wetland) **145**
mires 180–81
Mississippi River, floodplain wetland 153
mixed layer *see* boundary layer
mixing ratios 53, 54
Monet, Claude *52*
MOPITT (air pollution monitor) *94*
moraines *249*
mosquitoes 162
mosses 148, 150, 158, 161
mottling (wetland soils) **145**
moulins 239, *240*, *241*
Mount St Helens, Washington State, USA 279
multiyear ice **270**, 271
muskrat (*Ondatra zibethicus*) 168, *169*
Myrica gale (bog myrtle) 162

N

National Vegetation Classification system 180–82
natural gas *see* methane
net carbon accumulation 199
net primary production 187, 190
    effects of climate change 215
    in wetlands 199, 200, *201*
net solar radiation (Q_{NR}) 237, 238
Nettapus pulchellus (green pygmy goose) *176*
Nevado del Ruiz, Colombia 279
Nevado Huascaran, ice avalanches 278
Nigardsbreen, Norway 246
nilas 270
nitric oxide 193
    atmospheric source 57, 60
    natural sources *55*
nitrogen cycle
    'fixing' of atmospheric nitrogen 92, 150
    see also denitrification
nitrogen oxides *see* NO$_x$
nitrogen-fixing bacteria 150

nitrous oxide
    emission from wetlands 193
    natural sources *55*
nivation 262
NO$_x$ 75
    control of emissions 119–21
    emissions from fires 123
    reaction with hydroxyl radicals *76*
    role in ozone production 78–80, 110
    and smog formation 100–1, 104
    varying content in air 83, 91, 92–3
    see also nitric oxide; nitrous oxide
NO$_x$-limited regime 117, 118, 119
non-cryotic ground **256**
non-methane hydrocarbons 92
null cycles 78
number density 54
Nymphaea spp. 167

O

ocean circulation 273–4
ocean conveyor belt 273, 274
ocean currents *16*
octane number 102
Odden Ice Tongue 273
'Oetzi' (mummified body) 246
Oklahoma
    Mesonet 37
    tornado outbreak 41
Ondatra zibethicus (muskrat) 168, *169*
organic soils (wetland) **144**–5
oxidation 57–8, 146, 189, 194–5
oxygen
    excited atoms (O*) 75
    levels in wetlands 140–41
    photolysis *68*, *69*
    uv absorption 69
    see also ozone
oysters *176*
ozone 52, **67**
    air pollution by 96–123
        control 116–23
        effects on human health 112–13
        effects on plants 114–16
        regional perspective 109–11
        role of road transport 101–4
        role of thermal inversions 106–8
        trends in air quality 121–3
    in stratosphere *see* ozone layer
    in troposphere 71, 74–80, 96–8
        sources and sinks 97–8, 110
    varying content in air 96–7, 110–11
ozone layer 71–4

P

paludification *159*, **160**
Panatal wetland 154
pancake ice **270**
papyrus reed (*Cyperus papyrus*) 137
paraglacial environment 251, **260**
particulate organic carbon (**POC**) **188**, 199–201
    aerobic respiration of 188–91
    fermentation of 191–2
    leaching from wetland 199
    oxidation through nitrate reduction 192–3, *194*
parts per billion, ppbv **54**
parts per million, ppmv **54**
parts per trillion, pptv **54**
peat 144, **157**, 161, 178
peatlands 157–63, 177–8, 188, 189, 190
    carbon accumulation in 208–10
    carbon storage in 200, 201, 206–8
    effects of climate change 216
    frost cracking in *262*
    see also bogs; fens
perched wetlands 141, 143
periglacial environment **255**, 260
    effects of climate change 294–5
    processes and features of 260–64
Perito-Moreno glacier, Patagonia *269*
permafrost 255, 256–9
    effects of climate change 294–5, *296*
    hydrology 262–4
peroxy radicals 77, 78, 79, 80, 101, 104
phosphorus cycle, role of wetlands in 146–7
phosphorus sinks 173
photochemical processes **67**–9
    in stratosphere 71–4
    in troposphere 74–80
photochemical smog 52, 99–101, 104–5
    role of road transport 101–4
photodissociation 67
photolysis 67
    of atmospheric constituents 70–71
    of oxygen molecule *68*, *69*
photons 66
    absorption by atoms 68–9
photosynthesis 185–6
photovoltaic cells *123*
Phragmites spp. (reeds) *143*, 167, 181
Picea mariana (black spruce) 162
Pine Island Glacier, Antarctica 294, 301
pitcher plants 162, *163*
Planck constant 66

plants
    adaptations to wetlands 148–9, 150–51
    effects of ozone pollution on 114–15
    hydrocarbon production by 103, 105, 119
    of marshes 167–8, 178
    of peatland 161–2, 178
    of swamplands 170–71, *172*, 175, 178
    and water accumulation in wetlands 140
Plomo Glacier, Argentina 278
pneumatophores 175
POC 188
    see also particulate organic carbon
polar amplification (global warming) **287**
pollution *see* air pollution
polynyas 273
poplar (*Populus* spp.) 178
powder avalanches 280
prairie pothole marshes 166–7
primary pollutants 100, 101
proglacial environment **260**
prop roots 175
Puccinellia spp. 168

Q

quaking bogs 159

R

radar networks 38, 39, *40*
radiation inversions 107
radicals 63
    see also hydroperoxy radicals; hydroxyl radicals; methyl radicals; peroxy radicals
rainfall 7, 9
    hurricane 18, 27–9
rainforest swamps 17
Ramsar Convention 154–5
rate constants (*k*) **85**
rate equations 85
reaction cycles 77–9, 81–2
reaction mechanisms 62
reaction rates 84–5
rear-flank downdraft (storms) 33
redox status of soil **147**
reduction 57–8, 58, 189
residence times 83
    see also atmospheric lifetimes
respiration 186
    microbial 58
    see also aerobic respiration; anaerobic respiration
response time (glaciers) **246**
rhizomes (marsh plants) 167
rice paddies 152, 208, 210
    and greenhouse gases 214–15

Rio Santa valley, ice avalanches 278
rivers
    icings 263–4
    in periglacial environments 262–3
road transport
    emissions 92, 93, 100, 101–4
        control 119–22
rocks, frost weathering 260–61
Ronne Ice Shelf, Antarctica *268*
root exudates 187
Ross Ice Shelf, Antarctica 248
run-out distance (ice avalanches) **277**

S

Saas-Fee, Switzerland, avalanche defences *283*
Saffir–Simpson scale of hurricane intensity 18, **20**–21
Salicornia spp. (glasswort) *164*, 168
saline wetlands
    plant adaptations for 151
    see also mangrove swamps; saltmarshes
Salix spp. 178, 182
saltmarshes 163, 164, *165*, 166, 167, 181
    animals of 168, *169*
    plants of 168
salt water swamps *see* mangrove swamps
Sarracenia spp. (pitcher plants) 162, *163*
saturated hydrocarbons 103
Schönbein, Christian 67, 96, 97
Scirpus spp. 167
Scott Turnerbreen (glacier) *238*, *239*, *264*, *280*
sea ice 228, 266, 270–74
    and climate 272–3
    distribution 271, *272*
    effects of climate change 296, *297*
    formation 270
    and ocean circulation 273–4
sea-level, changes 292, 298, 302
sea-surface temperatures 9, *16*, 17
secondary pollutants 100
sedge family 162
sediment budgets 252
sediment yield 250–52, 263
segregation ice model of frost weathering *259*, **260**
sinks (trace atmospheric constituents) **55**
    surface 60–61
sitatunga antelope *169*
Skywarn project 38
slab avalanches 280
slush avalanches 280
smog 98–**99**, 107
    see also photochemical smog
snow, water drainage through 238–9

snow avalanches 280–83
snow cover 228
    effect of climate change 287–9
snowfall 11, 243
    effect of climate change 288
snowflakes, formation 228
soils
    mangrove 173
    marsh 165
    removing nitrogen from runoff 192–3, *194*
    wetland 144–7, 165, 173, 185
        aerobic zones in 150, *151*
        carbon storage in 204–7
solar energy *123*
solar radiation 65, 67
    net 237, 238
sources (trace atmospheric constituents) **55**
    surface 55–9
Spartina spp. (cordgrass) *166*, 168
Sphagnum spp. 158, 161, 181
spiral rainbands 18, 27, 29
stability of atmosphere 106–7
static sensitivity to climate change **234**, 302
sticky spots (glacier flow) **248**
stomatal pores 114
storms
    Burns' Day 1990 8, 10
    Christmas Eve 1998 8, 10, *11*
    October 1987 (England) 5, 7
stratification of oceans **270**
stratosphere 71, 75
    see also ozone layer
stratospheric dry intrusions 9, 10, *11*
subglacial drainage **240**, 241, 251
subsidence inversions 108, *109*
sulfate fertilizers 215
sulfate-reducing bacteria 192, 194, 195
sulfate reduction 194
sulfur dioxide
    natural sources 55, 56, *61*
    oxidation 59, 60–61
    reaction with hydroxyl radicals 76
sundews (*Drosera* spp.) 162, *163*
surface hoar on snow **281**
surges 19, 29
    glacial 277, 278
Svalbard
    Bayelva catchment 263
    cornice collapse *280*
    frost-weathered rock *260*, *261*
    glacial moraine *249*
    glaciers *238*, *239*, *240*
    icing *264*
swamps *143*, **157**, 170–78, 181, 189, 190

T

taliks 259, 263–4
tamarack (*Larix laricina*) 162
Taxodium spp. (cypress) 170, 175
terrestrialization 158–9, 166
thermal inversions 106–8, 109
thermohaline circulation 273, 274
three-way catalytic converter 120–**121**
throughflow water
    in marshes 166
    in wetlands 141, 142
Thuja occidentalis (white cedar) 170
Tollund Man *161*
toluene *102*, *103*, 104
tornadoes 29, **30**–41
    damage due to 40–41
    forecasting 32–3, 37–9, 41
    incidence, longevity and intensity 33–6
    origin and development 31–3
    'tornado alley' 33
trace gases 51, **53**–4, 83
    atmospheric lifetimes 85–6
        and transport processes 88
    energy required for photolysis *69*, 70
    reactions with hydroxyl radicals *76*
    surface sources 55–9
    see also air pollution
Trade Winds 87
trans-Alaska oil pipeline *256*, 277
trees, effect on carbon cycling 213–14, 216
tropical cyclones 12, 13–15
    see also hurricanes
tropical depressions 12
tropical storms 12
tropical swamps *143*
tropical waves 25
tropopause *71*, 75

troposphere *51*, 52, **71**
    air movements in 86–8
    ozone in *71*, 74–80, 96–8
        sources and sinks 97–8, 110
    photochemical processes in 74–80
tupelo (*Nyssa* spp.) 170, 175
turbulent heat fluxes 237
Tyndrum, Scotland, glacial moraines *249*
Typha spp. *166*, 167
typhoons 12

U

ultraviolet radiation 66, 67, 69, 70–72, 78
    absorption in atmosphere 72–4
unsaturated hydrocarbons 103, 104
upland environments 139

V

Vaccinium myrtillus (bilberry) 162
vehicle emissions 92, 100, 101, 103–4
vernal pools 179–80
vervet monkey (*Cercopithecus aethiops*) *176*
volcanic emissions 56
volcanoes 279

W

water
    conservation by wetland plants 151
    drainage through glaciers 239–41, *242*
    drainage through snow 238–9
    flow rates in wetlands 141–2
    quality in wetlands 142
    sources
        glacial meltwater 283–6
        in wetlands 139–42
water-equivalents 231
water-lilies 167
    pressurized gas flow in *149*

water-table, in wetlands 140, 141, 177
wavelength of radiation **65**
weather forecasting 6, 10–11, 18, 22–5, 32–3, 37–9
weather satellite images 6, *8*, 9, 10–11, 22–4, 37–8
West Antarctic Ice Sheet 298–302
wet deposition 60, 61
wet meadows 178, *179*
wetlands 137–8
    biota 147–51, 161–3, 167–8, 170–1, 172, 175–80
    buffer strips 193, *194*
    characteristics 139–53
        hydrology and water quality 139–44
    classification 180–82
    effects of drainage 213–14
    global distribution 152
    leaching from 198–9
    losses and protection 152–5
    restoration 215
    soils 144–7, 150, *151*, 165, 173, 185, 204–7
    sources of greenhouse gases 58, 146, 196–8, 210, 214–15, 216–17
    transitional communities 178–80
    see also carbon cycle (in wetlands); marshes; peatlands; rice paddies; swamps
white cedar (*Thuja occidentalis*) 170
willow (*Salix* spp.) 178, 182
winds 5, 9, 86
    cyclonic 14
    hurricane 17, 18–19, 20, 24–5, 27–8
    in periglacial environments 262
    tornadoes 30, 35
    see also hurricanes
wind crusts on snow 238, *239*
wind loading 12–13
woodlands 182
Wyeomyia smithii 162

0845 300 6090